T0318763

OUR EXTRACTIVE AGE

Our Extractive Age: Expressions of Violence and Resistance emphasizes how the spectrum of violence associated with natural resource extraction permeates contemporary collective life.

Chronicling the increasing rates of brutal suppression of local environmental and labor activists in rural and urban sites of extraction, this volume also foregrounds related violence in areas that we might not expect, such as infrastructural developments, protected areas for nature conservation, and even geoengineering in the name of carbon mitigation. Contributors argue that extractive violence is not an accident or side effect, but rather a core logic of the 21st-century planetary experience. Acknowledgment is made not only of the visible violence involved in the securitization of extractive enclaves, but also of the symbolic and structural violence that the governance, economics, and governmentality of extraction have produced. Extractive violence is shown not only to be a spectacular event, but an extended dynamic that can be silent, invisible, and gradual. The volume also recognizes that much of the new violence of extraction has become cloaked in the discourse of "green development," "green building," and efforts to mitigate the planetary environmental crisis through totalizing technologies. Ironically, green technologies and other contemporary efforts to tackle environmental ills often themselves depend on the continuance of social exploitation and the contaminating practices of non-renewable extraction. But as this volume shows, resistance is also as multi-scalar and heterogeneous as the violence that it inspires.

The book is essential reading for activists and for students and scholars of environmental politics, natural resource management, political ecology, sustainable development, and globalization.

Judith Shapiro is Chair of the Global Environmental Policy Program at the School of International Service at American University, USA. She is author/editor of numerous books, including *China Goes Green: Coercive Environmentalism for a Troubled Planet* (2020, with Yifei Li) and *China's Environmental Challenges* (2016).

John-Andrew McNeish is Professor of International Environment and Development Studies at the Norwegian University of Life Sciences, Norway. He is the editor of multiple books, including most recently *Sovereign Forces: Everyday Challenges to Environmental Governance in Latin America* (2021).

ROUTLEDGE STUDIES OF THE EXTRACTIVE INDUSTRIES AND SUSTAINABLE DEVELOPMENT

Mining and Sustainable Development
Current Issues
Edited by Sumit K. Lodhia

Africa's Mineral Fortune
The science and politics of mining and sustainable development
Edited by Saleem Ali, Kathryn Sturman and Nina Collins

Energy, Resource Extraction and Society
Impacts and Contested Futures
Edited by Anna Szolucha

Regime Stability, Social Insecurity and Bauxite Mining in Guinea
Developments Since the Mid-Twentieth Century
Penda Diallo

Local Experiences of Mining in Peru
Social and Spatial Transformations in the Andes
Gerardo Castillo Guzmán

Resource Extraction, Space and Resilience
International Perspectives
Juha Kotilainen

Our Extractive Age
Expressions of Violence and Resistance
Edited by Judith Shapiro and John-Andrew McNeish

For more information about this series, please visit:
https://www.routledge.com/series/REISD

OUR EXTRACTIVE AGE

Expressions of Violence and Resistance

*Edited by Judith Shapiro
and John-Andrew McNeish*

 Routledge
Taylor & Francis Group
LONDON AND NEW YORK

 earthscan
from Routledge

First published 2021
by Routledge
2 Park Square, Milton Park, Abingdon, Oxon OX14 4RN

and by Routledge
605 Third Avenue, New York, NY 10158

Routledge is an imprint of the Taylor & Francis Group, an informa business

British Library Cataloguing-in-Publication Data
A catalogue record for this book is available from the British Library

Library of Congress Cataloging-in-Publication Data
Names: Shapiro, Judith, 1953- author. | McNeish, John-Andrew, editor.
Title: Our extractive age : expressions of violence and resistance / edited by Judith Shapiro and John-Andrew McNeish.
Description: Abingdon, Oxon ; New York, NY : Routledge, 2021. |
Series: Routledge studies of the extractive industries | Includes bibliographical references and index. |
Identifiers: LCCN 2020054388 (print) | LCCN 2020054389 (ebook) |
ISBN 9780367650537 (hardback) | ISBN 9780367650520 (paperback) |
ISBN 9781003127611 (ebook)
Subjects: LCSH: Mineral industries--Moral and ethical aspects. | Natural resources--Management. | Environmental ethics. | Environmental accountability.
Classification: LCC HD9506.A2 O97 2021 (print) | LCC HD9506.A2 (ebook) | DDC 333.8--dc23
LC record available at https://lccn.loc.gov/2020054388
LC ebook record available at https://lccn.loc.gov/2020054389

ISBN: 978-0-367-65053-7 (hbk)
ISBN: 978-0-367-65052-0 (pbk)
ISBN: 978-1-003-12761-1 (ebk)

Typeset in Bembo
by Taylor & Francis Books

CONTENTS

ILLUSTRATIONS

Figure

Tables

Boxes

CONTRIBUTORS

John-Andrew McNeish. Professor of International Environment and Development Studies. Institute of International Environment and Development Studies (Noragric), Norwegian University of Life Sciences. McNeish is a social anthropologist with a focus on development and resource politics in Latin America. He is the author of multiple peer-reviewed articles and several co-edited volumes including *Contested Powers: The Politics of Energy and Development in Latin America; Flammable Societies: Studies on the Socio-Economics of Oil and Gas; Gender Justice and Legal Pluralities: Latin American and African Perspectives.* McNeish's latest book, *Sovereign Forces: Everyday Challenges to Environmental Governance in Latin America,* was published by Berghahn Books in 2021. McNeish is a member of several international research projects including the Norwegian Research Council funded projects: Riverine Rights: Exploring the Currents and Consequences of Legal Innovation on the Rights of Rivers and Green Curses and Violent Conflict: The Security Implications of the Renewable Energy Sector in Africa. https://orcid.org/0000-0001-6128-3997

Judith Shapiro. Chair, Global Environmental Policy Program and Director, Natural Resources and Sustainable Development Dual MA, American University, School of International Service. Dr. Shapiro is the author of *China Goes Green: Environmental Authoritarianism for a Troubled Planet* (with Yifei Li), *China's Environmental Challenges,* and *Mao's War against Nature* and the co-author of *Son of the Revolution, After the Nightmare* and other books on China. Her research focuses on the environmental implications of the rise of China, both domestically and throughout the globe. As one of the very first US citizens to live in China after US–Chinese relations were normalized in 1979, her career has spanned—and reflected—the post-Mao period from the post-Cultural Revolution to today's dramatic rise. https://orcid.org/0000-0002-4238-0027

Christopher Chagnon. PhD candidate in Development Studies in the Political, Societal, and Regional Change Doctoral Programme, University of Helsinki. Chagnon is affiliated with the Helsinki Institute of Sustainability Science (HELSUS) and works with the Global Extractivisms and Alternatives Initiative (EXALT). He is a member of the board of the Finnish Society for Development Research and has over a decade of work experience in development, business, and education across several continents, specializing in African-Chinese engagement. https://orcid.org/0000-0001-8127-4692

Francesco Durante. PhD candidate in the Political, Societal, and Regional Change Doctoral Programme in affiliation with the Aleksanteri Institute and Faculty of Social Sciences, University of Helsinki. Durante is a member of the thematic joint network led by the University of the Arctic and the Northern Research Forum on geopolitics and security. His research focuses on the governance of natural resource extraction in the Russian Arctic. https://orcid.org/0000-0002-2884-389X

Katharina Glaab. Associate Professor in Global Change and International Relations, Norwegian University of Life Sciences. Dr. Glaab has a PhD in Political Science/International Relations from the University of Münster. Previously she obtained a Magister in Political Science, Chinese Studies and history from the same university and studied Mandarin Chinese at Renmin University, China. She has held visiting fellowship positions at Jawaharlal Nehru University in Delhi, India, Tsinghua University China and Fridtjof Nansen Institute, Norway. Her fields of research are international relations theory, global environmental politics (especially genetically modified organisms and climate change), theories of power, and religion in global politics. https://orcid.org/0000-0002-0818-7522

Garrett Graddy-Lovelace. Associate Professor. American University, School of International Service. Dr. Graddy-Lovelace researches global environmental and agricultural policy and politics. A critical geographer, she draws upon political ecology and postcolonial studies in current research on agricultural biodiversity conservation, agrarian cooperatives, and domestic and global impacts of US farm policies. This includes community-based participatory action research with grassroots groups on Farm Bill reform, as well as ongoing research on Cuban–US agricultural relations. She is writing a book on agrobiodiversity conservation politics. https://orcid.org/0000-0002-6006-3498

Sophia E. Hagolani-Albov. PhD candidate in Interdisciplinary Environmental Studies, University of Helsinki. Hagolani-Albov is affiliated with the Faculty of Agriculture and Forestry and Helsinki Institute of Sustainability Science (HELSUS). She works as the project coordinator for the Global Extractivisms and Alternatives Initiative (EXALT) at the Faculty of Social Sciences. Sophia co-hosts the EXALT

Initiative podcast, a monthly conversation with academics, artists, and activists. http s://orcid.org/0000-0002-0958-524X

Saana Hokkanen. Graduate student, University of Helsinki. Hokkanen focuses on development studies and is interested in world-ecological theory, decolonial political ecology, and post-capitalist imaginaries. She has worked at the Helsinki Institute of Sustainability Science (HELSUS) and is part of the Global Extractivisms and Alternatives Initiative (EXALT). https://orcid.org/0000-0002-9050-969X

Victoria Kiechel. Faculty, American University, School of International Service. Kiechel is a practicing architect and native Washingtonian. Her courses in the American University's Global Environmental Politics Program include real-world, local projects in sustainable design. Vicky works for the Cadmus Group, LLC, an environmental consultancy. At Cadmus she has led research work for the Appalachian Regional Commission, sustainability consulting for the Smithsonian Institution, and consulting and review teams for projects seeking LEED certification. She worked for the US Green Building Council on LEED v.3 and advises the US Environmental Protection Agency's ENERGY STAR commercial and industrial branch. https://orcid.org/0000-0003-1535-2626

Markus Kröger. Associate Professor of Development Studies, University of Helsinki and Academy of Finland. Dr. Kröger is a member of the Helsinki Institute of Sustainability Science and one of the founding members of The Global Extractivisms and Alternatives research initiative (EXALT). He has written extensively on global natural resource politics, conflicts, and social resistance movements and their economic outcomes, especially in relation to iron ore mining and forestry. He is also an expert in political economy, development, and globalization in Latin America, India, and the Arctic. https://orcid.org/0000-0001-7324-4549

Will LaFleur. PhD candidate in the Political, Societal, and Regional Change Doctoral Programme, Faculty of Social Sciences, University of Helsinki. LaFleur is an affiliate of the Helsinki Institute of Sustainability Science (HELSUS). With a background in Anthropology and Education, his research engages the socio-sensory, phenomenological, and atmospheric contingencies of knowing and the making of resilience in local foodways practices amid global ecological crises and the COVID-19 pandemic. https://orcid.org/0000-0002-4427-7744

Yifei Li. Assistant Professor of Environmental Studies, NYU Shanghai. With support from the US National Science Foundation, the Rachel Carson Center for Environment and Society, and other extramural sources, Dr. Li's research examines the city and the environment in China, focusing on questions about bureaucracy, urban sustainability, and environmental governance. He is coauthor (with Judith Shapiro) of *China Goes Green: Coercive Environmentalism for a Troubled Planet*. His recent work appears in *Current Sociology, Environmental Sociology,*

Journal of Environmental Management, International Journal of Urban and Regional Research, and other scholarly outlets. He received his PhD and Master's degrees from the University of Wisconsin-Madison and Bachelor's from Fudan University. https://orcid.org/0000-0001-5156-0161

Philippe Le Billon. Professor, University of British Columbia, Geography and School of Public Policy and Global Affairs. Dr. Le Billon's work focuses on environment, development and security linkages and the political ecology of conflicts. He is co-founder of the *Environmental Peacebuilding Association.* A (co)author of about 100 refereed papers and book chapters, his books include *Oil* and *Wars of Plunder: Conflicts, Profits and the Politics of Resources.* https://orcid.org/0000-0002-4635-2998

Nick Middeldorp. Doctoral candidate, University of British Columbia, Geography. Middeldorp's work focuses on the tension between megaprojects (extractive, agribusiness, infrastructure), corruption, environment, and community needs and rights (land rights, indigenous rights). He has extensive experience in human rights investigations in Latin America. He is the author of *Derecho a la Consulta Previa, Libre e Informada: Una Mirada Crítica Desde los Pueblos Indígenas* [*Right to Free, Prior and Informed Consent: A Critical View from Indigenous Peoples*]. https://orcid.org/0000-0002-0204-108X

Simon Nicholson. Associate Professor, American University, School of International Service. Dr. Nicholson's work focuses on global environmental governance, global food politics, and the politics of emerging technologies, including climate-engineering (or "geoengineering") technologies. He is co-founder of the *Forum for Climate Engineering Assessment* and the *Institute for Carbon Removal Law and Policy.* As well as multiple peer reviewed articles Nicholson recently co-edited two important collections i.e. *Global Environmental Politics: From Person to Planet* and *New Earth Politics: Essays from the Anthropocene.* https://orcid.org/0000-0002-8081-5968

Whitney Richardson. Research Associate, Earth Law Center. Recent graduate, Norwegian University of Life Sciences, Master's degree in International Environmental Studies program. Richardson's work focuses on political ecology and intersectional environmental and ecological justice concerns, with an interest in environmental law, food sovereignty, human rights, nature's rights, and resistance movements. https://orcid.org/0000-0003-2538-4242

Kirsti Stuvøy. Associate Professor, Norwegian University of Life Sciences. Dr. Stuvøy is a political scientist who researches and teaches in the field of international relations and development studies. She has published on security theory, human security and methodology, war economy in Africa, Russian politics and civil society development, and Arctic governance. https://orcid.org/0000-0002-6671-0609

Paul Wapner. Professor, American University, School of International Service. Dr. Wapner's research focuses on global environmental politics, environmental thought, transnational activism, and environmental ethics. He is particularly concerned with understanding how societies can live through this historical moment of environmental intensification in ways that enhance human dignity, compassion, and justice, and come to respect and nurture the more-than-human world. He has published six books and numerous scholarly articles. His latest book is, *Is Wildness Over?* (Polity, 2020). Paul serves on the board of RE-volv, an organization that finances and installs solar for nonprofits, and teaches workshops on contemplative environmentalism at the Lama Foundation (www.earthlovego.org). https://orcid.org/0000-0002-6671-0609

Michael J. Watts. Class of 1963 Professor of Geography and Development Studies Emeritus at the University of California, Berkeley. A Guggenheim Fellow in 2003, Dr. Watts served as the Director of the Institute of International Studies from 1994–2004, Director of Social Science Matrix at Berkeley. He has written on political ecology, the political economy of development, agrarian transformations and political conflicts and has conducted research in Senegambia, Nigeria, India, Vietnam, and California. He has authored 20 books and over 300 articles, received grants and fellowships from the MacArthur Foundation, the National Science Foundation, and the Guggenheim Foundation, and is a member of RETORT, a Bay Area. Educated at University College London and the University of Michigan, held visiting appointments at the Smithsonian Institution and Universities of Bergen, Bologna, and New Delhi, and served as the Chair of the Board of Trustees of the Social Science Research Council (2010–15). He serves on a number of Boards of non-profit organizations including the Pacific Institute. https://orcid.org/0000-0002-1971-7269

INTRODUCTION

John-Andrew McNeish and Judith Shapiro

Our Extractive Age unpacks the complex character of an era in which human extraction and use of natural resources contribute to an environmental crisis of planetary proportions. This crisis has multiple complex features, but the most central of these—climate change, land and forest degradation, the loss of biodiversity, even the global pandemic—are directly linked to the physical impacts caused by the extraction of natural resources and the climate gas emissions caused by a growing human population's demands for energy, food, and consumer goods. Earth system scientists claim that through our rapacious interaction with the environment we have caused a sudden "state shift" of fundamental, irreversible change in the biosphere (Moore, 2016). Human pressures are pushing biospheric stability, climate, and biodiversity to the breaking point (Mace *et al.*, 2014; Dirzo *et al.*, 2014; Steffan et al., 2015). Multiple planetary boundaries are now being crossed, or soon will be (Rockström *et al.*, 2009). Moreover, natural and social scientists widely suggest that as a result of our damaging interaction with the planet, biosphere and geological time has been fundamentally transformed. While still controversial, a new conceptualization of geological time—one that includes humankind as a "major geological force"—has been proposed: we are living in the Anthropocene (Crutzen and Stoermer, 2000).

The contributors to this volume recognize the central role that resource extraction plays in defining both our time and the character of the present existential threat. A distinct contribution is the volume's planetary perspective and an understanding of extractive processes that moves past a narrower focus on non-renewable resources such as fossil fuels, minerals, and precious metals. The contributors argue that the pace and nature of extraction have dramatically accelerated and broadened. We argue that the expansion of extraction, aided by technological development, legal and illegal capital, and geo-political and national decision-making, is a truly global phenomenon now stretching from developing countries to the Poles, cyberspace, the Earth's atmosphere, and even outer space. Hyper-extraction is to be found in the unlikeliest of places.

Hydrocarbon extraction is linked to significant transformative events such as the Industrial Revolution, the two World Wars, the Cold War, and the more recent New Wars involving non-state as well as state actors (Kaldor, 1999). It has played—and continues to play—a defining role in 20th- and 21st-century global and national politics, economics, and society (Mitchell, 2011). However, as we make evident in this volume, fossil fuel extraction was exploited in parallel with a much wider range of thermodynamic and calorific-rich resources, and in many cases it helped to harness them. Indeed, the extraction of other energetic resources such as biomass, somatic resources (human slaves), atomic power, industrial exploitation of hydropower, and industrial-scale agriculture have all played significant roles in the accelerated globalization of the state system and capitalism. The fossil fuel economy cannot account for all of the human influence on the climate and planet (Malm, 2016). Moreover, although carbon dioxide emissions are important, there are other greenhouse gases, including methane, nitrous dioxide, ozone, and sulphur hexafluoride, that also have important social histories linked to human adaptation and modern development. Chapters in this volume chart some of those wider social and extractive histories.

A major contribution of this volume is its exploration of the violence that surrounds, and is caused by, natural resource extraction. The volume emphasizes how the spectrum of violence—from dramatic/direct to slow/hidden—permeates contemporary collective life. In addition to increased rates of brutal suppression of local environmental and labor activists in rural and urban sites of extraction, we observe and consider a multiplicity of related violence/s in areas we might not expect, such as infrastructural developments, protected areas for nature conservation, and even geoengineering in the name of carbon mitigation. Acknowledgement is made not only of the visible violence involved in the securitization of extractive enclaves, but also of the symbolic and structural violence that the governance, eco-nomics, and governmentality of extraction have produced. Extractive violence is shown not only to be a spectacular event, but an extended dynamic that can be silent, invisible and gradual, a process of long dyings (Nixon, 2011). As a result, the volume goes further than earlier reductionist analyses that emphasizes a *resource curse*. We also recognize that much of the new violence of extraction has become cloaked in the discourse of "green development," "green building," and efforts to mitigate the planetary environmental crisis through totalizing technologies. As well as recog-nizing the contextual specificities of the violent political economy and ecologies of resource extraction, we suggest, in line with Mbembe (2019), that the predation of natural resources forms part of a *necro-political complex*, as we explain further below.

Perhaps most importantly, this book is more than a new reading of the visible physical traces and impacts of resource extraction. The authors argue that extractive violence is not an accident or side effect, but rather a core logic of the 21st-century global experience. Contributors make the case that our world can no longer be defined merely as late-capitalist, postmodern, neoliberal, or authoritarian. Rather, *hyper-extractivism* is a defining dynamic and mentality of our era. An extractivist logic justifies the violence of removal and exploitation that are hallmarks of our hyper-extractive age. Ironically, green technologies and other contemporary efforts to tackle environmental ills often themselves depend on the continuance of social exploitation and the contaminating practices of non-renewable extraction.

We also emphasize that extractivism is the result of a particular ontological assemblage. Throughout human history, ideas of civilization, empire, sovereignty, accumulation, *terra nullius*, capital, and modernity have become layered and intertwined to form a rationale for intensifications of both social and planetary exploitation. In our moment, this assemblage has created particular expressions, contestations, and logic. However, these are met with contestation and resistance. In this volume, contributors see the power of ontologies as expressions of power, but they also acknowledge countervailing acts and ideas of extrActivism (Willow, 2018). Multiple communities across the world express political ontologies that not only confront the displacement and destruction caused by resource extraction, but also militantly and legally resist, interact with, and contest *extractivist onto-logic* claims as to the necessity of earth and life removal.

These points fall within three key cross-cutting themes: 1) the universalization of sacrifice zones; 2) extractive necro-politics; and 3) political geo-ontologies. We now move to a deeper examination of these themes before concluding this introductory chapter with an overview of the volume's structure and a short account of the foundational story of the book itself.

When Sacrifice Zones Become Universal

Nicholson comments in this volume that "An extractivist mindset or pervading set of understandings opens the whole world to human exploitation, justifying taking with too little regard for the environmental and social consequences." In this chapter and in the other contributions to the book, authors powerfully highlight this insight by exploring the historical and geographical reach and forms of extractive activities.

This volume explores multiple forms and locations of resource extraction. In addition to the usual suspects of oil and mining (see chapters by Watts; Richardson and McNeish; Le Billon and Middeldorp), chapters highlight the extractive activities and logics of the building trade (Kiechel); industrial agriculture (Graddy-Lovelace); food, tourism, and talent industries (Li and Shapiro); information and data industries (Chagnon *et al.*) and geo-engineering (Nicholson). In a chapter theorizing the meaning of extraction and extractivisms, Durante *et al.* provide an important sketch of the historical relationship between these terms and an expanding scholarship in political economy and political ecology critical of resource extraction across various fields of activity. Although the violence and environmental impacts of mining remain of central concern, Durante *et al.* highlight that literatures have also developed emphasizing the extractive nature of activities in the agriculture and forestry sectors.

In extending earlier understandings of the scale and character of extraction, this volume adopts a theoretical and empirical position different from that of some earlier authors. Gudynas (2018) has, for example, influentially maintained that an expansion of the concept of extractivism beyond the realm of natural resources—to finance, or all forms of development, for instance—is detrimental to the analytical and descriptive power of the concept, and thus undermines the search for alternatives. Chagnon *et al.* emphasize in this volume, however, that the concept of "extractivism in fact rests upon a universalizing 'natural law' in which the exploitation of 'nature' features as an ontological prerequisite to the forms that European modernity developed over the last 500 years." Moreover, they acknowledge in line with other authors (Mezzadra and Neilsen, 2017) that "new forms of financial and digital processes facilitate the expansion of resource extraction in the global economic system." The digitization of finance and data renders these sectors of the global economy dependent on one another in increasingly complex ways. As Chagnon *et al.* suggest, an emphasis on the "existence and prominence of less visible and tangible thrusts" aligns with what Dunlap and Jakobsen (2020, p. 6) have termed "total extractivism" and its "deployment of violent technologies aiming at integrating an reconfiguring the earth." For Dunlap (2019), this now involves the imposition of industrial-scale wind and other renewable energy projects in the name of clean power and green capitalism.

The global extent and geo-political significance of extraction and extractivism are also made evident in this book. Chapters detail specific contexts of extraction in Latin America, Africa, and Asia and emphasize their positioning within historic and contemporary economic development and geo-politics. Emphasis is placed on the role of resource extraction as a key dynamic of previous capitalist *primitive accumulation*, and present *accumulation by dispossession* (Graddy-Lovelace). Resource extraction is observed here as playing a central role in nationalism and contemporary racialized capitalism (e.g. Richardson and McNeish on Colombia; Watts on Nigeria) and in regional and transnational ambitions as well as new imperial ambitions, importantly including the vast Belt and Road programme of the Chinese state (Li and Shapiro). Our authors recognize that technological advances, both in terms of information technologies and extractive technologies, have played a significant role in increasing the speed, volume, and scale of extraction.

Although geologies and resource geographies are constant, the producers and politics of extractive resources have changed dramatically (Magrin and Perrier Bruslé, 2011). Owing to the continued growth of the global economy and growing number of emerging economies, the demand for natural resources continues to increase. Taking into account all the materials that are extracted, resource extraction has more than doubled in the last 25 years. Estimates further demonstrate that material extraction went into high gear in 2000, owing to expanding demands from emerging powers. Indeed, many of the new powers added significantly to foreign investment and direct involvement in extractive projects as well as related mega-development and infrastructure projects beyond their borders. China's Belt and Road Initiative is the most significant of these efforts. Responding to increasing demands for commodities, international commodity market prices spiked for almost a decade before falling again in 2012.

As Chagnon *et al.* observe in this book, whereas in previous eras resource extraction might have remained unseen, playing out at the frontiers in marginalized spaces, technological advances allow a window into the extractivist activities taking place. Given the widespread access to information technologies, social media, and the active campaigning of environmental and human rights organizations, consumers can no longer avoid awareness of the social and environmental consequences of the manufacturing and extractive practices involved in the production of the goods they purchase. As Watts indicates, technological advances also allow new zones of exploitation to be opened, new resource forms to be exploited, value chains to be expanded, and new players—both legal and illegal—to enter the resource-extraction business. As Glaab and Stuvøy observe in their chapter, knowledge has achieved the status of a raw material that can be traded. This development is evident, for example, in the notion of "biopiracy," whereby the knowledge of local communities is commercially exploited, or biological resources are patented without adequate compensation. Novel forms of extractivism are also seen, in the work of Li and Shapiro, to extend to cultural appropriation and to the commandeering of human talent.

"Sacrifice zones," which can be understood as geographic areas impaired by environmental damage or economic disinvestment (Lerner, 2010), are no longer hidden from the view of those who benefit from the consumption of goods produced from extracted raw materials. As Wapner importantly makes clear, environmental harm primarily affects those living on the frontlines of extraction, but these frontlines are changing. As infrastructural technologies reconstitute the frontiers of extractive enclaves, production sites have shifted into areas previously out of reach of industry (e.g. drilling and mining in isolated and challenging environments such as Amazon jungles, the high Andes, Sub-Saharan Africa, and Greenland and other Arctic zones). They have moved ever closer to major populations (e.g. sites of petrochemical production or the construction of wind, geo-thermal, and solar parks on the doorsteps of cities around the world). And they have expanded into areas previously delimited for national parks and nature conservation.

As Watts masterfully demonstrates in this volume, a mapping of the visible and invisible networks leading from extractive frontiers—exemplified in his chapter by the Arctic, Nigeria, and Mexico—across the face of the planet defies any previous restricted spatial or scalar notion. Watts emphasizes the speed, intensity, and energy of contemporary extractive systems, but also the friction, disorder, and layered sovereignties of the flows of minerals, materials, and capital. Indeed, with the value chain of extraction moving ever further out (we even see projects under way to mine the far side of the moon and asteroids in outer space), and ever closer (to protected species and middle-class urban communities), it now appears that nothing is to be spared. The sacrifice zone has not only become planetary but universal, with the possible exception of the lived spaces of the extremely wealthy. Our age is not only extractive, but hyper-extractive.

Wapner, in this volume, recognizes the disappearing wilderness and warns us of the divisions that have been produced by narratives of global collapse and threshold politics. These forms of thinking, he argues, enable the privileged to displace environmental harms onto others and to sidestep direct experience of the dangers they predict. This allows the privileged to postpone their environmental reckoning and thus divert attention from global environmental decline. Wapner (2020) also suggests in a recent book that this politics of postponement needs to be disrupted through rewilding—a process through which we must relinquish the fantasy of mastery over the natural world.

Extractive Necro-Politics

The link between extraction and violence has been a major theme of political economy and political ecology for some time. Key works in political economy have contributed an important analysis of the relationships between resource-based conflict and geo-politics, the structure of the international economy, and national-level political and economic competition over rents and territory (Homer-Dixon, 2001; Harvey, 2009; Mitchell, 2011; Ross, 2012). The political economy of resource conflict also explores the linkages between the opening of new extractive frontiers and civil war (Auty, 1993; Humphreys et al., 2007). Despite the frequent expectations of wealth produced by new discoveries, the resource curse literature argues that the exploitation of "point" resources commonly generates low levels of economic growth and a series of adverse effects on governance, including authoritarianism, militarization, regional secessionism, chronically unstable governments, and high levels of conflict (Lynn-Karl, 1997; Collier, 2000).

Despite widespread acceptance of the resource curse concept both in academic and policymaking circles, significant critique is now made of its common rational-actor and behaviorist assumptions, particularly the view that extraction necessarily represents an opportunity for elites and the weaponized to maximize their advantages and wealth. This analysis of extractive violence is now countered by a growing body of studies (McNeish and Logan, 2012). Instead, these studies highlight the roles of history and social structure in guiding or catalyzing the direction of political actions (Rosser, 2006; Stevens and Dietche, 2008; Omeje, 2008). While some writers highlight a heightened level of conflict resulting from either naturally occurring or politically induced resource scarcity (Kahl, 2006), a contrasting body of literature suggests that such conditions can also lead toward increased cooperation (Wolf et al., 2003). There is also a growing understanding that the study of extractive violence has been positioned too often at the national and international levels. This has fed reductionist assumptions regarding the inevitability of civil war or the paradox of plenty in resource-rich societies (Lynn-Karl, 1997; Collier, 2010). Moreover, it has left understudied the relationship between sub-national political dynamics and extraction.

In contrast to the costs of extraction emphasized by political economy, political ecology has emphasized the close relationship between environmental costs and impacts on local life and culture, from corruption to changing patterns of consumption (Peluso and Watts, 2001; Le Billon, 2001; Le Billon and Bridge, 2012). While a "political ecology of the subsoil" is still in its early days as compared to political economy (Bebbington and Bury, 2013), a growing literature recognizes the need to study the emergence of extractive conflicts over the distribution of costs and benefits associated with the subsoil. For example, water can become unavailable or contaminated, while few of the supposed financial benefits of its use in extractive processes are accrued in the community (Kirsch, 2014; Perreault, 2017).

Another important strand to the political-ecological study of extractive conflicts considers the way that natural resources are differentially valued and understood in different contexts (McNeish *et al.*, 2015). There is a growing awareness that environmental goods such as water, land, and soil cannot be understood in merely physical terms. For example, the concept of "waterscape" recognizes water as a "socio-natural entity" (Loftus, 2009) rather than as something to be theorized purely in material terms. Boyer's (2017) recent proposals for *energo-power* as a means to express an alternative genealogy of modern power further contribute to a bio-political take on extractive violence, arguing that our very bodies are intertwined with the pipes and ductwork of energy installations and the logic and expression of extractive politics. The environmental impacts of extractive activities are also increasingly recognized in political ecology as leading to the formation of environmental movements in defense of rural livelihoods and resources (Li, 2015; Bury and Bebbington, 2015), and visible expressions of an "environmentalism of the poor" (Guha and Martinez-Alier, 1997).

This rich bibliography, only partly surveyed here, provides an important foundation for this volume. However, there is much room to build on this to understand the expressions of violence in a hyper-extractive age. Volume contributors draw on the analytical benefits of both political economy and political ecology by employing multiple scales of analysis, economic and cultural understandings of how value is differentially created, and the possibility of combining structure and agency in decoding power relations. Here we also suggest the need to understand violence not only as physical action but also as a force with a plurality of expressions and consequences.

Kiechel, for example, in her chapter on the built environment, comments, "In a seeming paradox, construction involves destruction—not only of raw material stocks, but often of local economies excluded from benefit, of the socio-spatial fabric of neighborhoods, and of construction workers themselves." The built environment, she suggests, whether beneficial or oppressive in its social and ecological effects, owes its very existence to extractive actions. In highlighting these connections Kiechel provides a powerful new twist to previous studies that highlight the links between technology and asymmetries in global exchange and that uncover the relationships between ecology and power. In an earlier comment on *imperial thermo-dynamics*, for example, Hornborg (2001) argued that we should reconceptualize "the machine"—or industrial technomass—as a species of power and a problem of culture. As technological devices multiply exponentially in a vain attempt to make life "efficient," "luxurious," and "productive," Hornborg argues that on planet Earth everything is a zero-sum game, and that one person's gain is always another's loss. Kiechel's extension of technomass from machines to the built environment not only updates the civilizationary nature of this zero-sum game, but its violent reconfiguring of humans and nature, animate and inanimate life, or what Hornborg terms biomass.

Our volume explores the multiple dimensions, scales, and local contexts of this global reconfiguration. Evidence and discussion of the violence against nature caused by resource extraction is visible in all contributions to the book. We concur with other studies that highlight that oil and petrochemical production has led to dumping millions of barrels of a cocktail of chemicals, drilling fluids, and formation water into our seas, rivers, and forests. Mining practices, including those for rare earths, have caused river courses to change and heavy metals and chemicals used in processing to leach into drinking water and aquifers. Persistent toxic leaks, periodic catastrophic spills, large-scale mining, and oil-exploitation projects play important roles in opening areas of sensitive biodiversity and human population to industrial development. Extractive activities involve the construction of supporting infrastructures such as roads, pipelines, hydro-electric dams, pylons and cable networks, ports and storage facilities (see Watts for a detailed characterization of these infrastructures in the oil assemblage). Many of these projects have their own direct costs for the environment and encourage problematic destabilization and displacement of human populations.

The visible violence against nature is also being waged on human populations and individuals who dare to stand in the way of extractive development. In the chapters by Le Billon and Middeldorp, Graddy-Lovelace, Watts, and Richardson and McNeish, we see the extent of the physical violence waged against land defenders by the actors involved in resource extraction. In doing so, we mirror in our academic work the reports produced by journalists and human rights organizations that highlight a rising trend of violence against land defenders across the world. For the Past three years the *The Guardian* newspaper and the international human rights non-governmental organization Global Witness have worked together to form and update a global database that attempts to record and map what they call "a murder epidemic." In its 2019 report, *Enemies of the State?*, Global Witness highlighted that on average more than three "land defenders" (civil society leaders, human rights activists, indigenous and peasant leaders) were killed every week in 2018. These attacks occurred in relation to extractive industries such as mining, logging, and agri-business. The report also reveals how countless more people were threatened, arrested, or thrown in jail for opposing the governments or companies seeking to profit from their land. Our work extends this picture by pointing to the relationships that exist between legal and illegal actors, companies, and politicians involved in the business of resource extraction. It also discusses the manner in which these relationships background the manipulation and circumvention of laws, regulations, and consultations meant to govern the sector, as seen in particular in the chapters by Graddy-Lovelace and Le Billon and Middeldorp.

In their chapter, Glaab and Stuvøy study the nexus between extractivism and violence. Their work argues for a need to push beyond obvious violent impacts and their common normative explanations. In common with many of the other chapters, their work emphasizes the structural nature of extractive violence at different scales e.g. its character as horizontal and vertical (Watts); its grounding in histories of structural racism and criminalization (Graddy-Lovelace); its positioning within capitalist and now multipolar geo-politics (Li and Shapiro). However, their theoretical analysis goes even further. Arguing that "violence is ambiguous and transcends the local/global divide," Glaab and Stuvøy suggest that researching violence in extractivism brings into discussion the multi-scalar and hetero-temporal character of violence. Instead of thinking of scales as separate, they stress the need for a focus on entanglements (Tsing, 2005). The global is in this perspective not a separate scale, but part of scale-making processes. Such a perspective moves a focus away from the directly observable to what is commonly unseen, or what they more precisely define as "site effects"—violence as built into structure, symbolism, and space; and the temporalities of violence on humans and the environment as both fast and slow, both immediate and incremental.

Significantly, Glaab and Stuvøy recognize that the degree, character, and manifestation of extractivism have changed in the neoliberal age. This signposts the need for a more comprehensive, or suitably entangled, meta-narrative of our extractive age and the place within it of resource extraction and related violence. Watts's chapter makes a particularly significant contribution to such a meta-narrative. His chapter here on extractive value formation in the global oil assemblage not only lays bare the complex licit constellation of science, technology, rent-seeking, and financial speculation, but also its hidden deep reliance on the illicit and criminal.

Building on our colleagues' observations and analysis we also suggest here that Mbembe's (2019) deep, and necessarily complex, characterization of necro-politics could represent a timely meta-narrative in this regard. Mbembe's concept of necro-politics is a "decolonial" meditation on the current epoch—one defined by a sense of global realignment, an inward turn, and the reorganization of space and being between the living and the dead. Acknowledging the renewed rise of the extreme political right and the camps imprisoning migrants, Mbembe posits that we live in an epoch defined by a politics of enmity and separation. If globalization and neoliberalism were said to shrink the world, Mbembe now sees a backlash of retrenchment and borderization. Democracy, he suggests, has begun to embrace its dark side, or nocturnal body, based on the historical fuel of the desires, fears, affects, relations, and violence that drove colonialism. Necro-politics entails the "subjugation of life to the power of death" (Mbembe, 2003, pp. 39–40). To this end, a necro-economy encourages as one of its central features the predation of natural resources in which violence on humans and nature are justified, and populations are displaced and eliminated in the causes of consumer- or energy-protection and security.

Political Geo-Ontologies

Durante *et al.* contribute to this volume an exploration of the etymological evolution of the concept of extractivism and the developmental trajectory that has served to inform the underlying and overarching extractivist logic. In line with the intentions of many of the other contributions, they suggest that "this could be described as an extractivist mindset, or ontology, which has its particular expressions, practices, and understandings in different extractive sectors, which have become global (global extractivisms)." The chapters by Li and Shapiro, and by Nicholson, are particularly important in revealing the global scale and expression of an extractivist onto-logic, even in connection to supposedly green fixes.

With the example of the Chinese Belt and Road Programme, Li and Shapiro highlight how the extractive mindset can inspire ambitions with massive proportions. Moreover, they importantly characterize how the pattern of social and ecological destruction of the classic extractive economy—the transfer of key resources to benefit the people and economy of the destination to the detriment of the people and environment of the origin—not only spills over into a wider range of economic areas, but also takes on a seemingly more benign form that appears not to be extractive at first glance. Their work reveals less well-known aspects of extractivism, including securing food supply chains, commodifying cultural heritage for mass tourism, and appropriating human talent and intellectual property. From a different geo-political positioning, but still with clearly global and extractive ontologic intent, Nicholson describes the politics and deployment of U.S. and European geo-engineering strategies. Although carbon-removal technologies are ostensibly aimed at cleansing the air and helping to stabilize the climate by removing the dangerous byproducts of our production and use of fossil fuels, Nicholson argues that carbon-removal schemes operate according to an extractivist logic. This is indicated by the requirements and outcomes of the geoengineering techofix, including massive land-use changes and multiple interventions in the geology and atmosphere of the planet. These are outgrowths from the fossil fuel industries of a vast new industrial infrastructure requiring further vast expenditures of mineral and energy wealth.

Our volume recognizes that while an extractive ontologic is hegemonic in current geo-political and economic decision-making, it has met considerable resistance and contestation. The chapters by Richardson and McNeish, Graddy-Lovelace, and Wapner narrate histories of extrACTIVISM i.e. activism opposing the impacts of resource extraction (Willow, 2018). Importantly, our work demonstrates that extrACTIVISM does not necessarily imply acts of protest and political militancy. Richardson and McNeish highlight in Colombia the recent use of the courts, and an alliance between local indigenous and afro-descendant communities and specialists in the national legal system, to combat illegal-mining and secure the environmental protection of the Atrato River. Graddy-Lovelace gives an account of the efforts by agrarian movements around the world to proactively harness legal and political channels to seek justice for those who have been killed and to defeat efforts aimed at the criminalization of land and environmental defenders. She highlights the successful extrACTIVISM leading to the 2020 signing of the Escazú Agreement (Regional Agreement on Access to Information, Public Participation and Justice in Environmental Matters) by 22 Latin American and Caribbean countries. Wapner makes a further connection between the actions of the USA and international environmental justice movements and extrACTIVISM.

Importantly, these acts of extraACTIVISM are driven by ontologies that challenge an extractive onto-logic. Indeed, it is evident that multiple local communities and international networks envision and embody a different relationship between people and the planet. Although with varied histories, characteristics, and nuances of expression, there is a common mobilization of beliefs and sensibilities about our common human connection with the environment and nature. While our chapters only touch on this theme, we can indirectly contribute to the theorization and empirical study by a growing number of philosophers and social scientists who emphasize the "more than human" ontological turn. In this perspective, the environment is intrinsically entangled and co-evolving with society. Agency is also no longer seen as the sole privilege of human consciousness.

Latour (2014; 2017), who is one of the leading leaders of this intellectual turn, highlights that global warming and climate change threaten our existence and force us to acknowledge that the earth is agential in its own right. He argues that if we bear in mind the current ecological crisis, we must devise a new theory of agency for recognizing the active role of nonhumans, both organic and inorganic. As "more than human" thinking has gained increasing currency, new efforts have also been made to test its value in practice. A series of writers now emphasize the manner in which ontologies have been made political, or the basis of a cosmo-politics (e.g. de la Cadena, 2010; Escobar, 2015; Blaser, 2016). Wapner's call for rewilding and its connection to environmental justice is an expression of this. Importantly, Povinelli (2016) suggests the terms geontology and geontopower as a means of capturing the intensifying visibility of the interaction between components of nonlife (geos) and being (ontology) in late liberal governance and economics. A clear example of this is the trend towards ecocentric or "earth law" made visible in the legal case of the Atrato River described by Richardson and McNeish in this volume.

Volume Structure

Part 1, *Theorizing Violence in An Extractive Age,* provides a theoretical grounding in the themes of extraction and violence, and of the connections between them. In *Extraction and Extractivisms: Definitions and Concepts,* Durante, Kröger, and LaFleur provide an intellectual history of the scholarship of extractivism; in *Politics of Violence in Extractivism: Space, Time and Normativity,* Stuvøy and Glaab refine understandings of "violence," and in *Thresholds of Injustice: Challenging the Politics of Environmental Postponement,* Wapner argues that the hyper-extractivism of the current age has revealed and disrupted the displacement of environmental harm onto the vulnerable and created a wave of resistance.

Part 2, *Exacerbated Violence at the Local Level*, explores the intensification of violence in classic realms of extraction. In *Empowerment or Imposition? Prior Consultation, Indigenous Peoples and Extractive Violence*, Le Billon and Middledorp examine the practice of prior consultation in advancing extractive projects in contrast to its envisioned ideal of enforcing compliance with Indigenous and environmental human rights. In *Criminalization of Agrarian Movements and the Escazu Agreement: Leveraging Law and Violence against Land Defenders in Latin America and the World*, Graddy-Lovelace contextualizes recent violence against agrarian movements within the divergences and convergences of historical peasant movements, neoliberalized environmentalist movements, and contemporary resurgences of agrarian and legal mobilization. In *Building Boom: Deconstructing Violence and Other Social Consequences of Extraction in the Built Environment*, Kiechel documents the ways in which the act of construction depends on violence and extraction, and reveals the embedded social and social justice impacts of multiple aspects of a building project's life-cycle.

Part 3, *New Ways of Thinking about Extractivism* reveals new contexts of extraction, expressions of extractivism, and extrACTIVIST innovations. In *Rethinking Extractivism on China's Belt and Road: Food, Tourism and Talent*, Li and Shapiro highlight the role that a quickly globalizing China is playing in catalyzing and intensifying novel forms of extraction and violence. In *Granting Rights to Rivers in Colombia: Significance for extrACTIVISM and Governance*, Richardson and McNeish consider the significance of legal cases that recognize the personhood rights of rivers as a means to control illegal mining and as an innovative form of extrACTIVISM. In *Extraction at Your Fingertips*, Chagnon, Hagolani-Albov, and Hokkanen provide an analysis of a complex web of extractivisms where digital and data extractivism intersect with natural resource extractivisms in their underlying logic and processes. Finally, in *Carbon Removal and the Dangers of Extractivism*, Nicholson shows how the concept of *hyper-extractivism* can help us understand and guard against problematic potentials in large-scale carbon removal activities at the level of the planetary atmosphere.

Part 4, the final section, is devoted to a major chapter by Michael J. Watts, *Hyper-Extractivism and the Global Oil Assemblage: Visible and Invisible Networks in Frontier Spaces*. Focusing primarily on the oil industry in the Arctic, Nigeria, and Mexico, Watts details the planetary nature of extractive capitalism and reveals its reliance on intertwined legal and illegal logics and actions. The chapter epitomizes the multi-scalar and complex analysis that all of the contributors argue is needed to understand the violence embedded in the hyper-extractivism of our age, and thus it deserves space of its own at the volume's finale.

The Story of the Book

This volume is the product of a multi-year research collaboration between the Department of International Environment and Development Studies, Norwegian University of Life Sciences and the Global Environmental Politics program at American University's School of International Service, with the support of the Norwegian Agency for International Cooperation and Quality Enhancement in

Higher Education (DIKU).. Scholars from the two institutions who work on environment and development identified the intensification of extraction and concomitant violence as a key element of our age. They identified their multi-disciplinary training as a core strength for a common research agenda, and invited scholars from other institutions to join them, particularly including scholars with extensive prior work on extraction and extractivism. These include Le Billon, Middeldorp,and Watts, as well as scholars belonging to the Global Extractivisms and Alternatives Project (EXALT) at the University of Helsinki (i.e. Chagnon, Hagolini-Albov, Hokkanen, Durant, LaFleur and Kröger).

The scholars represented in the volume come from many different disciplinary backgrounds: they are linked to geography, political ecology, global environmental politics, development and resource economics, international relations, architecture, environmental law, regional studies, and philosophy. They include senior scholars, junior scholars, graduate students, and project associates. They come originally from the USA, UK, Norway, Finland, Germany, Holland, France, Italy, Canada, China, and New Zealand. Examples and cases span Latin America, Africa, Europe, the USA, China, the Middle East, and the Arctic. We hope that the resulting common research project will deepen and extend understanding of multi-scalar extractive processes; we hope that it will offer fresh insights into the dynamics of both the overt and hidden violence of such extraction; and we hope that it will point a way forward in addressing the new forms of violence that characterize our hyper-extractive age.

The editors and contributors wish to express gratitude to the Norwegian Agency for International Cooperation and Quality Enhancement in Higher Education (DIKU) for their support of this collaborative project and to Professor Katharina Glaab for spearheading the administration of the grant. We also thank Jacqueline Kessler, a graduate student at American University's Global Environmental Policy program, whose editorial support helped make this project a delight.

References

Auty, R. (1993) 'Natural Resources and Civil Strife: A Two-Stage Process', *Geopolitics*, 9 (1), pp. 21–49.

Bebbington, A. and Bury, J. (2013) *Subterranean Struggles: New Dynamics of Mining, Oil, and Gas in Latin America*. Austin, TX: Texas University Press.

Blaser, M. (2016) 'Is another cosmopolitics possible?', *Cultural Anthropology*, 31 (4), pp. 545–570.

Collier, P. (2010) *Plundered Planet: Why We Must and How We Can Manage Nature for Global Prosperity*. Oxford: Oxford University Press.

Crutzen, P. and Stoermer, E. (2000) 'The Anthropocene', *IGBP Newsletter*, 41, pp. 17–18.

De la Cadena, M. (2010) 'Indigenous Cosmo-politics in the Andes: Conceptual Reflections Beyond "Politics"', *Cultural Anthropology*, 25 (2), pp. 334–370.

Dirzo, R., Young, H., Galetti, M., Ceballos, G., Isaac, N., and Collen, B. (2014) 'Defaunation in the Anthropocene', *Science*, 345, pp. 401–406.

Dunlap, A. (2020) *Renewing Destruction: Wind Energy Development, Conflict and Resistance in a Latin American Context*. Lanham, MD: Rowman & Littlefield.

Dunlap, A. and Jakobsen, J. (2020) *The Violent Technologies of Extraction: Political ecology, critical agrarian studies and the capitalist worldeater*. London: Palgrave Pivot.

Escobar, A. (2015) 'Thinking-feeling with the Earth: Territorial Struggles and the Ontological Dimension of the Epistemologies of the South', *Revista de Antropología Iberoamericana*, 11 (1), pp. 11–32.

Gudynas, E. (2018) Extractivisms: tendencies and consequences. In Munck, R and Wise, D (eds). *Reframing Latin American Development*. Abingdon: Routledge.

Guha, R. and Martínez-Alier, J. (1997) *Varieties of Environmentalism: Essays North and South*. London: Earthscan.

Harvey, D. (2004) 'The New Imperialism: Accumulation by Dispossession', *Socialist Register*, 40.

Homer-Dixon, T. (2001) *Environment, Scarcity and Violence*. Princeton, NJ: Princeton University Press.

Hornborg, A. (2001) *The Power of the Machine: Global Inequalities of Economy, Technology and the Environment*. Lanham, MD: AltaMira Press.

Humphreys, M., Sachs, J., and Stiglitz, J. (2007) *Escaping the Resource Curse*. Columbia University Press.

Kahl, C. (2006) *States, Scarcity, and Civil Strife in the Developing World*. Princeton, NJ: Princeton University Press.

Kaldor, M. (1999) *New and Old Wars: Organised Violence in a Global Era*. Cambridge: Polity Press.

Kirsch, S. (2014) *Mining Capitalism. The Relationship Between Corporations and their Critics*. Durham, NC: Duke University Press.

Latour, B. (2014) 'Anthropology at the Time of the Anthropocene: A personal view of what is to be studied'. Distinguished lecture American Association of Anthropologists.

Latour, B. (2017) *Facing Gaia: Eight Lectures on the New Climatic Regime*. Cambridge: Polity Press.

Le Billon, P. (2001) 'The political ecology of war: natural resources and armed conflicts', *Political Geography*, 20 (5).

Le Billon, P. and Bridge G. (2012) *Oil*. Hoboken, NJ: John Wiley & Sons.

Lerner, S. (2010) *Sacrifice Zones: The Front Lines of Toxic Chemical Exposure in the United States*. Cambridge, MA: MIT Press.

Li, F. (2015) *Unearthing Conflict: Corporate Mining, Activism and Expertise in Peru*. Durham, NC: Duke University Press.

Loftus, A. (2009) 'Rethinking Political Ecologies of Water', *Third World Quarterly*, 30 (5), pp. 953–968.

Lynn-Karl, T. (1997) *The Paradox of Plenty: Oil Booms and Petro-States*. Los Angeles, CA and London: California University Press.

Mace, G., Reyer, B., Alkemadec, R., Biggse, R., ChapinIII, S., and Cornelle, S. (2014) 'Approaches to Defining a Planetary Boundary for Biodiversity', *Global Environmental Change*, 28, pp. 289–297.

Magrin, G. and Perrier Bruslé, L. (2011) 'Nouvelles géographies des activités extractives', *EchoGéo*, 17, pp. 2–11.

Malm, A. (2016) *Fossil Capital: The Rise of Steam Power and the Roots of Global Warming*. London: Verso Publications.

Mbembe, A. (2003) 'Necropolitics', *Public Culture*, 15 (1), pp. 11–40.

Mbembe, A. (2019) *Necropolitics*. Durham, NC: Duke University Press.

McNeish, J. and Logan, O. (2012) *Flammable Societies: Studies on the Socio-Economics of Oil and Gas*. London: Pluto Press.

McNeish, J., Borchgrevink, A., and Logan, O. (2015) *Contested Powers*. London and New York, NY: Zed Books.

Mezzadra, S. and Nielsen, B. (2017) 'On the multiple frontiers of extraction: Excavating contemporary capitalism', *Culture Studies*, 31(2–3), pp. 185–204.

Mitchell, T. (2011) *Carbon Democracy: Political Power in the Age of Oil*. London: Verso Publications.

Moore, J. (ed.) (2016) *Anthropocene or Capitalocene? Nature, History, and the Crisis of Capitalism*. Oakland, CA: PM Press.

Nixon, R. (2011) *Slow Violence and the Environmentalism of the Poor*. Cambridge, MA: Harvard University Press.

Omeje, K. (2008) *Extractive Economies and Conflict in the Global South: Multi-Regional Perspectives on Rentier Politics*. Aldershot: Ashgate.

Peluso, N. and Watts, M. (2001) *Violent Environments*. Ithaca, NJ: Cornell University Press.

Perrault, T. (2017) 'Governing from the ground up? Translocal networks and the politics of environmental suffering in Bolivia'. In Horowitz, L. and Watts, M. (eds.). *Grassroots Environmental Governance: Community Engagements with Industry*. London: Routledge, pp. 103–125.

Povinelli, E. (2016) *Geontologies: A Requiem to Late Liberalism*. Duke University Press.

Rockström, J., Steffen, W., Noone, K., Persson, Å., Chapin II, S., Lambin, E., and Lenton, T. (2009) 'Planetary Boundaries', *Ecology and Society*, 14 (2), pp. 299–310.

Ross, M. (2012) *The Oil Curse: How Petroleum shapes the development of nations*. Princeton, NJ: Princeton University Press.

Rosser, A. (2006) 'The Political Economy of the Resource Curse: A Literature Survey', I Working Papers, 268.

Steffen, W., Broadgate, W., Deutsch, L., Gaffney, O., and Ludwig, C. (2015). 'The Trajectory of the Anthropocene', *Anthropocene Review*, 2 (1), pp. 81–98.

Stevens, P. and Dietche, E. (2008), 'Resource Curse: An Analysis of Causes, Experiences and Possible Ways Forward', *Energy Policy*, 36 (8), pp. 56–65.

Tsing, A. (2005) *Friction: An Ethnography of Global Connections*. Princeton, NJ: Princeton University Press.

Wapner, P. (2020) *Is Wilderness Over?* Cambridge: Polity Press.

Willow, A. (2018) *Understanding ExtrACTIVISM: Culture and Power in Natural Resource Disputes*. Abingdon: Routledge.

Wolf, A., Stahl, K., and Macomber, M. (2003) 'Conflict and Cooperation Within International River Basins: The Importance of Institutional Capacity', *Water Resources Update*, 125 (June): 31–40.

PART 1

Theorizing Violence in an Extractive Age

1

EXTRACTION AND EXTRACTIVISMS

Definitions and Concepts

Francesco Durante, Markus Kröger, and William LaFleur

The first section is translated from the Portuguese, from an interview with Aldira Munduruku conducted by Markus Kröger in November 2019 on the shores of Tapajós river in a village inhabited by the Munduruku people, south of Itaituba in the Amazon Basin, Pará, Brazil:

MARKUS: From the time you got here, to this day, has life changed? For example, are there more fish, or fewer fish, more trouble, or less trouble, what was it like before and what is it like now?

ALDIRA: Every year things change, you know? So the climate is also changing, the sun comes very hot, and the river is also not filling correctly, from time to time it fills, then dries, fills, dries…so, every year it's been changing. Also, the lack of fish. At times there is not much fish. Game too, there is no more game behind the village. Because, you know, there are many access roads [illegally built inside the Amazon rainforest areas of the Munduruku people]. We saw them and the warriors got lost because of so many access roads that the Pariuás [non-indigenous people] are making, right, the acai palm-heart cutters, the loggers. So, my husband goes often to hunt on this side, you know, and almost every time he brings nothing. This is the way we live, and then comes the hunger. Then, fish not so many, right? Fishing, every time he goes fishing, he brings fish, even if they are small.

MARKUS: How do you see your children's future, do you think they're going to live here, or do you think they're going to have to move from here, how are they going to be when they're grown?

ALDIRA: I still have hopes that our area will be demarcated. Bolsonaro's government is always bringing bad projects to us, death projects as we call them. But we're going to face him until…until death. Until we get the demarcation. And my hope is that my children will be happy with the demarcation, right? And we're going to be feeling at peace. And I have hopes that my children will always live here.

Introduction

Extractivism characterizes the modern era. We define extractivism in this publication as *a particular way of thinking and the properties and practices organized towards the goal of maximizing benefit through extraction, which brings in its wake violence and destruction.* Extractivism plays out particularly brutally at resource frontiers, invisible to the majority of the distant users of the commodities appropriated under this frontier-logic (Moore, 2015). Yet globalization has made the effects—physical, social, and mental—of the ever-intensifying extractivisms more visible, as they increase in scale and scope to maintain the global rush into modernity (Kröger, 2015). Ignorance about the tolls of extractivism and an increase in hyper-extractive activity is no longer an excuse. Nonetheless, the violence played out against humans and other living beings, as well as against lived environments on the multiple frontiers of extractivisms need to be further scrutinized (Acosta, 2013; Taylor, 2015; Gudynas, 2015; McNeish, 2018; de la Cadena and Blaser, 2018; Svampa, 2019; Kröger and Nygren, 2020).

The modern era has seen a rise in the scope and scale of extractivist violence. A major driving factor has been the global expansion of extractive activities by traditional powers and rising economic powers, such as China (see Li and Shapiro in this volume). However, technological advances have played a significant role in transforming ontologies, practices, and spiritual, reciprocal, or sacredness-based relations with the environment and the planet (Merchant, 1983). In addition, these technological advances support an increased volume of extraction (Gudynas, 2015; Dunlap and Jakobsen, 2020) and allow a window into the extractivist activities taking place. In previous eras these activities might have remained unseen, playing out at frontiers in marginalized spaces (Peluso and Watts, 2001; Arboleda, 2020).

However, the logic of extractivism is still firmly in place. The violent logic of taking resources—without reciprocity, without stewardship—has gained traction in the past two decades, despite an increase in on-the-ground resistance and some localized regulatory attempts to hamper its operations and impacts (Jalbert *et al.*, 2017; Willow, 2018; Kröger, 2013; 2020a). In addition to the evident push for natural resource extraction, the underlying logic of extractivism is increasingly revealed to be a fundamental driving force of capitalism—as well as of other modern world-systems (Szelényi and Mihályi, 2020). In fact, the extractivist logic, operating through depletion, has been in operation for thousands of years (e.g. over-logging, deforestation, etc.), as empires have been built and capital amassed for wealthy families, enterprises, and colonizing powers (Frank and Gills, 1993; Perlin, 2005). While empires have been resisted by local communities for thousands of years, less has been written about this because history tends to be written by the winners, the established "civilizations," and states. This kind of resistance based on rooted dwelling and anti-state attitudes is still visible, as e.g. Scott (2017), de la Cadena and Blaser (2018), and Kröger (2020a) have elucidated ethnographically. As non-modernist framings stemming from these communities have proliferated (Kröger, 2013), sectors of the global economic system that have not historically been directly associated with the concept of extractivism, such as the financial and digital sectors, are now increasingly being understood as "extractive," "colonial," and a feature of contemporary capitalism(s) (Thatcher *et al.*, 2016; Gago and Mezzadra, 2017; Mezzadra and Neilson, 2017; Couldry and Meijas, 2019; Sadowski, 2019; Dunlap and Jakobsen, 2020). These new arenas of extractivism are multi-faceted and change rapidly as the technological capability and creativity for their myriad uses (and abuses) continues to evolve, as we see in chapters in this volume by Chagnon *et al.*, Li and Shapiro, and Nicholson.

In this chapter, we explore the etymological evolution of the concept of extractivism and the developmental trajectory that has served to inform the underlying and overarching extractivist logic. This could be described as the extractivist mindset, or ontology, which has its particular expressions, practices, and understandings in different extractive sectors, which have become global (global extractivisms). As a result, we aim to cast light on the ontological underpinnings that inform extractivist logic, or as we refer to it here, an extractivist "onto-logic" that underwrites the machinations of much globalized economic activity, from natural resources to the digital and data infrastructures on which the world is increasingly dependent. The final section brings the paper full circle, linking the opening vignette with a discussion on resistance, including the way that this has been approached through the concept of extrACTIVISMS as developed by Willow (2018) and, in this volume, by Wapner and by Richardson and McNeish, highlighting resistances to the lived material consequences of this hegemonic onto-logic. We conclude that extractivist logics are inextricably bound up with colonialism, capitalism, and other configurations of modernity, and that distinct modes of violence are associated with different extractivist spheres.

Definitions of Extractivism and Extraction

The term extractivism derives from the Latin American concept of *"extractivìsmo,"* which originally emerged in the 1970s to describe developments in the mining and oil export sectors (Gudynas, 2018). The word originates from the Latin verb *"extrahĕre"* which is a combination of *"ex-"* meaning "from" and *"trahĕre"* meaning "draw." Thus, *extrahĕre* quite directly means to 'draw from' (Willow, 2018). One of the most widely used definitions of extractivism in the academic literature relates extractivism solely to natural resources, "appropriation of natural resources in large volumes and/or high intensity, where half or more are exported as raw materials, without industrial processing or with limited processing" (Gudynas, 2018, p. 62).

Extractivism is often categorized as a feature, imperative, or characteristic. For example, it is described as "a mode" (Acosta, 2013, p. 62), or "a particular mode of capitalist accumulation" (Teràn Mantovani, 2016, p. 257: *"un particular modo de acumulación capitalista"*), a "structural feature of capitalism as a world economy" (Machado Aràoz, 2013, p. 131: *"un rasgo estructural del capitalismo como economía-mundo"*), and as the "imperative driving the global capitalist economy" (Dunlap and Jakobsen, 2020, p. 6). Furthermore, in its version of neo-extractivism, the concept is labelled as "a way of appropriating nature" (Svampa, 2019, p. 6) and an "economic…(and) development model" (Brand *et al.*, 2016, pp. 133, 131) or solely as a "development model" (Svampa, 2019, p. 6).

The concept of extractivism is related to and builds on a long prior tradition of political ecology and political economy critical of resource extraction, especially in Latin America, and focused particularly on excessive and highly conflictive mining expansion in the Andes region (see Bebbington and Bury, 2013). The terms extraction and extractivism stand in an ambiguous yet symbiotic relation (Kröger, 2020b). There are particular literatures for the global analysis of different extractive sectors, as well as key actors and dynamics, such as the roles of social movements, states, and corporations (Kröger, 2020c). Specific literatures on agrarian or agro-extractivism further specify the terms and offer analytical tools to use the concepts for analyzing recent transformations, especially in the Latin American countryside, through political economy and political ecology (McKay, 2017; Alonso-Fradejas, 2018). Meanwhile, forestry extractivism in the form of monoculture tree plantations is a constantly growing trend, pursued under the umbrella label of a so-called bioeconomy (Kröger, 2013; 2016), with carbon sequestration and other claims hiding the actual circumstances of rising pollution and deaths caused by such extractivist expansions (Ehrnström-Fuentes and Kröger, 2018; Kröger and Ehrnström-Fuentes, 2020).

More recently, authors from different disciplines with a wide-spanning scope of foci have tried to comprehend the essence of extractivism in order to expand its analytical use to other applications, such as in the financial sector (Gago and Mezzadra, 2017; Mezzadra and Neilson, 2017) and digital environments (Sadowski, 2019). In the case of the digital environment, data extraction consists of information being "taken without meaningful consent and fair compensation" (Sadowski, 2019, p. 7). As for extractivism, Gago and Mezzadra (2017) follow the definition of Acosta (2015), "extraction of huge volumes of natural resources, which are not at all or only very partially processed and are mainly for export according to the demand of central countries" (translated in Gago and Mezzadra, 2017, p. 576). Insights can be drawn from these theories on extraction to delve into the meaning of extractivism.

To be able to deconstruct the word "extraction," we make a quick exploration of the verb "extract" and its current use in American and British English dictionaries. The Merriam-Webster's Dictionary[1] provides the following pertinent definitions of the transitive verb "extract": 1 a: *to draw forth;* 1 b: *to pull or take out forcibly,* 1 c: *to obtain by much effort from someone unwilling.* These definitions ascribe to the verb the qualities of strength and effort, especially addressed against a non-cooperative counterpart. Therefore, one can discern here the violent nature of extraction. The former feature is described by Gago and Mezzadra regarding finance as "an accumulation of drawing rights on the wealth to be produced in future" (Gago and Mezzadra, 2017, p. 583), as well as extraction as "forced removal" (Mezzadra and Neilson, 2017, p. 188). Finally, the aspect of unwillingness is found in Sadowski concerning data being "taken without meaningful consent" (2019, p. 7).

A second set of definitions reveals an additional interesting feature: 2 a: *to withdraw (something, such as a juice or a constituent element) by physical or chemical process,* 2 b: *to treat with a solvent so as to remove a soluble substance,* 3: *to separate (a metal) from an ore.* The Oxford English Dictionary[2] reports the same meanings, adding an extra nuance in its definition 2 a: *to take from something of which the thing taken was a part.*

These last definitions describe actions that as practiced on the ground often result in irreversible transformations that radically change the target of extraction, such as the landscape and the environment and often also the socio-economic and ecological relations between and within populations and landscapes. In other words, such transformation results in major modifications, as through the loss of lives and spaces to live, which in turn affects the inhabitants of a given territory. Mezzadra and Neilson describe this element as "processes that cut through patterns of human cooperation and social activity" (2017, p. 194). Ye *et al.* (2020) provide a deeper analysis of capital accumulation, value creation, and the political economic and agrarian dimensions of extractivism, noting that typically these are actions "where value generation is necessarily temporary and generally followed by barrenness and an inability to sustainably reproduce livelihoods in the affected habitat" (p. 155).

Concerning the word extractivism, the attention to the noun suffix "-ism" sheds light on the relation between extraction and extractivism, conferring pre-eminence to the former, yet providing foundations and clarifying the latter. Looking to the Latin languages—as the concept of *extractivismo* was developed in a Spanish-speaking context—we can find more nuance in the definition compared to the English definition of "-ism." The Spanish dictionary[3] reveals that "-ismo" primarily gives the meaning of a doctrine, system, school, or movement. Then it provides attitude, tendency, and quality. In the Dictionary of Physical Sciences in Italian[4] (which sometimes preserves more original meaning in words derived from Latin), "-ismo" denotes abstract concepts such as a way of thinking or properties of a thing, or a bundle of things organized towards a certain goal (as a mechanism). Based on this definition, in contrast to "extraction," we see that the concept of "extractivism" should be used to denote, in an abstract way, *a particular way of thinking and the properties and practices organized towards the goal of maximizing benefit through extraction.* This would allow for wider comparisons through "extractivism" as a heuristic device and an underlying logic, allowing for comparability across quite similar instances of the same process between different contexts, while retaining the possibility for specific instances to be explored in detail.

The specific act of extraction, then, is different from extractivism, which can be understood as a constellation of logics or drivers, but the first is a prerequisite to talk about the second. The scoping of dictionaries above allows one to see the different nuances of extractivism, as either a feature of something (a political regime), a doctrine, a theory (e.g. economic or social), an attitude, or a disposition, such as in ontology. Extractivism also denotes being organized towards a goal, these goals varying from development to profit, and even to changes in ideology or mindsets.

To conclude this etymological excursus, it is worth noting that the concepts of extraction and extractivism, so far exposed, present also their respective homonyms yet antithetical notions. These are extrACTION (Jalbert *et al.*, 2017) and extrACTIVISM (Willow, 2018), denoting *in toto* active rejection of and resistance to the extractivist logic. As part of the new wave of environmental movements, this activism incarnates environmental justice by an *in loco* resistance and contestation against large-scale extractive activities, standing firmly against the predominant extractivist mindset (Wapner, this volume). In sum, the definitions and the etymology of the term extractivism shed light on the intrinsically violent nature of extraction, providing us with two *foci* of inquiry. The aspect of strength opens up the realm of power relations, while the unwillingness of the affected subjects introduces issues of freedom. If these two aspects are taken together, they can effectively launch extractivism into new spheres of analysis.

Universalizing Exploitation as Natural

In an attempt to conceptualize the violence of modern extractivist practices, it is necessary to trace the seeds of this onto-logic as it has operated through the centuries and forms the functional core of the modern world-system and world-ecology (see Moore, 2015). What is now referred to as extractivism did not spontaneously emerge in European colonial times, nor does it characterize only one economic system (capitalism), but indeed permeates other modern iterations—socialism and its variants included (Gudynas, 2018). Scholars have shown that modern extractivist practices began to take root, or at least gather significant momentum, around 500 years ago (Wallerstein, 1974; Mintz, 1986; Escobar, 1995; Acosta, 2013; Gudynas, 2015; Moore, 2015; Willow, 2018). Thus, extractivist practices are inextricably entangled with European colonialism, the development of the modern world system, and the Enlightenment and scientific revolution (Merchant, 1983). Scholars such as Moore (2015) and Escobar (2016) argue that the economic forces and imaginaries creating the contemporary world system started to become dominant during the *longue durée* of the 16th century and thereafter became hegemonic among states of all ideologies, permeating the epistemological culture of most governments and international politics. Extraction and anthropocentric appropriation, which are ecologically destructive and aimed at building empires, have even deeper roots, with world systems analysts pointing to a 5,000-year history of imperial capital demolishing environments (Frank and Gills, 1993).

A clear example of anthropocentric appropriation is the long and wide-ranging history of deforestation (Perlin, 2005). However, what differentiates the ancient deforestations and other extractions from the past 500 years of extractivism is the scale, and the greater domination of certain mindsets, by the advancement of modern technology alongside political and military power. These have gradually wiped out prior and co-existing, nurturing, regenerative, and sacredness-based understandings and ontologies of the earth (Merchant, 1983). Yet Merchant (2013) traces the foundation of modernity's relations of the domination of nature to Greek and Christian narratives which helped to pave the way for the modern colonial-capitalist era. Similarly, Harvey emphasizes how the idea of the "domination of nature" has strongly influenced both scientific writing and the popular imagination since the Enlightenment era (2014, p. 247). More precisely, the historian Andrew Fitzmaurice (2007), writing on the origins of the concept of *terra nullius,* provides a lucid account of this conception of the world as it was maintained through the Ancient, Christian, and Enlightenment histories of Europe—at first philosophical and later legal—and which holds the key to the ontological basis for a logic of extractivism.

The ontological basis for the logic of extractivism was first promulgated by the ancient Greeks as "natural law" before being codified by the Romans as the "law of first taker" (or *ferae bestiae,* the law of wild beasts), and eventually reified as *res nullius,* or "a thing belonging to no one" under international law in the late 19th century (Fitzmaurice, 2007). The genealogy of the natural law concept is significant on at least two accounts. First, the law asserts that property, and therefore humanity, is established through the *exploitation* of the potential of things in the physical world. Where the exploited thing in question has no (human) owner, it comes under ownership of the exploiting party (Fitzmaurice, 2007). It is the notion of exploitation in natural law—dependent on assumptions about property and the dominance of humans over "nature"—where the seed of this extractivist onto-logic begins to express itself. Second, the interpretation of the *ferae bestiae* was central to debates in Europe on the justification of colonial dispossession and domination. For example, Francisco de Vitoria, a theologian at the School of Salamanca in the 16th century, used *ferae bestiae* to argue against Spanish colonial plunder (Fitzmaurice, 2007; de Vitoria, 2010). However, the English inverted the interpretation of this law (e.g. John Locke's *Second Treatise of Government,* Chapter V), reasoning that by not having properly exploited nature—that is, through their labor—the native populations of the Americas (and later Aboriginal Australians) had not established their humanity, and therefore, did not hold just dominion over the lands on which the Europeans first encountered them (Fitzmaurice, 2007). The colonists had only to exploit the nature (and people) of their newfound lands through their labor in order to take ownership of them. Thus, exploitation was first carried out through direct physical violence so as to carve out lands for the new immigrants to "properly" exploit, followed by transformations of these landscapes that more closely resembled those they had left across the Atlantic. Extractivist violence was thus the central tool and mindset in consolidating and legitimizing Western colonial dominance.

This extractivist mindset paved the way for centuries of violence and destruction against indigenous communities and ecosystems. It is no stretch to see that this reinterpretation of natural law via *ferae bestiae* required additional forms of subjugation, including the racialization of non-Europeans, which was added to the already-operative linear notion of human development as progressing from savage to civilized. This rationale marked a turning point in the modern colonial project and set the stage for a world economy that increasingly and more intensively hinged upon an extractivist conception of world-making, as filtered down through the centuries from ancient Greece and Christianity. By the end of the 17th century, there was a gathering ethos of gaining mastery over nature through technology and the new science of mechanics, which supported a Western worldview of exploitation, subduing prior notions of a "nurturing earth mother" (Merchant, 1983, p. 116). It is notable that this process turned global only within the past 500 years, marking the era of global extractivisms. The onto-logic of exploitations-cum-extractivisms established a prerequisite for the advent of the modern world-system and appears pervasive in all forms that the system now takes, having intensified even more in the past century. We find an increasing number of scholars utilizing extraction and extractivism to understand new forms of violent regimes that have emerged only in the 20th and 21st century, ranging from neoliberal to progressive governments' macro-developmental projects (e.g. Gudynas, 2015; Svampa, 2019).

Futures for Resistance, extrACTIVISMS, and Violence

Extractivist activity and its enablers are increasingly being contested on the ground, particularly by those whose ontological understandings of the world are, or have become, antithetical to the extractivist onto-logic, and who face extractivist violence and threat of extinction by extractions. Understanding how these resistances play out and their effects on extractivist practices locally, regionally, and globally, is a key task for researchers and extrACTIVISTS (Willow, 2018) wishing to analyze, critique, or intervene. Wapner (in this volume) argues that extrACTIVISM is a new phenomenon in the line of environmental justice struggles starting in North America. Willow (2018) offers examples of how extrACTIVISM takes place in particular places, where populations—such as the indigenous people resisting oil pipelines in North America—are fundamentally resisting the increased extraction at these sites while simultaneously being opposed to the logic, global impacts, and connotations of these extractivist expansions. Kröger (2013; 2020a) analyzes how this kind of resistance to extractivism has been able to succeed in discontinuing or slowing down the expansion of extractivisms of global industrial forestry and mining in Brazil and India. He emphasizes that there are five key resistance strategies that resistance to global extractivisms should use, as well as some strategies that they should avoid, if, for instance, the goal of the resistance is not simply to allow for a modified expansion of extractions or receive compensation, but to block, discontinue, and reverse investment projects that are destructive for the planet and the localities. These key strategies support the fostering of contentious agency and the establishment of resistance actors that vary in shape (including non-governmental organizationss, networks, difference-scale actors, social movements, indigenous groups, trade unions, environmentalists, different classes, and mixes of these) and share the quality of resisting primarily via physically disruptive and innovative protest tactics such as roadblocks, blocking entrances to would-be extraction sites, bodily resistance, cutting down eucalyptus plantations, or other tactics which are not violent toward humans. They also include "non-violent" strategies such as the use of lawsuits and even intervention by activist government agencies. Key to these resistances is the retention of alternative physical, social, and symbolic territories and territorialities. They rely on tight and distinct place-based affinities to places rendered as non-commensurable and non-commodifiable— practices that challenge the onto-epistemic logic of extractivism. Besides protesting, the other necessary strategy that Kröger (2020a) identifies is the avoidance of making private deals directly with companies in private politics, such as stakeholder dialogues, which do not seem to work for the benefit of extrACTIVISTS. There are other strategies and factors that need to be considered when assessing the role of resistance to extractivisms in different sectors, polities, and contexts, in order to understand and analyze the complexities and dynamics in global extractivist politics: Kröger (2020d) offers methodological guidelines on how to analyze these factors through multi-sited political ethnography and Qualitative Comparative Analysis. More detailed, systematic, globally spanning analyses across and within different global extractivisms and their resistance are needed.

While identifying the manifestations of violence produced through extractivist activities is paramount, it is not always an easy task. For example, when the violence that originates in extractive practices plays out in ways far removed from their locus, as in the case of digital extractivisms, as Chagnon *et al.* show in their chapter in this volume. Nonetheless, the identification of how violence plays out is crucial for devising activist strategies and organizing research in ways that augment more effective resistance, anticipatory actions, or media strategies. Taken from an ontological perspective, the presence of extractivist logics is more ubiquitous than meets the eye, but for this reason allows for the possibility of more effective diagnosis and action. A future research agenda, for which this volume boasts a collection of key examples, should strive to communicate and collaborate with extrACTIVISMS as they unfold and emerge both in established and new resource frontiers.

Conclusion

This chapter has outlined how extraction and extractivism have been used as concepts and how they can be used in reference to violence. We provided an etymological analysis of the Latin roots of the key concepts and distinguished different historical epochs, timelines, and roots of global extractivisms as an ontological feature of the modern world-ecology, as nested in a longer succession of world systems. Although these dynamics began to gain traction about 500 years ago, ontologically extractivist mindsets and practices already existed—depleting, destroying, and deforesting lived environments for thousands of years before the modern world-system—as is visible in the building of empires and ancient civilizations. However, in our age, the extractivist thrust has become ever more widespread, violent, and global in scale and pace. The line between tangible and intangible realms of extraction has become blurred as financial speculation and markets that are digitized and run by algorithms have spread. There are precursors to this current state-of-affairs in bio-piracy and intellectual theft, as well as the longer necessity of mapping and codifying lands, resources, and the possibilities of exploitation and extraction in capital's commensurability project (see Moore, 2015). These expansions have direct and indirect links to extractions based on appropriative accumulation, both in human and other-than-human natures.

The different arenas of extractivist logic need to be further explored by providing broader, overarching assessments that seek to unveil the complex webs of extractivisms in this era. In this endeavor, we suggest an inclusive definition of extractivism as *a particular way of thinking and the properties and practices organized towards the goal of maximizing benefit through extraction, which brings in its wake violence and destruction.*

Notes

1 "extract." *Merriam-Webster.com*. 2020. https://www.merriam-webster.com
2 "extract, v." *Oxford English Dictionary Online*. 2020. Oxford University Press. https://www-oed-com.libproxy.helsinki.fi/view/Entry/67080?rskey=BaUkPe&result=3
3 "-ismo." *Diccionario de la lengua española*. 2020. Real Academia Español. https://dle.rae.es
4 "-ismo." *Dizionario delle Scienze Fisiche*. 1996. Treccani. http://www.treccani.it/enciclopedia/ismo_ %28Dizionario-delle-Scienze-Fisiche %29/

References

Acosta, A. (2013) 'Extractivism and neoextractivism: Two sides of the same curse', *Beyond development: alternative visions from Latin America*, 1, pp. 61–86.

Acosta, A. (2015) 'Después del saqueo: Caminos hacia el posextractivismo, Perspectivas, Análisis y Comentarios Políticos', *América Latina*, 1, pp. 12–17. Available at: https://mx.boell.org/sites/default/files/perspectivas_1_version_online.pdf.

Alonso-Fradejas, A. (2018) 'The rise of agro-extractive capitalism: Insights from Guatemala in the early 21st century', PhD Thesis, Erasmus University Rotterdam, International Institute of Social Studies at The Hague, Netherlands.

Arboleda, M. (2020) *Planetary Mine: Territories of Extraction under Late Capitalism*. London: Verso Books.

Bebbington, A. and Bury, J. (2013) *Subterranean Struggles: New Dynamics of Mining, Oil, and Gas in Latin America*. Austin, TX: University of Texas Press.

Brand, U., Dietz, K., and Lang, M. (2016) 'Neo-Extractivism in Latin America. One Side of a New Phase of Global Capitalist Dynamics', *Ciencia Política*, 11 (21), pp. 125–159.

Couldry, N. and Mejias, U. (2019) *The Costs Of Connection: How Data Is Colonizing Human Life And Appropriating It For Capitalism*. Stanford, CA: Stanford University Press.

De la Cadena, M. and Blaser, B. (2018) *A World of Many Worlds*. Durham, NC: Duke University Press.

De Vitoria, F. (2010) *Political writings*. A. Pagden and J. Lawrance (eds). Cambridge: Cambridge University Press.

Dunlap, A. and Jakobsen, J. (2020) *The Violent Technologies of Extraction: Political Ecology, Critical Agrarian Studies and the Capitalist Worldeater*. London: Palgrave MacMillan.

Ehrnström-Fuentes, M. and Kröger, M. (2018) 'Birthing extractivism: The role of the state in forestry politics and development in Uruguay', *Journal of Rural Studies*, 57, pp. 197–208.

Escobar, A. (1995) *Encountering development: The making and unmaking of the Third World*. Princeton, NJ: Princeton University Press.

Escobar, A. (2016) 'Thinking-feeling with the Earth: Territorial Struggles and the Ontological Dimension of the Epistemologies of the South', *AIBR. Revista de Antropología Iberoamericana*, 11 (1), pp. 11–32.

Fitzmaurice, A. (2007) 'The genealogy of terra nullius', *Australian Historical Studies*, 38 (129), pp. 1–15.

Frank, A. and Gills, B. (eds.) (1993) *The World System: Five Hundred Years or Five Thousand?* London: Routledge.

Gago, V. and Mezzadra, S. (2017) 'A critique of the extractive operations of capital: Toward an expanded concept of extractivism', *Rethinking Marxism*, 29 (4), pp. 574–591.

Gudynas, E. (2015) *Extractivismos: Ecología, economía y política de un modo de entender el desarrollo y la naturaleza.* 1st ed. Cochabamba: Centro de Documentación e Información Bolivia.

Gudynas, E. (2018) 'Extractivisms: Tendencies and Consequences'. In Munck, R. and *Wise*, R.D. (eds.) *Reframing Latin American Development.* New York, NY: Routledge.

Harvey, D. (2014) *Seventeen Contradictions and the End of Capitalism.* London: Profile Books.

Jalbert, K., Willow, A., Casagrande, D., and Paladino, S. (eds.) (2017) *ExtrACTION: Impacts, Engagements, and Alternative Futures.* London: Routledge.

Kröger, M. (2013) *Contentious Agency and Natural Resource Politics.* London: Routledge.

Kröger, M. (2015) 'Spatial Causalities in Resource Rushes: Notes from the Finnish Mining Boom', *Journal of Agrarian Change*, 16 (4), pp. 543–570.

Kröger, M. (2016) 'The political economy of 'flex trees': a preliminary analysis', *The Journal of Peasant Studies*, 43 (4), pp. 886–909.

Kröger, M. (2020a) *Iron Will: Global Extractivism and Mining Resistance in Brazil and India.* Ann Arbor, MI: University of Michigan Press.

Kröger, M. (2020b) *Politics of Extraction: Theories and New Concepts for Critical Analysis.* Oxford Bibliographies Online in International Relations.

Kröger, M. (2020c) *Natural Resources, Energy Politics, and Environmental Consequences.* Oxford Bibliographies Online in International Relations.

Kröger, M. (2020d) *Studying Complex Interactions and Outcomes Through Qualitative Comparative Analysis: A Practical Guide to Comparative Case Studies and Ethnographic Data Analysis.* London: Routledge.

Kröger, M. and Ehrnström-Fuentes, M. (2020) 'Forestry Extractivism: Uruguay'. In McKay, B.M., Alonso-Fradejas, A., and Ezquerro-Cañete, A. (eds.) *Agrarian Extractivism in Latin America.* London: Routledge.

Kröger, M. and Nygren, A. (2020) 'Shifting Frontier Dynamics in Latin America', *Journal of Agrarian Change.* 20 (3), pp. 364–386.

Locke, J. (2010) Second Treatise of Government. Project Gutenberg. Available at: https://www.gutenberg.org/files/7370/7370-h/7370-h.htm.

Machado Araóz, H. (2013) 'Crisis ecológica, conflictos socioambientales y orden neocolonial: Las paradojas de NuestrAmérica en las fronteras del extractivismo', *REBELA—Revista Brasileira de Estudos Latino-Americanos*, 3 (2), pp. 118–155.

McKay, B.M. (2017) 'Agrarian Extractivism in Bolivia', *World Development*, 97, pp. 199–211.

McNeish, J.A. (2018) 'Resource Extraction and Conflict in Latin America'. *Colombia Internacional* (93), pp. 3–16. Available at: https://dx.doi.org/10.7440/colombiaint93.2018.01.

Merchant, C. (1983) '*Mining the Earth's Womb*'. In Rothschild, J. (ed.) *Machina Ex Dea: Feminist Perspectives on Technology.* Oxford: Pergamon Press.

Merchant, C. (2013) *Reinventing Eden: The Fate of Nature in Western Culture.* New York, NY: Routledge.

Mezzadra, S. and Neilson B. (2017) 'On the multiple frontiers of extraction: Excavating contemporary capitalism', *Cultural Studies*, 31(2–3), pp. 185–204.

Mintz, S.W. (1986) *Sweetness and power: The place of sugar in modern history*. New York, NY: Penguin Books.

Moore, J.W. (2015) *Capitalism in the Web of Life: Ecology and the Accumulation of Capital*. London: Verso Books.

Peluso, N.L. and Watts, M. (eds.) (2001) *Violent Environments*. Ithaca, NY: Cornell University Press.

Perlin, J. (2005) *A Forest Journey: The Story of Wood and Civilization*. New York, NY: The Countryman Press.

Sadowski, J. (2019) 'When data is capital: Datafication, accumulation, and extraction', *Big Data & Society*, 6 (1), pp. 1–12.

Scott, J.C., (2017) *Against the Grain: A Deep History of the Earliest States*. New Haven, CT: Yale University Press.

Shapiro, J. (2018) 'As China Goes, So Goes the Planet: The Environmental Implications of the Rise of China'. In Kütting, G. and Herman, K. (eds.) *Global Environmental Politics*. Routledge, London.

Svampa, M. (2019) *Neo-extractivism in Latin America Socio-environmental Conflicts, the Territorial Turn, and New Political Narratives*. Cambridge and New York, NY: Cambridge University Press.

Szelényi, I. and Mihályi, P. (2020) *Varieties of post-communist capitalism: A comparative analysis of Russia, Central Europe and China, Studies in critical social sciences*. Leiden, Boston, MA: Brill.

Taylor, M. (2015) *The Political Ecology of Climate Change Adaptation: Livelihoods, Agrarian Change and the Conflicts of Development*. London: Routledge.

Teràn Mantovani, E. (2016) 'Las Nuevas Fronteras de Las Commodities En Venezuela: Extractivismo, Crisis Histórica y Disputas Territoriales', *Ciencia Política*, 11 (21), pp. 251–285.

Thatcher, J., O'Sullivan, D., and Mahmoudi, D. (2016) 'Data colonialism through accumulation by dispossession: New metaphors for daily data', *Environment and Planning D: Society and Space*, 34 (6), pp. 990–1006.

Wallerstein, I. (1974) *The Modern World-System* I. New York: Academic Press.

Willow, A.J. (2018) *Understanding ExtrACTIVISM: Culture and Power in Nature Resource Disputes*. London: Routledge.

Ye, J., van der Ploeg, J.D., Schneider, S., and Shanin, T., (2020) 'The incursions of extractivism: Moving from dispersed places to global capitalism', *The Journal of Peasant Studies*, 47 (1), pp. 155–183.

2

THE POLITICS OF VIOLENCE IN EXTRACTIVISM

Space, Time, and Normativity

Katharina Glaab and Kirsti Stuvøy

Introduction

Extractivism is central for the working of the global capitalist system in the age of the Anthropocene. The never-ending extraction of resources for profit provides fuel for a global economy anticipated to be able to grow endlessly. Resource extraction is often linked to violence, as it creates conflicts about access, control, and use of nature (Martín, 2017). The global dispersion of violence in relation to extractivist practices is recognizably complex: Reports about environmental activists being killed and indigenous groups being expelled from their traditional land as economic activities expand demonstrate the severity of the violence (Global Witness, 2015). The expansion of extractivist practices has led to environmental destruction, social inequality, and hardship for many communities, with yet unforeseeable consequences. Research on the link between violence and extractivism has shown various forms of violence connected to extractivist practices and carefully identified nuances across contexts (Tyner and Inwood, 2014; Springer and Le Billon, 2016). In this volume, a guiding theme is that extractive violence is a core logic of the 21st century global order—expressed in the title, *Our Extractive Age*. This chapter engages with that global perspective and aims to explore the interconnections of various manifestations of violence with global structures and processes. With this agenda our focus is on the following conundrum: how can we carefully attend to nuances in particular contexts and simultaneously make sense of global dimensions that permeate violence in the hyper-extractivist age?

In order to explore this local/global perspective, we suggest taking a step back from studies of empirical manifestations of violence to address this question conceptually (Springer, 2011; see also Tyner and Inwood, 2014). With this approach we open more fundamental questions about what characterizes violence in extractivism, and how conceptions of violence limit or enable us to see and understand the global embeddedness of violence. We argue that conceptualizations of violence show different ontological qualities and epistemological scale, leading to different foci in research on actors, structures, and processes that ultimately define what research makes visible, as well as what remains invisible. In this chapter, we develop a framework that emphasizes the spatial and temporal properties of violence in extractivism. As the basis for developing this framework, we elaborate on the everyday as a lens through which we examine the global (Enloe, 1989; Tsing, 2005; Guillaume and Huysmans, 2019). This, we argue, allows for an epistemological position that is open to a multitude of modalities of violence in extractivism and simultaneously addresses global entanglements.

The chapter proceeds as follows: In a first step, we discuss different manifestations of violence and their conceptual contribution to our understanding of violence in extractivism. Drawing on literature across various fields, including conflict research, historical sociology, geography, anthropology, and international relations, we develop our main argument that the global embeddedness of seemingly local violence can be approached through a study of the "everyday." In a second step, we add to this discussion by focusing conceptually on the spatial and temporal dimensions of violent practices in extractivism. For analytical clarity, space and time are addressed separately in this chapter. The section "violent politics of space" examines spatial properties of violence in extractivism across local/global interconnections, focusing on violence as boundary-drawing activity and as inscribed in places, with particular spatial effects that we approach with the term "site effects" (Bourdieu, 1999). The section "violent politics of time" focuses on the temporal properties of violence in extractivism and emphasizes the fast and slow paces of violence and its temporal dispersion. These sections are followed by a concluding discussion of the impact of spatio-temporal conceptualizations of violence on research practices. We bring forth normative concerns in research on violence in extractivism pertaining to our scholarly responsibility with regard to what we make visible and what we leave invisible when we approach the study of violence in extractivism.

The Nexus of Violence and Extractivism

The discussion of violence and its different forms and manifestations is debated within multiple disciplines. These debates have developed and modified concepts and research focus in reaction to changing forms of conflict and warfare. In the 1990s the "new wars" debate (Kaldor, 2001; Münkler, 2005) moved away from state-led violence and highlighted the blurring of boundaries between state and non-state actors, focused on what is done for economic and political motives, and emphasized changes in modes of warfare and how wars are financed. As part of the efforts to analyze the newness of these wars, attention was directed towards the role of the economy and economic motives for war. The resource-curse thesis tied in to these debates and linked certain resource-rich countries to endemic violence owing to increased internal instability (Homer-Dixon, 1999; Collier, 2000). This reductionist lens was soon challenged by perspectives that addressed the financing of the "new wars" and how the various actors and activities involved in the war economy defined conflict dynamics in terms of legitimacy. Questions were asked about the functions of violence as means of creating access to economic profit for the purpose of sustaining particular social orders and forms of domination and hierarchy, rather than necessarily to end wars (Le Billon, 1999; Keen, 2000; Kaldor, 2001). These debates on the entanglement of economics, politics, power, and violence in the post-cold war era directed attention to various kinds of resources, e.g. oil, diamonds, and cobalt, and how they defined and amplified conflict dynamics, sustaining particular violent social orders that made violence appear endemic (Bakonyi and Stuvøy, 2005; Bakonyi et al., 2006). A particular focus on the micropolitics of violence emerged (e.g. Schlichte, 2009), addressing how violent entrepreneurs have expanded their legitimacy within armed groups and within larger society.

The emphasis of conflict research on answering questions about how the power relations of a particular (violent) social order matter to how resource extraction is enabled locally served to develop a relational approach to violence. Focusing on micropolitics, this approach investigates the social and historical contexts that create conditions for the emergence of violent action. In this historical sociological perspective on violence the processual character of violence is underscored (Malesevic, 2010; 2017). Focusing on how violence is constituted relationally, this perspective directs attention to how violence is embedded in structures and processes and shaped by ideas of what is considered legitimate and illegitimate action in a particular context. The approach to micro-dynamics of violence characteristically studies specific acts of violence and how they are committed, with attention to what actors are involved and how violence is perceived and understood (Verweijen, 2020). Relational approaches to violence have also been applied in anthropology through socio-cultural perspectives that emphasize the relationships among victim, perpetrator, and witness. Attention has also been directed at the connections among economic, political, and institutional dimensions and the everyday experiences of violence among individuals and families (Accomazzo, 2012). Across these different relational approaches to violence, the interconnection between the micro-dynamics of violence and broader global processes has emerged as a core issue. For example, economic activities in violent conflicts are incorporated into the world economy through black market operations of conflict minerals; global development and shifts of energy demands drive resource extraction locally.

Expanding on the interconnection between micro-dynamics of violence and global processes, we find important insights in the spatial turn in violence research (Fregonese, 2012; Tyner and Inwood, 2014). Emerging from the field of political geography, questions about how violence is shaped by space and how violence affects space underscore the relational constitution of space and violence (Massey, 1996). While adding to and expanding the relational approach, a recurrent position across these scholarly contributions on space and violence is that there is a need to advance the theorization of violence itself (Blomley, 2003; Springer, 2009, 2011; Springer and Le Billon, 2016). For example, Tyner and Inwood insist that "violence, in short, is given its own materiality: it simply exists" (2014, p. 773). This is evident in how, from realist perspectives, violence is considered a natural expression of human nature, whose violent aggression is taken for granted and exposed in competition to survive within a global anarchic system. With this approach, human aggression appears "natural" and violence an ever-present aspect of human life, part of humankind's past history, present and future. Theorising violence challenges this idea of violence as pre-discursive and defined simply by its own material presence.[1] Furthermore, Tyner and Inwood underscore violence as produced discursively in ongoing political struggles, and they conceptualize violence as "the outcome of the political" (ibid). Such concern with violence and meaning-making is, however, criticized for deflecting attention from the "kill chain," that is, the factors that produce direct, physical violence in a particular context, unless it is demonstrated empirically how such discourses feed the production of violence (Verweijen, 2020). There is thus a normative contestation over which forms of violence should be given priority in research.

The title of this chapter—the politics of violence in extractivism—reflects our understanding that political processes are relevant to the understanding of violence and underscore that it matters how violence is given meaning. We depart from the understanding of violence as a social activity and are concerned with the scholarly *meaning-making* connected to violent practices. This includes the contestations of meaning of violence and the existence of competing discourses on violence. Furthermore, we understand violence in a processual and relational way and not as an attribute of an individual or a group. We are therefore concerned with violence as it unfolds over space and time and are interested beyond its manifestations in its emergence, aftermaths, or effects. Importantly, it matters to the issue of violence in extractivism how the local and global interact and are given meaning, directing our attention to aspects of scale. The typical social science approach is to think of scales, or levels, as separate. For example, global governance institutions are discussed as a separate "global" arena detached from the everyday realities of its citizens. However, with the turn to the everyday in international relations scholarship, there is growing attention to how analysis of global processes and practices can be grounded in particular localities. This pertains, for example, to questions of how "global" norms are translated into particular local contexts (Zwingel, 2012; Zimmermann, 2017), or how everyday lives are entangled in global power relations (Enloe, 1989; Sylvester, 1994; Wibben, 2011). Instead of thinking of scales as separate, our focus is on their entanglements. We draw here on the work of Tsing (2005; 2015), who, in her research on how value is produced from nature, draws attention to the everyday and how it is embedded in global norms, processes, and practices. The global is in this perspective not a separate scale, but part of scale-making processes. This perspective underscores that we need to ask questions about and research what the global is (processes, institutions, practices, world views) and how it manifests in particular local contexts–or how "local" violence is embedded in global processes and structures and whether and how these can be violent, too. Ontologically we thus follow Tsing in her relational approach to scale, expressed in the concern for scale-making projects in which global projections are linked with regional and local scale-making projects, in which the "global scale" either could succeed strongly, tentatively, or partly (Tsing, 2005).

The concern with scale-making links to the understanding of the extractive political economy as multiscalar in nature (McNeish, 2018). We draw attention here to two specific dynamics that we argue characterize extractivism and require our attention when it comes to violence: *First*, and most importantly, extractivism is inseparable from the global spread of the capitalist paradigm. Extractivism is embedded within dynamics of the globalized capitalist system. Resource-intensive lifestyles have led to a growing demand for and use of raw materials. Increasing financialization and privatization capture all areas of life, including nature. Within this context, extractivism appears as a consequential logic within a global market. In this extractivist reasoning, raw materials are discursively presented as a necessity, e.g. the lifeblood of a nation, which legitimizes exceptional measures that are often detrimental to the environment and people. In addition, the drive for accumulation of surplus wealth pushes extraction to ever-new territories and physical frontiers. *Second*, and related to the point above, the degree, character, and manifestation of extractivism have changed in the neoliberal age. While mining, drilling, and harvesting are activities that have been associated for a long time with exploitation, "the notion of extractivism [...] remains associated with a narrow and literal sense of extraction" (Mezzadra and Neilson, 2017, p. 185). Within the capitalist-extractivist logic, however, we can observe a turn away from these "traditional" material manifestations of extraction towards the extraction of data, information, and knowledge. Knowledge has achieved the status of a new raw material that can be traded. This development is, for example, evident in the notion of "biopiracy," where the knowledge of local communities is commercially exploited, or biological resources are patented without adequate compensation. As we see in this book's chapters, moreover, novel forms of extractivism extend to the built environment, to cultural appropriation, to the commandeering of talent, and even to the reverse-extraction dynamics of global geoengineering.

In the context of these two logics and the diversity of empirical modes and forms of extractivism, the link between violence and extractivism can be very direct. This is the case when conflict and contestations of extractivism lead to physical harm or even to people's deaths. There is thus a direct and physical form of violence connected to extractivism, but violence also takes more subtle and indirect forms. In order to expand on the multitude of manifestations of violence in extractivism, we propose to emphasize space and time dimensions as pathways to develop a more encompassing understanding of violent practices in resource extractivism. For analytical purposes, we present them stepwise, first emphasizing spatial relations and subsequently temporality with regard to violence in extractivism.

Violent Politics of Space

Extractivism is embedded in power relations that reflect different kinds of systemic inequalities and economic rationalities. Yet, aiming to identify how violence is interwoven with extractivism directs attention to the everyday lives of people involved in or affected by extractivist practices and processes, and to the structures and institutions that permeate these everyday lives. Therefore, we utilize the "everyday" as an epistemological point of departure to study violence in extractivism, to approach the global as entangled in the everyday, and attend to the multiplicity of meanings, practices, and relations that thereby come into focus (cf. Guillaume and Huysmans, 2019). Thus, in this perspective, particular extractivist locations constitute entry-points for asking questions about the global in regard to violence in extractivism. We draw here on Massey (2004), who views locations as nodes for examining multiple interconnections and power geometries, towards which we have responsibility (as scholars). Below we outline three dynamics defined by spatial relations that we see as significant to violence in extractivism.

Violence Sits in Places and Can Be Informed by "Elsewheres"

Violence is usually perceived as localized in particular places, linked for example to violent conflicts over resources or extractivist projects. We recognize the embeddedness of such localized experiences in global power relations thus linked to "elsewheres." Drawing on Springer (2011), who notes that violence is perceived as an embodied, localized experience, we argue that this perception should not direct our thinking in terms of seeing place only as defined merely by local and contextual factors. Rather, as elaborated above, we need to recognize that (extractivist) places are (globally) interconnected. A spatial approach to violence in extractivism thus needs both to identify the contextual histories, actor-constellations, and institutions that caused the violence and to ask questions about global entanglements across places. Springer's point is that violence sits in places in terms of how we perceive its manifestations, yet that this violence is co-constituted by and mediated through a "wider experience of space." A relevant example of that is how neoliberalism manifests. Conceptualized as "an assemblage of rationalities, strategies, technologies, and techniques" (Springer, 2011, p. 95), neoliberalism is a governmentality that ensures and enables rule at a distance. It disciplines actors, institutions, and individuals to conform and "subjectivate" to market norms of competition and individual entrepreneurship. Applied to the Global South, this governing is a form of neo-colonialism and the subjection to market norms is at times complemented with the direct use of violence that "maintain the interest of an internationalised global elite" (Springer, 2011, p. 95). The argument that emerges from this spatial approach to violence culminates in a notion of extraction, which "provides a means to map and join struggles that unfold in seemingly distant and unrelated landscapes" (Mezzadra and Neilson, 2017, p. 187).

The global logic of market and capital relations thus interlinks various sites of extractivism and the violence that happens in these places. Drawing on the work of Mbembe, the historical link between governing authority in the form of European imperialism and extractivism can serve as an illustration (Mezzadra and Neilson, 2017). The slave trade can be understood in extractive terms when identifying the close connections between forced labour and extractive practices. This connection was central to the continuation of extractivism through the centuries of colonialism and imperialism, creating, and reproducing coercive and fear-based cultures as part of extractivist practices. Today, the histories of these governing practices enabling extraction, and how they were justified, maintain continuity as new extractive sites are opened, and not without controversy: As Mezzadra and Neilson point out, "The violence of this opening often manifests in controversies surrounding property and land rights" (2017, p. 192). Such controversies are often pronounced when the identification and exploitation of an extractive project intersect with territorial usage by indigenous people and their traditions and interests in cultural and economic reproduction. Such conflicts evoke multiple layers of property relations concerning how access to and use of the land has traditionally been practiced and how it has often resulted in dispossession and displacement. Thus, global modes of production in one part of the world have links with extractivist and violent practices in others through governing practices and rationalities. This leads us to our next point, concerning how violence in extractivism is linked to *boundary-drawing*.

Violence Creates Order by Creating Spatial Boundaries

When a certain location is designated for an extractivist project, this is usually the result of a process that determines something as "extractable." Representation-making practices delineate and create boundaries between "extractable," and "non-extractable" places and objects and can imply the exclusion and marginalization of certain people and their ideas from such places. This exposes how power and hierarchy are at play in representation-making practices and shows that spatial representations are best viewed in terms of power and politics, and, importantly, in terms of how they produce inequalities.

For example, certain conceptions of extractivism produce differences between "extractivist" countries and "industrialized" countries (Martín, 2017), and also reproduce dichotomies of victims and beneficiaries (Gago and Mezzadra, 2018). Once a certain place is constructed as "extractable," it takes away the possibility of alternative uses of the land on which the extractivist practices take place. If extractivism denies certain individuals or groups access to land and thus the use of this land for alternative usage, it has severe impacts on some peoples' livelihoods. The effects of such spatial reordering are institutionalized through practices such as spatial mapping, land titling, and registration. Clearly, the law is a powerful boundary-making tool with regard to claims to property and land use that are essential for resource extraction (Blomley, 2003). For example, a government has the power to issue a lease to an individual if it deems this in the interest of the state. Such state action could trigger a compensation payment to people dispossessed of their land. As the revenues serve the owner, who is often located far away from the extraction site, those who receive compensation could be dealing with environmental damage. This demonstrates how the spatial boundaries that are set by property regimes can allow "the violence of extractivism to proceed" (Mezzadra and Neilson, 2017, p. 192).

The making of spatial boundaries also has consequences for what violence is deemed legitimate and what is not: When the power to delineate spaces and assign property lies with the state, violence is typically considered legitimate, although this legitimacy needs to be recognized by people whose belief in its righteousness determines that legitimacy (Blomley, 2003, p. 133). Violence therefore plays an integral role in the legitimation, foundation, and operation of property regimes that are essential to extractivism (Blomley, 2003), and they require violence as a last resort for the enforcement of its rules. We postulate that core questions about violence in extractivism include how spatial representations and boundary-drawing make a difference to an extractivist operation, how they shape its modalities, and how they create violent effects. This requires analytical attention to the spatializing powers of governments, (para-)militaries, businesses, and other contextually specific actors. With the everyday as a point of view, we are not limited to elite actors but draw attention also to how violence in extractivism is enacted through boundary-drawing that affects trends in inclusion and exclusion of women, workers, citizens, or individuals belonging to subaltern groups (Guillaume and Huysmans, 2019). The spatial effects of such violence are in focus of our third spatial dynamic of violence in extractivism.

Violent Site Effects

The concept of "site effects" captures the idea that various dimensions of power, once inscribed in space, shape the everyday (Bourdieu, 1999, 2018; Wacquant, 2018). For example, poor housing conditions, lack of basic infrastructure, and, we can add, environmental degradation, are all forms of dispossession inscribed in space. People might experience the coercive effects of these conditions as bodily harm, which exemplifies how power inscribed in space is part of the production of precarious lives. Such violent site effects are, however, neither necessarily intentional nor caused by a subject (a perpetrator), but part of what Marx called the "silent coercion of economic circumstances" (1885/1988, p. 76). Instead, violent site effects draw attention to violence as built into the structure, and are part of the (social) order and mode of operation that defines extractivist practices.

Violence can encompass both structural and symbolic forms. Structural violence directs attention to systemic constraints, including unequal access to resources and lack of legal protection or political power, that are usually taken for granted but by their very existence perpetrate violence (examples are poverty, racism, and colonialism) (cf. Accomazzo, 2012). Structural violence operates indirectly and prevents people from realizing their full potential (Galtung, 1969). Similarly, the concept of symbolic violence (Bourdieu, 1977/1991) shows how structured arrangements uphold particular social orders, often with the complicity of its participants or members. Within a social order, there is always a power asymmetry in which some people or groups remain in positions of dominance, while others remain in non-dominant positions. These power relations are sustained as certain kinds of practices are valued more highly than others and people often unwillingly accept the power upholding these relations. Symbolic violence is thus the violence that is exercised upon an agent with his or her complicity. This kind of violence is habitual, non-intentional, and indirect, and is produced in the everyday as an effect of a symbolic order that structures society. In regard to violence in extractivism, these concepts emphasize how extractivist practices are embedded in social structures of racism, classism, and nationalism and how the logic of extraction is often taken for granted and sustained within a capitalist social order. Within such orders, symbolic violence defines what is to be critiqued or problematized and what is defined as naturalized and thus not to be questioned. It allows the continuation of practices with violent effects.

The concept of "violent site effects" thus encapsulates the interconnections between spatial arrangements and socio-economic inequalities and directs attention to the often quite violent effects of spatial relations of power. For example, the divisions between extractivist countries and non-extractivist countries, and between extractivist regions and non-extractivist regions, create particular (global) economic geographies. Unequal access to resources generates spatial disparities which underlie uneven development and inequality. Focusing on the post-socialist space, Golubchikov *et al.* have elaborated on these interconnections and underscored that with particular contextual histories such as the socialist heritage (legacy), "it is easy to overlook the more fundamental nature of capitalism, including its systemic propensity to produce inequalities–no matter what original spatialities and legacies it colonises" (Golubchikov *et al.*, 2014, p. 618). Once difference is inscribed spatially, by becoming a structure affecting access to infrastructure, among other things, this can reproduce dominance and marginalization. This is where the site effects operate and become violent when threatening people's livelihoods and ability to survive. Violent site effects are thus the effects that occur when violence happens and materializes in spatial differences.

Conceptually, violent site effects direct attention to how extractivist practices, as a particular utilization of space, affect people's physical conditions. These include health deprivation, risk of death, and the systematic reproduction of certain extractivist places as deprived of alternative futures. A change of material structure as required for extractivism can be drastic and durable, not easily undone. This leads to our next step, in which we emphasize the temporal dimensions of violence in extractivism.

Violent Politics of Time

The act of violence is intimately linked to our understanding of its temporal dispersion, ruptures, and continuities, and its effects through time. The relation between violence and time has been less theorized, however, than the relation between violence and space. But violence does not only transcend time (and space). How we think about violence also relates to how we imagine, experience, remember or forget violence–in other words, how the flow of time shapes our understandings of and assumptions about violence and its effects in relation to extractivism. Therefore, in the following, we discuss three assumptions about the temporalities of violence that are crucial for our understanding and evaluation of violence in extractivism.

Violence Is Fast and Slow

Direct physical actions of harm and force, such as the destruction of livelihoods, the forceful relocation of people, or the killing of environmental protesters in order to establish sites for extractive industries, often appear to be the most obvious and alarming signs of violence in the age of hyper-extractivism. Especially in times when media attention is set towards the spectacular in politics and society, the documentation of and focus on immediate and exceptional forms of violence seem to be more important than ever.[2] While the materiality of violence in extractivism needs our urgent scholarly attention, the focus on immediate violence also demands critical scrutiny about society and the media's preoccupation with the political spectacle (Edelman, 1988; Nixon, 2011). In addition, this time dimension underscores important questions about the ontological status of violence, as alluded to above. Actions of physical harm and force are deemed to be violent as they are visible and ultimately "knowable" to the observer. This violence is often termed to be "fast," as it is time-bound and fixated on certain violent events at a specific point in time. While its immediacy makes it available to the researcher to be exposed, analyzed, and explained, it concomitantly sustains scholarly attention towards the material and physical expressions of violence.

The focus on fast violence is, however, limited on an ontological level as it assumes that violence is imperceptible when it is not directly visible. If we take seriously, as Tyner and Inwood (2014, p. 771) claim, that "violence has no material reality," as noted above, there is a need to rethink the ontological status of violence in terms of time. For this, we can turn to Nixon, who famously reminds us that violence is not only "a contest ... over space, or bodies, or labor, or resources, but also over time" (2011, p. 8). In contrast to "fast" violence, he argues that we need to rethink our understanding of violence and look at the unspectacular and some-times invisible forms of violence. Nixon shifts our attention to "slow" violence— "a violence that occurs gradually and out of sight, a violence of delayed destruction that is dispersed across time and space, an attritional violence that is typically not viewed as violence at all" (Nixon, 2011, p. 2). Just like structural violence or the conception of violent "site effects," slow violence highlights the invisible and therefore often silenced violence such as the violence of poverty, inequality, or racism. The focus on the temporal dimensions of violence emphasizes the delayed effects of violence that in extractivism are often related to the "conflictual inter-penetration of industrial and environmental temporalities" (Adam, 1998, p. 56).

Climate change, toxins, and growing inequalities are examples of "long dyings" (Nixon, 2011, p. 2) and less visible harm inflicted on people and environment over time. This violence unfolds slowly over long periods and, in contrast to "fast" violence, leads to invisible and therefore also unaccounted "disposable casualties" (2011, p. 13). Building on Rachel Carson's ground-breaking thinking about death by indirection, Nixon's concept of "slow violence" challenges us to rethink the causes and effects of violence. Nixon argues, for instance, that displacement is not just space-bound but also time-bound, when places become uninhabitable for residents in the future owing to environmental costs that make places "irretrievable." In the case of extractivism, the environmental impacts for groundwater, biodiversity, or climate of the building of extractive sites and related infrastructure could become visible only after several decades. Hence, the extraction of materials that are potentially destructive for climate and environment might also affect opportunities and life choices of future generations. This perspective on time is explored further in the next section.

Violence Disperses Throughout Time

The focus on the spectacle of violence ascribes to violence a certain "presence" and allows the possibility of representation in the now. Nixon takes issue with this as he argues that the relative invisibility of delayed destruction and casualties poses a representational challenge. Spectacular violence can be observed, documented, and remembered and therewith also facilitates immediate responses. In the case of extractivism, burning oilfields, killings of people, or forced displacements are more likely to be reported and lead to calls for action. This can encompass legal responses, the criminalization of violence, or reconciliation processes in the aftermath of the event. In contrast, the invisible and temporarily dispersed violence of effects on the environment and people due to extractivist practices is often not represented. How something can be addressed, remembered, and acted upon when it is not represented (in the now and as part of the obsession with presentism) becomes a pertinent question.

Lundborg's work (2012; 2016) ties into this debate. He criticizes conceptualizing violence as a singular event with a clear beginning and ending. He productively reminds us of Deleuze's differentiation between the "pure event" and the "historical event." The historical event alludes to an exceptional intervention that clearly presents a disruption of time and separates a "before" from an "after." Violence is here perceived as rupture, as a radical unsettling and disrupting of the present and its preconceived path (Lundborg, 2016). In contrast, the pure event does not refer to a particular moment in time and place. Instead of "being," the pure event alludes to the ambiguous process of "becoming" which displaces the presence of violence as something that moves into the past and future at the same time (Lundborg, 2012). This differentiation is useful for our understanding of violent extractivism as it makes visible how violence is represented differently over time and with what implications. The temporal borders that the historical event creates require immediate responses and often exceptional measures, but do not take into account "long-term processes and large-scale patterns" (Lundborg, 2016, p. 115). Understanding violence "as an event with a definite beginning and end, duration and conclusion, contributes to privileging instantiations of violence like war and armed conflict over more diffuse, ongoing, structural forms of violence" (Lundborg, 2016, p. 126). In contrast, persecution, rebuilding, or compensation around future losses is difficult to establish, owing to lack of representation. It is ultimately this inaction built on doubt that creates a fatal yet invisible form of violence that is relevant to extractivism.

The discussion around the temporal representation of violence also taps into important debates within postcolonial scholarship that advocates for a temporal refor- mulation of violence (see, for instance, Agathangelou and Killian, 2017b). If we take seriously the claim that the present forms of violence are temporarily dispersed, one needs to reconsider the focus on representation of violence within a linear, chron- ological timeline. Linear understandings of time make temporal distinctions between the modern and pre-modern world that entail a teleological orientation towards pro- gress and an end of violence. The ontologization of the Western world as modern and secular (Agathangelou and Killian, 2017a) in contrast to the violent and irrational pre- modern world, indicates a clear understanding of temporal progress. Against this temporal linearity, which separates "'before' from 'after' and introduce[s] us to a 'new beginning'" (Lundborg, 2012, p. 1), stands the postcolonial suggestion of temporal multiplicity, or what Chakrabarty (2007) termed "heterotemporality." Hetero- temporality challenges the notion of a universal history which narrates and makes capitalism global and resists dominant meta-narratives by articulating time in plural ways. Accordingly, next to political time we can find a multiplicity of spiritual and religious temporalities, where violence seems to be inevitable when these temporalities clash (Agathangelou and Killian, 2017a).

Acknowledging the temporarily dispersed character of violence shifts our atten- tion towards the history of extractivism and the inability to define violence as a singular act in a singular present. Not only are extractivist practices often deeply entangled with violent histories of colonialism, as many of the world's resources are situated in the Global South, but new forms of extractivism in the Global North are similarly entangled with the historical rise of capitalism and the violence that is exercised through the establishment of industrial and economic temporalities. Taking seriously the postcolonial claim that plural temporalities mingle, as expres- sed in the term heterotemporality, helps us to see how extractivist practices often build on a teleological meaning of time where extraction is part of an under- standing of teleological progress and change towards becoming a "modern" and "developed" country. Within such temporal dimensions of modernization and development, we can also expand on temporalities in relation to agency as expressed through a focus on subjectivities. This is the focus of our final step.

Violence Constitutes Temporal Subjectivities and Strategies

Thinking beyond linear time enables us to understand the exclusion of many sub- ject positions and possibilities of action (Hom, 2018). From the perspective of heterotemporality and the coexistence of multiple (political, spiritual, religious) temporalities, it can be acknowledged how perceptions of time enable, inhibit, or conflict with extractivist practices. Extractivism is deeply bound up with modern modes of production. Acceleration shapes the world by speeding up production and consumption processes and ultimately links the good life of some with the suffering of distant others. This furthers the complicity of violence and capitalism by amplifying structural inequality and exploitation (Glezos, 2013). But the capitalist logic also prioritizes other temporal goals and focuses on short-term gains at the expense of long-term societal and environmental effects. Bourdieu showed how different temporal logics affect the ability of actors to plan. In his ethnographic studies of Algeria, he distinguished between traditional time consciousness on the one hand and the time characteristics of the spirit of capitalism on the other (Atkinson, 2018). Accordingly, time-consciousness of the Algerian peasant is rooted in the circadian rhythms and routines of the workday and the holidays given by the ritual calendar. Recurrent past experiences and the patterns of the everyday form ideas of the future, i.e. an expectation of the forthcoming. According to Bourdieu, this does not enable one to concern oneself with longer time spans or provide the ability to "colonize the future."

This conflicting difference of temporalities and how they are acted upon becomes clear when we see how violence constitutes different subjectivities or what Fanon calls a "time lag, or difference of rhythm" (Fanon, 2002 [1963], p. 106). Fanon reminds us that violence has deep effects on individuals. Violent experiences have a particular effect on the psyche and become internalized dispositions that operate as guides to action (Fanon, 2007 [1952]). While this violence is embodied in immediate suffering and pain, the ongoing bodily existence also affects further motivation and choices (Frazer and Hutchings, 2008). Subjects are caught between backward and forward temporalities (Solomon, 2014) between the-what-has-been and the-not-yet-there. Therefore, the past becomes part of the present and similarly informs imaginations of futures. For example, there might be a discrepancy in "rhythm" and strategizing of the future between poor and rich people. While those struggling to make ends meet probably orient toward the short-term, those without immediate survival needs have the luxury to build long-term projects. In extractivism, these different rhythms influence the temporal strategies of those affected by extractivist practices. They further define how someone will experience the violence of extractivist actions and processes and inform whether the reaction will be resistance to, or acceptance of, everyday suffering.

Conclusion: On the Normativity of Violent Extractivism Research

In this chapter we have reflected on the questions, "what characterizes violence in extractivism?" and "how can we conceptualize the global dimensions that permeate violence in the hyper-extractivist age while simultaneously attending to nuances in particular contexts?" We proposed a spatio-temporal framework to study violence in extractivism that can serve as a conceptual map to expand the discussion on the politics of researching violence in extractivism. In this conclusion, we want to foreground the scholarly responsibility with regard to what we make visible—and what we leave invisible—when we approach the study of violence in extractivism.

The conceptualization of violence can, as we have underscored, be normatively restricting in terms of what is seen or not seen as violence. For example, an often-underlying assumption in research on violence and extractivism is that extractivist practices necessarily lead to violent conflicts. An opposite view, equally often evoked throughout history, is an optimistic view of resource extraction as contributing to economic and national development, reflecting the idea of extractivism as a developmental model. These equally one-sided, dichotomous attributes reflect an ontological bias towards studying violence as the exercise of "power over" something or someone without paying attention to more subtle forms of violence that are invisible, "slow," or indirect. In addition, the assumption directs attention to the physicality of violence, that which we can directly observe. However, violence, when restricted to a focus on visible, discursively framed forms of violence, comes with normative restrictions.

In this chapter we have shown that violence is ambiguous and transcends the local/global divide. We have argued that researching violence in extractivism brings into discussion the multi-scalar and hetero-temporal character of violence, which is embedded in a particular approach to the global, not as a separate scale but as process and becoming of scale-making projects (Tsing, 2005). Our conceptual discussion reflected on the categories with which we think and our presumptions regarding the situatedness of knowledge. Hence, the ontological quality and epistemological scale that we assign to violence and extractivism define what we make visible and what we leave in the shadow. Wolff (Martín, 2017, for example, points out that the challenge to violence in extractivism can address very different things, including specific conflicts over particular extractive projects, contestation of the development model behind extractivism, and contested perceptions about what is acceptable with regard to extractivism and "post-extractivist" alternatives. The potential for agency, or alternative thinking and ways of development, underscores possible normative consequences implied in research on violence in extractivism.

In other words, it matters how we approach and elevate violence in extractivism, whether extraordinary and exceptional instances of violence in extractivism draw our attention, or whether the ordinary, almost unseen violence is addressed. However, studying what is silenced and unseen is not easy. Just as slow violence is hard to observe, so are the reactions to it. As this violence is temporarily dispersed, too, the response is at best not immediate and at worst not happening at all. Fast and slow violence have a different heft and as scholars we need to go beyond the immediate event and shift our attention to the violence that is slowly creeping upon us. A particular normative challenge is representational, as Nixon points out, in slow violence the "long dyings are underrepresented" (Nixon, 2011, pp. 2–3). This again poses questions about the role of the scholar in research, our responsibility in representation-making practices, and the underlying normativity of research when it comes to addressing these uncomfortable questions.

We have argued that conceptualizing the space and time dimensions of violence is helpful for research on the violence in extractivism, as it leads to more reflexivity about our scholarly responsibility when invoking a term such as violence. This conceptualization is built on debates about what violence renders visible and invisible. It implies an ethical responsibility of scholars to engage critically with the conceptualization of violence, because our conceptual and theoretical choices have implications for how the world can be understood and ultimately imagined to be changed. As Martín (2017, p. 22) warns, "by disregarding the multidimensionality and complexity of the phenomenon, we run the risk of losing the transformational and enlightening power of criticism." As researchers, we therefore need to do more than take a critical stance towards extractivism and reflect more deeply upon our own positionality and conceptual starting points.

Notes

1 Following Tyner and Inwood (2014, p. 774), the point here is not to suggest that violence has no materiality—which it certainly does when we consider, for example, the documentation of violent attacks on environmental activists or homicide rates across cities in the world. The point is rather to address violence as embedded in relations and to give this priority in theorization.
2 See, for instance, *The Guardian*'s project on "The Defenders" to document the deaths of activists around the world who fight for the environment. www.theguardian.com/envir onment/series/the-defenders (retrieved February 14, 2019).

References

Accomazzo, S. (2012) 'Anthropology of Violence: Historical and Current Theories, Concepts, and Debates in Physical and Socio-cultural Anthropology', *Journal of Human Behavior in the Social Environment*, 22 (5), pp. 535–552.

Adam, B. (1998) *Timescapes of Modernity. The Environment and Invisible Hazards*. London: Routledge.

Agathangelou, A.M. and K.D. Killian. (2017a) '*Introduction. Of Time and Temporality in World Politics*' in Agathangelou, A.M. and Killian, K.D. (eds.) *Time, Temporality and Violence in International Relations. (De)fatalizing the Present, Forging Radical Alternatives*. New York, NY: Routledge, pp. 1–22.

Agathangelou, A.M., and K.D. Killian. (2017b) *Time, Temporality and Violence in International Relations. (De)fatalizing the Present, Forging Radical Alternatives*. New York, NY: Routledge.

Atkinson, W. (2018) 'Time for Bourdieu: Insights and oversights', *Time and Society*, pp. 1–20.

Bakonyi, J. and Stuvøy, K. (2005) 'Violence & Social Order Beyond the State: Somalia and Angola', *Review of African Political Economy*, 104 (5), pp. 359–382.

Bakonyi, J., Hensell, S., and Siegelberg, J. (2006) *Gewaltordnungen bewaffneter Gruppen. Ökonomie und Herrschaft nichtstaatlicher Akteure in den Kriegen der Gegenwart*. Baden-Baden: Nomos Verlagsgesellschaft.

Blomley, N. (2003) 'Law, Property, and the Geography of Violence: The Frontier, the Survey and the Grid', *Annals of the Association of American Geography*, 93 (1), pp. 121–141.

Bourdieu, P. (1977/1991). 'On Symbolic Power' in Bourdieu, P. (ed.) in *Language & Symbolic Power*. Cambridge: Polity Press.

Bourdieu, P. (1999) *The Weight of the World: Social Suffering in Contemporary Society*. Oxford: Polity.

Bourdieu, P. (2018) 'Social Space and the Genesis of Appropriated Physical Space', *International Journal of Urban and Regional Research*, pp. 106–114.

Chakrabarty, D. (2007) *Provincializing Europe. Postcolonial Thought and Historical Difference*. Princeton, NJ: Princeton University Press.

Collier, P. (2000) 'Doing Well out of War. An Economic Perspective' in Berdal, M. and Malone, D. (eds.) in *Greed and Grievance. Economic Agendas in Civil Wars*. Boulder, CO: Lynne Rienner, pp. 91–111.

Edelman, M. (1988) *Constructing the Political Spectacle*. Chicago, IL: University of Chicago Press.

Enloe, C. (1989) *Bananas, Beaches & Bases: Making Feminist Sense of International Politics*. London: Pandora Press.

Fanon, F. (1952/2007) *Black Skin, White Masks*. New York, NY: Grove Press.

Fanon, F. (1963/2002) *The Wretched of the Earth*. New York, NY: Grove Press.

Frazer, E. and Hutchings, K. (2008) 'On Politics and Violence. Arendt Contra Fanon', *Contemporary Political Theory*, 7 (1), pp. 90–108.

Fregonese, S. (2012) 'Urban Geopolitics 8 Years on. Hybrid Sovereignties, the Everyday, and Geographies of Peace', *Geography Compass*, 6 (5), pp. 290–303.

Gago, V. and Mezzadra, S. (2018) 'A Critique of the Extractive Operations of Capital: Toward an Expanded Concept of Extractivism', *Rethinking Marxism*, 29 (4).

Galtung, J. (1969) 'Violence, Peace, and Peace Research', *Journal of Peace Research*, 6 (3), pp. 167–191.

Glezos, S. (2013) *The Politics of Speed*. London: Routledge.

Global Witness. (2015) *Annual Report 2015*. Available at: www.globalwitness.org.

Golubchikov, O., Badyina, A., and Makhrova, A. (2014) 'The Hybrid Spatialities of Transition: Capitalism, Legacy and Uneven Urban Economic Restructuring', *Urban Studies*, 51 (4), pp. 617–633.

Guillaume, X. and Huysmans, J. (2019) 'The concept of 'the everyday': Ephemeral politics and the abundance of life', *Cooperation & Conflict*, 54 (2), pp. 278–296.

Hom, A.R. (2018) 'Silent Order: The Temporal Turn in Critical International Relations', *Millennium: Journal of International Studies*, 46 (3), pp. 303–330.

Homer-Dixon, T.F. (1999) *Environment, Scarcity, and Violence*. Princeton, NJ: Princeton University Press.

Kaldor, M. (2001; reprint of the 1999 ed.). *New & Old Wars. Organised Violence in a Global Era*. Cambridge: Polity Press.

Keen, D. (2000) *The Economic Functions of Violence in Civil Wars*. New York, NY: Oxford University Press.

Le Billon, P. (1999) '*A Land Cursed By Its Wealth? Angola's War Economy 1975–99*' in *Research in Progress*, 23. The United Nations University/World Institute for Development Economics Research.

Lundborg, T. (2012) *Politics of the Event. Time, Movement, Becoming*. London: Routledge.

Lundborg, T. (2016) 'The Limits of Historical Sociology. Temporal Borders and the Reproduction of the 'Modern' Political Present', *European Journal of International Relations*, 22 (1), pp. 99–121.

Malesevic, S. (2010) *The Sociology of War and Violence*. Cambridge and New York, NY: Cambridge University Press.

Malesevic, S. (2017) *The Rise of Organised Brutality. A Historical Sociology of Violence*. Cambridge: Cambridge University Press.

Martín, F. (2017) 'Reimagining Extractivism: Insights from Spatial Theory' in Engels , B. and Dietz, K. (eds.) *Contested Extractivism, Society and State: Struggles over Mining and Land*. London: Palgrave Macmillan, pp. 21–44.

Marx, K. (1885/1988) *Das Kapital. Kritik der politischen Ökonomie Band. 1: Der Produktionsprozeß des Kapitals, Karl Marx—Friedrich Engels—Werke*, Band 23. Berlin: Dietz.

Massey, D. (1996) 'Space/Power, Identity/Difference: Tensions in the City' in Merrifield, A. and Swyngedouw, E. (eds.) *The Urbanisation of Injustice*. London: Lawrence & Wishart, pp. 100–117.

Massey, D. (2004) 'Geographies of responsibility', *Geografiska Annaler B*, 86 (1), pp. 5–18.

McNeish, J. (2018) 'Resource Extraction and Conflict in Latin America', *Colombia Internacional*, 93, pp. 3–16.

Mezzadra, S. and Neilson, B. (2017) 'On the multiple frontiers of extraction: Excavating contemporary capitalism', *Cultural Studies*, 31 (2–3) pp. 185–204.

Münkler, H. (2005) *The New Wars*. Cambridge: Polity Press.

Nixon, R. (2011) *Slow Violence and the Environmentalism of the Poor*. Cambridge, MA, and London: Harvard University Press.

Schlichte, K. (2009) *In the Shadow of Violence: The Politics of Armed roups*. Frankfurt: Campus Verlag.

Solomon, T. (2014) 'Time and Subjectivity in World Politics', *International Studies Quarterly*, 58 (4), pp. 671–681.

Springer, S. (2009) 'Culture of Violence or Violent Orientalism? Neoliberalisation and Imagining the 'Savage Other' in Post-Transitional Cambodia', *Transactions of the Institute of British Geographers*, 34, pp. 305–319.

Springer, S. (2011) 'Violence Sits in Places? Cultural Practice, Neoliberal Rationalism, and Virulent Imaginative Geographies'. *Political Geography*, 30 (2): 90–98.

Springer, S. and Le Billon, P. (2016) 'Violence and Space: An Introduction to the Geographies of Violence', *Political Geography*, 52 (1–3), pp. 521–533.

Sylvester, C. (1994) *Feminist Theory and International Relations in a Postmodern Era*. Cambridge: Cambridge University Press.

Tsing, A.L. (2005) *Friction: An Ethnography of Global Connection*. Princeton, NJ: Princeton University Press.

Tsing, A.L. (2015) *The Mushroom at the End of the World. On the Possibility of Lie in Capitalist Ruins*. Princeton, NJ and Oxford: Princeton University Press.

Tyner, J.A., and Inwood, J. (2014) 'Violence as fetish: Geography, Marxism, and dialectic', *Progress in Human Geography*, 38 (6), pp. 771–784.

Verweijen, J. (2020) 'A microdynamics approach to geographies of violence: Mapping the kill chain in militarized conservation areas', *Political Geography*, 79.

Wacquant, L. (2018) 'Bourdieu Comes to Town: Pertinence, Principles, Applications', *International Journal of Urban and Regional Research*, pp. 90–105.

Wibben, A.T.R. (2011) Feminist Security Studies. A Narrative Approach. Edited by J. Peter Burgess, *PRIO New Security Studies*. London and New York, NY: Routledge.

Zimmermann, L. (2017) More for Less: The Interactive Translation of Global Norms in Postconflict Guatemala', *International Studies Quarterly*, 61 (4), pp. 774–785.

Zwingel, S. (2012) 'How Do Norms Travel? Theorizing International Women's Rights in Transnational Perspective', *International Studies Quarterly*, 56 (1).

3

THRESHOLDS OF INJUSTICE

Challenging the Politics of Environmental Postponement

Paul Wapner

Introduction

The modern environmental movement arose in a moment of panic. Throughout the 1960s and 1970s, "prophets of doom" warned, literally, of the end of the world. They based their alarm on the fact that the Earth has biophysical limits and that industrial society was pushing up against critical ecological thresholds. Exploding population, growing affluence, unparalleled technological innovation, and mass consumerism, while signs of human wellbeing, were nonetheless ripping at the ecological fabric of the Earth and undermining the life-support system of the planet. If left unchecked, they would quickly deplete critical resources, overfill essential sinks, and compromise ecosystem functioning. Many scientists and activists at the time predicted planetary collapse within decades (Ehrlich, 1970; Commoner, 1971; Falk, 1972; Meadows *et al.*, 1972; Shabecoff, 2003).

Anti-environmentalists remind people about the exaggerated claims made by earlier environmentalists. They point out that, despite dire warnings, "we" are still here. The planet's organic infrastructure seems to be holding, and civilization seems to be thriving. To be sure, many species have disappeared, multiple forms of pollution plague the world, and many ecological hotspots have been severely despoiled. Nonetheless, there have been no planet-wide famines or extinctions and, while humans might have altered the carbon, nitrogen, and hydrological cycles in dramatic ways, none of these has yet cracked in a manner to induce planetary ecosystem failure. The overshoot that many foresaw simply did not happen. As libertarian Ronald Bailey put it, "The prophets of doom were not simply wrong but spectacularly wrong" (2000).

How does one make sense of earlier environmental predictions? The question is not simply of historical interest. Contemporary environmentalists continue to see biophysical limits and presage ecological disaster. They recognize impending calamity—especially in the context of climate change and loss of biological diversity but also with regard to toxic chemicals, non-degradable plastics, freshwater scarcity, and deforestation. They share the view that the Earth's ecosystems can handle only so much assault and that current human practices are pushing the limits. To the degree that earlier warnings have come to naught, however, one can legitimately question current concerns—and this is, indeed, what is actually happening. Today, climate skeptics point to exaggerated forecasts of the 1970s to ridicule contemporary climate alarm while other critics highlight earlier warnings about mass starvation, depletion of oil supplies, or deadly air pollution to cast doubt on contemporary environmental concerns (see, for instance, Michaels and Maue, 2018; Ebell and Milloy, 2019). As William Faulkner famously wrote, "The past is never dead. It's not even past" (2012, p. 73). Reflecting on the character and fate of earlier prognoses is necessary to understand the fate of contemporary environmental affairs.

The most powerful rejoinder to critics of earlier predictions is that forecasts of the 1960s and 1970s were not wrong *per se* but merely inaccurate in terms of timeframe. The Earth's biophysical infrastructure might seem to be holding, but arguably not for long. Species are disappearing at unprecedented rates (United Nations, 2019); atmospheric carbon concentrations are higher than they have been in three million years (Mingle, 2020, p. 50), a mere fifteen percent of forests around the world remain intact (Scranton, 2015), and 90 percent of fish stocks around the world are overharvested (Kituyi, 2018). Moreover, many other indicators suggest that planetary boundaries are being pressed as never before (Steffen *et al.*, 2015). All of this might not add up to a single, apocalyptic punch, but this could be merely a matter of time. Earlier environmentalists suffered from what Richard Falk calls "premature specificity" (Falk, 1975, p. 1002). They lent too much precision and temporal explicitness to their extrapolations. In this sense, their timelines might be inaccurate, but their overall insights about planetary collapse remain relevant. As many environmentalists would say, if you are doubtful about the veracity of earlier warnings about planetary wellbeing, just wait.

In this chapter, I want to offer a second, complementary response to critics of earlier environmental warnings. This second reply has less to do with the accuracy of previous predictions and more with the politics those predictions have come to recommend. By invoking scenarios such as the "end of civilization," "planetary collapse," "massive die-offs," and "human extinction," scientists and activists unwittingly created a visibility problem. They cast a gaze towards the planet as a whole and offered planetary dismemberment as the primary criterion for significant environmental harm. In doing so, they blinded themselves and others to how environmental harm actually "lands" in the world. They therewith unwittingly offered a politics that could capitalize on such blindness. To put it more specifically, by focusing on the globe, earlier environmentalists conceptualized environmental destruction at such a high level of abstraction that they bleached out the lived experience of those on the frontlines of environmental degradation and consequently politicized determination of what constitutes genuine environmental harm and how societies respond. By focusing on global thresholds rather than on-the-ground hardship, politicians and others in power are able to mitigate the urgency of environmental threats and delay meaningful action. They can engage in a "politics of postponement."

In the following section, I explain the politics of postponement. I show how ecological pressures lodge into particular bodies and minds and how political power both structures such hardship and uses it to extend global limits into the future. At the heart of the analysis is a focus on "we." When earlier environmentalists advanced a narrative of global collapse, they lumped everyone together; they spoke of a single humanity that faced extinction or at least a unitary environmental experience. The legacy of that narrative makes possible a type of political elasticity that forestalls (for many) coming into direct contact with or even noticing the crisis proportion of environmental destruction. More pointedly, the narrative encourages uneven distribution of suffering such that, when resources dwindle, sinks are depleted, or general environmental conditions worsen, those best able to avoid pain do so by shifting hardship onto the backs of others. They build protective walls, design complex financial shell games, lengthen and bend commodity chains, or simply enjoy the fruits of material consumption and comfort while exporting the harm involved to the less privileged. In this way, thresholds become political tools of "epistemic injustice" (Fricker, 2009). They hide unfairness while seemingly trying to reveal global realities. In short, the 1960s and 1970s narrative allows the more powerful to place others at the frontier of environmental degradation while nonetheless using a language of "we" to measure and categorize genuine environmental harm. Such practice fuels a politics of postponement.

A politics of postponement hides not only specific environmental harms but obscures the hyper-extractivism taking place today. As this volume explains, hyper-extractivism involves the pervasive use of exploitation and violence to wrestle more resources from the Earth and more labor from already overtaxed workers in ever-shorter spans of time. To the degree that environmentalists continue to draw attention to the globe itself and eclipse the lived experience of those on the frontlines of climate disruption, environmental toxicity, freshwater scarcity, and so forth, they normalize hyper-extractivism. To be sure, many environmentalists now include challenges of injustice in their activism, and they have reconceptualized their singular focus on global wellbeing. But this has gone only so far. The concept and empirical intensification of extractivism invite renewed scrutiny of the overall frame of much environmentalism. They welcome a chance to reconceptualize environmental politics in the Anthropocene.

This chapter proceeds in the following manner. In the next section, I explain the development of threshold thinking. I trace a line of thought that stretches from Thomas Malthus to the Intergovernmental Panel on Climate Change (IPCC) and describe how such thinking establishes a general framework that distracts from concerns of justice, exploitation, and violence. In the second section, I explain how threshold thinking enables the privileged to displace environmental harms onto others and thus to sidestep the direct experience of the kind of dangers they predict. This move allows the privileged to postpone their own environmental reckoning and thus steal attention from global environmental decline. In the third section, I relate how the Environmental Justice (EJ) movement has tried to disrupt a politics of postponement by bringing disproportional environmental suffering into high relief. This section not only explains EJ's efforts but also points out some unintended consequences of linking environmental concerns with social justice and how this has emboldened the powerful to continue practicing a politics of postponement. In the fourth section, I describe the emergence of a new wave of the EJ movement, namely, extrACTIVISM. ExtrACTIVISM represents a politics that avoids concern for global thresholds and focuses primarily on resisting the extraction of resources on particular lands and on empowering local communities to gain greater control over their environmental fates. I present extrACTIVISM not as an answer to environmental assaults but as a distinct effort to reframe environmental harm in ways that can disrupt the politics of postponement. In the concluding section, I summarize the article's main argument—that global threshold thinking and disproportional environmental pain enable elites to postpone environmental action—and I discuss the stakes for hyper-extractivism and the Anthropocene.

The Legacy of Postponement

Most people locate the beginnings of environmental apocalypticism with Reverend Thomas Malthus. Living in the late eighteenth century as the Industrial Revolution was gathering increasing speed, Malthus worried about population increase. At the time, fewer than a billion people lived on Earth, yet they were multiplying quickly as agricultural advancements produced greater amounts of food. Instead of using abundance to enhance the quality of life, food supplies simply encouraged more population growth—a condition known as the "Malthusian trap." The problem, as Malthus saw it, is that, once sparked, population would grow geometrically while food production could grow only arithmetically, and thus it would be impossible continuously to feed the world's population. At some point, human numbers would outpace food supply, resulting in hunger, disease, and war. In this way, population growth would lead to ecological overshoot. At the heart of Malthus's views rests an understanding of a finite Earth. Food production is tied to the Earth's productivity. Malthus saw food availability then as determinate. "The power of population is indefinitely greater than the power in the earth to produce subsistence for man [sic]" (Malthus, 1789/2007, p. 5). Earth, in other words, possesses biophysical limits. If mankind crosses Earth's biophysical limits it will face hardship and, in the extreme, planetary collapse. (For a nuanced reading of Malthus that distinguishes "limits" from "scarcity," see Kallis, 2019.)

As is now widely recognized, Malthus's predictions were wrong—or at least unrealized in his time. Malthus failed to appreciate that, while human numbers undoubtedly would skyrocket, so could food production. Using innovative technologies, including better crop breeding and eventually the Green Revolution, people found ways of accelerating agricultural productivity. Indeed, in many but not all parts of the world, food production has kept pace with and even outpaced human population growth since Malthus's time—even if many question if this can continue in the face of climate change, urbanization, and lack of investment (Elferink and Schierhorn, 2016). As the existence of food "mountains" suggest, hunger, conflict, and pestilence no longer result from food shortages but rather from unequal distribution. Malthus misjudged human ingenuity, but equally he misjudged humanity's tolerance for hardship and extreme inequality.

Modern environmentalists of the 1960s and 1970s picked up on Malthus' essential understanding (Kallis, 2019). To them, Malthus saw the big picture but got the details wrong. A few thinkers maintained that population would someday outpace food supply (Brown, 1974). Others thought that Malthus misidentified the critical factor. To them, food supply might well keep up with population growth, but the same cannot be said for freshwater, clean air, fertile soil, fish, timber, or critical minerals (Meadows et al., 1972). These undergird the planetary ecosystem and, unlike food production, have more determinate physical limits. By 1970, world population had quadrupled since Malthus's time (reaching close to 4 billion); world GDP increased from a few billion dollars to nearly $27 trillion (Roser, 2018); and consumerism became a mass phenomenon increasingly global in scope. Together, these increases pressed against the Earth's ability to produce resources and absorb waste as never before. Humanity might have dodged the food bullet, but it could not avoid crippling the less pliant dimensions of the planet's ecosystem.

Warnings came in a variety of forms. Paul Ehrlich's 1968 book, *The Population Bomb*, sounded a neo-Malthusian alarm around growing human numbers. Ehrlich argued that the 1970s and 1980s would be a dark era of resource scarcity and widespread famine wherein 'hundreds of millions of people will starve to death' (Ehrlich, 1970, p. xl). The Club of Rome's 1972 book, *The Limits to Growth*, offered a related prediction by identifying growing numbers, affluence, and consumption as incompatible with the planet's biophysical parameters. The authors warned that if current trends continued, the 'most probable result will be a sudden and uncontrollable decline in both population and industrial capacity' (Meadows et al., 1972, p. 23). Richard Falk's 1972 book, *This Endangered Planet*, added an important dimension to such admonitions by explaining how the war system and a competitive nation-state system exacerbate both resource extraction and the generation of waste, even as they fail to provide an appropriate political response. If left unreformed, the international state system would guarantee planetary environmental overload. Indeed, one of two scenarios Falk that foresaw was a twenty-first century of desperation followed by annihilation (1972). These volumes highlight a movement of thought that identified humanity as a growing, predatory, and voracious species seemingly unstoppable in its quest to grab resources and pump out waste, therewith compromising Earth's ecosystem functionality. Like Malthus, they envisioned a set of brittle thresholds that, once crossed, would usher in planetary decline and much hardship. The Earth's fundamental organic infrastructure was not forever pliant and forgiving. At some point it would reveal itself as inviolable. When such a threshold is reached, humanity itself faces widespread hardship and even extinction.

Despite Malthus's blind-spot and the premature specificity expressed by thinkers of the 1960s and 1970s, much contemporary environmentalism remains Malthusian. To many, the Earth is still finite and will, sooner or later, buckle, as a combination of population and consumptive pressures cripples the planet's regenerative biophysical capacity. Indeed, since the 1960s and 1970s almost every indicator of environmental harm has worsened, and many of these have implications for global unsustainability (Steffen *et al.*, 2009; 2015). This should come as no surprise, as the fundamental drivers that earlier environmentalists identified have not gone away but rather have intensified and globalized. Today, there are almost eight billion people on Earth, and although global population growth is decelerating, it continues to increase at 1.09 percent per year. In addition, the capitalist economy penetrates almost every niche of economic life, leading to intensified commodification and an unending practice of extractivism. Furthermore, changed technological capacity enables people to commandeer resources and emit waste at an increasingly faster pace, while mass consumerism, which equates value and happiness with material accumulation, has gone global as an act of secular faith (Assadourian, 2015). It certainly seems that if Malthus's time is ever to come, it is now.

Most environmentalists see the Malthusian apocalypse expressed in the form of climate change. The buildup of greenhouse gases is oversaturating the Earth's absorptive capacity. Heatwaves, wildfires, droughts, rising sea levels, intensified storms, and coastal flooding have become the new normal and, very soon, these will punch through the Earth's last remaining ecosystem bulwarks. Scientists are clearly fighting on the barricades. As a recent Intergovernmental Panel on Climate Change (IPCC) report underlines, temperatures should not exceed 1.5 degrees Celsius (°C) and certainly should go no higher than 2°C—the internationally agreed upper limit of acceptable global average temperature change (IPCC, 2018). Beyond 1.5°C and certainly 2°C, the planet bakes. Positive feedback loops—associated with the planet's albedo effect, release of methane from melting permafrost, thermal expansion of the oceans, and so forth—kick in, resulting in runaway climate change. The IPCC report makes clear that the world must cut emissions by 45 percent below 2010 levels by 2030 to stay within the 1.5°C limit to have a realistic chance of avoiding catastrophic climate change (2018). However, no concrete political plan exists within or across states to approach such necessarily aggressive mitigation. Indeed, the world is expected to surpass the 1.5° Cdegree limit by 2030 (McKibben, 2019, p. 15). Seen in this way, additional carbon emissions will soon hit the seemingly most important planetary threshold; they will become the metaphorical straw that breaks the Earth's back. After centuries of guessing, environmentalists have finally identified the golden eco-systemic marker. Carbon dioxide and other greenhouse gases stand as the limiting factors. Malthus is back, and his warning is on steroids—the best scientific evidence suggests that this time it is for real.

Thresholds

There is no doubt that climate change endangers the Earth as a living, functioning ecosystem. However, it is worth questioning the discursive frame within which the climate apocalypse is being expressed. The intellectual thread from Malthus to the IPCC envisions an inelastic Earth with brittle thresholds, whose demise will endanger humanity as a whole. Lost from such an understanding—or at least underplayed—is the role of *political elasticity* in shaping how the "end" literally unfolds. Lost is an incisive look at how politics tends to direct the fangs of climate calamity, so that they rip first and most violently into the lives of the underprivileged while shielding the powerful from coming into catastrophic contact with climate change. By lumping everyone together and envisioning a unitary environmental experience, the Malthusian legacy occludes the possibility of noticing the "end" as many currently experience it.

Consider the extractivist pattern that gives rise to climate change. Unearthing fossil fuels involves not only hard work to wrest oil, gas, and coal out of the Earth, but also the negative health effects on people who live near coal mines, oil refineries, and hydraulic fracturing facilities. Pollution from such operations is linked to asthma, emphysema, and heart disease, as well as toxic exposure to petroleum hydrocarbons (Israel, 2012). These costs of a carbon economy are unevenly distributed and often hidden from those "living across town" who enjoy the advantages produced by such hardship. At work is a dynamic wherein some directly benefit from the wonders of fossil fuels while others pay the price. To be sure, miners, refiners, and others employed in the carbon extraction industry gain economically by having jobs, and a job can go a long way towards offsetting certain hardships. But this should not blind one to the broader pattern of fossil fuel production. As Bullard and others powerfully argue, work in dirty industries is a form of economic blackmail. Many cannot afford or otherwise lack the ability to leave their jobs despite health and environmental concerns (Bullard, 2000; Pellow and Brulle, 2005; Lerner, 2010). At the production end of climate change, the economically strapped then serve as the absorbers of ecological harm yet their pain fails to register as catastrophic since it does not eventuate in planetary collapse. It is as if global endangerment represents a single planetary moment instead of an unraveling process that etches itself onto other people's skin.

Something similar happens at the opposite end of the carbon economy, as the costs of burning fossil fuels disproportionately fall on the marginalized. Heatwaves, storm surges, wildfires, and flooding in and of themselves might not distinguish the rich and politically connected from the poor and politically sidelined, but the politics involved ensure that they afflict the latter more than the former. For instance, living on fragile lands and in deficient housing, cut off from many social services, usually the last to receive aid, and immediately vulnerable to income disruption or loss of work, the poor cannot easily escape or recover from climate-related calamities. They stand naked, as it were, in the eye of climate intensification. This is particularly troubling, considering that they contribute the least to the problem. For example, Nepal, a country of almost three million people, generates almost all its energy through hydroelectric power and biomass; its per capita energy use is tiny compared to most other nations; and it houses no significant fossil fuel industry. Yet it remains one of the most vulnerable countries to climate change (USAID, 2012). This is largely due to its extreme poverty, which makes it difficult to adjust to vulnerabilities associated with its topography (steep mountains in the north make it particularly vulnerable to the flooding and landslides often associated with glacier melt) and to its system of rain-fed agriculture (that is defenseless against sustained drought). Over the past few years, Nepal has been hit particularly hard by landslides and mountain flooding, owing to erratic and intensified rains in the north and sweltering heat and droughts in the southern plains that many associate with climate change. Lacking the means to escape such conditions or recover from extreme hardship, many Nepalese bear the brunt of climate intensification. They stand on the receiving end but, far from international limelight, suffer largely on their own and thus obscure evidence of climate catastrophe.

Such obscuration and experiential inequality get lost in the threshold thinking that informs most climate assessments. The 2018 IPCC report makes clear that a rise of 1.5°C will send many ecosystems into feedback loops and make it difficult to stop coral reef die-off, coastal flooding, and an ice-free Arctic Ocean (2018). At 1.5°C and certainly 2°C, runaway climate change will be unavoidable. As mentioned, whole ecosystems will buckle. And yet, what does 1.5°C mean to the victims of hurricanes Katrina or Sandy, the typhoons of Tembin or Mangkhut, or the 2020 wildfires in Australia? For that matter, what does a threshold mean in general for those who have already been victimized by climate devastation? The planet as a whole might seem like the most important measure of climate calamity, but it does not monopolize climate misfortune or erase the upheaval that many already experience. It does not fully capture what the "end" can mean and, more troubling, the political dynamics of climate intensification.

So far, in every significant climate-related tragedy, the wealthy and otherwise more powerful have fared better than their counterparts. For instance, the poor were the least able to leave New Orleans during Hurricane Katrina, lacked sufficient insurance coverage, and were among the last to receive aid during and directly after the hurricane. In fact, strong evidence suggests that they were deliberately neglected during the actual hurricane and have been taken advantage of in the reconstruction (Dyson, 2007; Adams, 2013). This was also the case with Hurricane Sandy. The least able lacked the means to escape the storm and many of them faced significant hurdles seeking compensation (Huang, 2012; Sellers, 2017). In the Philippines, recent typhoons disproportionately victimized those living in vulnerable structures and on fragile lands. Most deaths were associated with collapsing buildings and landslides (Beech, 2018). Even wildfires get refracted through the lens of social stratification. Although the 2018 Woosley fire ripped through Malibu—one of the most affluent neighborhoods around Los Angeles—its effects damaged some more than others. The few remaining middle-class families in the area have not only lost their homes, but many also lack adequate insurance, sufficient savings to rebuild, and a reliable social safety net to ensure that they can ride out the hardship. Indeed, for some in this situation, temporary homelessness or unemployment can set off a spiral of downward mobility and poverty. This is very different from the experience of the more comfortably off Malibu residents who were able to jet away from the catastrophe and, although they lost their homes, possessed the means to permanently relocate or rebuild. Some of the super-wealthy even hired private firefighters to save their properties (White, 2018). This does not mean, of course, that hardship is ever easy to endure or that some people's pain is less important than others. However, it does point to the nexus between inequality and climate change. According to a 2016 United Nations report:

> Large inequalities in access to physical and financial assets; unequal access to quality health services, education and employment; and inequality with respect to voice and political representation aggravate the exposure and vulnerability of large population groups to climate hazards
>
> *(Islam and Winkel, 2016, p. 1)*

This should be no surprise; it makes sense that one's ability to respond to hardship rests partly on one's socio-economic status, gender, skin color, and other stratifying characteristics. Threshold thinking, however, erases such distinctions.

Threshold thinking also, by consequence, encourages a certain kind of politics. It enables the practice of displacing environmental harm. In this sense, people rarely solve environmental problems so much as export them to others. They send them across space, time, and species. For instance, as already mentioned, when it comes to climate change, people transfer the harm of extraction and exposure to those living "downstream"—to the poor and politically weak who contribute the least to the buildup of greenhouse gases but lack the means to avoid the pain of climate disruption. Likewise, they export the pain of extreme climate intensification to future generations—another category of people who did not cause the problem and who are essentially politically voiceless. Finally, climate politics ends up shifting much climate intensity onto other species as plants and animals often absorb the brunt of climate disruption. To cite one example, upwards of half a billion species died in the 2020 Australian wildfires (British Broadcasting Corporation, 2020). At work is a political shell game of enjoying the benefits of fossil fuels while shifting the harms involved.

Displacement politics is part of the politics of postponement introduced earlier in this chapter. If others are on the frontlines, if the privileged have the ability to dodge environmental pain by building fortresses or otherwise having the means to adjust, if climate and other environmental calamities always seem to happen "elsewhere," if, in other words, the experience of environmental harm falls disproportionately on the least able, then it is easy not to notice and thus delay addressing environmental challenges. In fact, the disaggregation of environmental victimhood necessitates deferment. As the adage goes, "a problem postponed is partially solved." Deflecting environmental degradation is a perfect tool of postponement. It shifts the burden and thus hides reality and consequently mitigates a sense of urgency. This is why so many predictions—in the 1960s and 1970s as well as today—can easily be labeled irrational and overblown. So long as the privileged can shift environmental assault, they can dispel any sense of imminent doom.

The Promise and Peril of Confronting a Politics of Postponement

Comparing degrees of oppression and levels of pain is fraught with difficulty. One cannot ethically declare that one form of atrocity is worse than another or that one person's hardship is more agonizing than someone else's. Suffering is part of the human experience, and, sadly, it seems that so are persecution and subjugation. In this context, there is nothing new about the inequities that accompany environmental displacement, nor do they represent unprecedented torment or misery. The extractivism going on today—of people, of other creatures, and of the Earth itself—is part of a long chain of extortion and distress often associated with colonial legacies, including "internal colonialism" (Willow, 2018, p. 8). However, it is precisely its familiarity and normalized quality that makes it so pernicious.

Today, as some theorists have pointed out, we live in a "post-ecological" age (Bluhdorn and Walsh, 2007). This means that, while environmentalism has gone mainstream, and widespread public opinion accepts the need to change established values, lifestyles, and social practices to address environmental dilemmas, there is a paradoxical unwillingness or inability to do so. Current institutions continually reproduce the causes of environmental degradation and even intensify them; if anything has changed, it is the adaptation of our politics to unsustainability itself. As Bluhdorn has demonstrated in an impressive body of work (2001; 2007), the world goes through the motions of seeking environmental alternatives but never generates the willingness to carry them out. Instead, states, businesses, and civil society practice what he calls a "politics of unsustainability." They have devised a way to organize power so that societies can lumber through climate intensification, freshwater scarcity, increasing toxification, and biodiversity loss. They have, in Bluhdorn's words, perfected the desire to "sustain the unsustainable" (2007). This is possible precisely because of the elasticity of environmental dislocation. Unsustainability rests on the backs of those on the receiving end of environmental displacement.

The EJ movement has tried to disrupt the politics of unsustainability. It has pointed out how poor neighborhoods receive the brunt of dirty industries and suffer the worst environmental conditions and how weaker nations become the hazard-plagued workhorses and dumping grounds of the global economy. The EJ movement has delineated the racist, classist, and gendered dimensions of such injustice and offered the strategic insight that any environmental analysis must take into account the role of the poor and marginalized, and that environmental groups must ally with social justice organizations (see, for instance, Bullard, 2000; Agyeman, 2005; Taylor, 2014; Detraz, 2016; Gaard, 2019). Indeed, today, many climate activists understand themselves as part of a climate justice movement. Groups that are focused on a panoply of environmental issues recognize that mass mobilization rests on making common cause with social justice politics. Today, one can no longer separate environmental and justice issues. And yet, the alliance has not demonstrably enhanced environmentalism's power to induce change. It might have given the movement more legitimacy and has certainly added to environmentalism's analytic understanding of how power operates when it comes to resources, pollution sinks, and decisions around extraction siting. However, the politics of unsustainability has largely subsumed EJ within its hegemonic grasp. Indeed, to the degree that environmental harm is now seen as simply another dimension of injustice, it has become dangerously normalized as part of a wider normalization of injustice in general.

Not only has a concern for social justice failed to sharpen environmentalism's critique; it has arguably blunted the movement's efforts. It has done so in two ways. First, the perennial character of injustice implicitly encourages a wallpapering of lines of conflict. To the degree that environmental harm is simply another instance of the powerful lording over the weak, a response is increasingly, "What else is new?" This has been a problem of many social movements in that, by expanding campaign foci, they dilute the intensity of their critiques. The inclusion of social justice has, to a degree, muffled the cutting edge of environmentalism criticism.

Second and perhaps more importantly, linking social justice with environmental concerns has, paradoxically, created new lines of conflict insofar as it has allowed various states to associate environmentalists with other, more enduring and more threatening forms of dissent. Today, because environmentalists aim to correct long-standing social injustices, many governments have come to see environmental activists as threats to the state itself. This has led to labeling environmentalists as national security threats and thus aggressively cracking down on their activities.

Environmentalists have become regular participants in movements for social justice. They played a significant role in the Occupy Movement and continue to be involved in demonstrations at G20 summits, anti-mining protests in countries including South Africa, Peru, Chile, Bolivia, and Colombia, and anti-globalization campaigns around the world (Rowe and Carroll, 2014; 2015). This association has allowed governments to criminalize environmentalist efforts. This has been especially the case since 9/11. In a post-9/11 world, various states have securitized dissent and environmentalists have been swept up as accomplice targets (Dauvernge and LeBaron, 2014). Today, states are using excessive force to repress public protest. Tactics include clubbing, pepper-spraying, tear-gassing, and military-style law and order methods of crowd control. Moreover, states are using surveillance, paramilitary policing, and infiltration to disrupt all kinds of dissent, including and, in many cases especially, environmentalist organizing. This has resulted not only in chilling environmental opposition—captured most dramatically in labelling of environmentalists as "ecoterrorists" or, in the case of the Tibetans of China, "separatists"—but also in deploying direct violence against environmentalists. According to Global Witness, between 2002 and 2013 a total of 908 environmental activists were killed for protesting or merely questioning corrupt land use practices (Lakhani, 2014). Such incidents show no sign of abating. Indeed, Global Witness identified 2017 as the deadliest year for environmentalists, documenting over 200 murders involving resistance to mining and logging, or standing up for land rights (Zachos, 2018). As Peter Dauvergne and Genevieve LeBaron make clear, these numbers are almost certainly an underestimate,a s many states rarely report such killings or identify them under the misleading labels of "accidents," "muggings," or "missing persons" (2014, p. 63).

Furthermore, within social justice/environmental activism, some states have specifically identified environmentalists as deserving of heightened scrutiny. For instance, the Harper government in Canada listed "eco-extremists" as a key threat in its 2012 anti-terrorist strategy, and in the USA the Federal Bureau of Investigation has, since 2004, identified eco-extremists as among the top domestic terrorist threats (Dauvergne and LeBaron, 2014). Today, as part of a surge in nationalist, populist politics, environmentalists are coming under increasing fire. Brazilian President Jair Bolsonaro has opened the Amazon for business even more than his predecessors and is encouraging cattle rangers to arm themselves against indigenous people and environmentalists working to protect the forests (Branford *et al.*, 2018). Likewise, President Donald Trump labelled environmentalists as terrorists and, together with former Interior Secretary Ryan Zinke, blamed them for the 2018 wildfires in California (Kasler and Sabalow, 2018) and castigated them for challenging climate skepticism (Waldman, 2018). Indeed, the US Department of Homeland Security listed activists who protest oil pipelines as "extremists," together with white supremacists and mass murderers. For instance, a recent intelligence bulletin evaluating domestic terrorist threats identified a group known as the "Valve Turners," who have nonviolently disrupted pipelines by turning off intake valves, as "suspected environmental rights extremists" worthy of surveillance and prosecutorial excess (Federman, 2020) and nonviolent activists protesting siting a plastics plant in a predominantly black neighborhood have been accused of terrorism (Brown, 2020). In short, as environmentalists have recognized the broader social justice dimensions of their work, they have run into the national security machine. As a result, they have suffered intimidation, organizational-fracturing, and even assassination and murder.

The normalization of violence against environmentalists represents a modern-day effort to advance the extractivist agenda and the politics of unsustainability. As activists react to structural and proximate environmental injustices—seeking to erect "dikes" for those living downstream—it is unsurprising that the powers-that-be strike back. The logic of the system requires that harm shift outward into the lives of the less fortunate. This is the nature of extractivism and the face of contemporary hyper-extractivism. It thus tries to smother resistance. In this sense, the EJ movement is fighting an uphill battle. As it works to resist options for dispersing hazards to the less fortunate, the more powerful crack down with greater effort. Hyper-extractivism kicks into high gear. Sustaining the unsustainable thus fuels the politics of postponement.

ExtrACTIVISM

The EJ movement has evolved since it first emerged in the 1980s. Initially a Western-based movement focused on how pollution and environmental degradation disproportionately affect people of color and low-income communities, EJ has grown to be a worldwide movement made up of various groups focused on the intersection between environmental protection and civil, social, and cultural rights. In some places, EJ emphasizes equal rights to environmental protection under the law and targets law-making and regulatory policy as means to achieve environmental justice. In other places, EJ adopts a grassroots orientation wherein activists work at the frontlines of environmental exploitation while trying literally to stop environmental assaults and to protect the dignity and livelihoods of victimized communities. This latter dimension has grown significantly—or at least gained greater public prominence—over the past few decades and is altering the political character of the EJ movement as a whole.

One area of grassroots EJ deserves particular attention in the context of this chapter. Over the past few decades, numerous countries have pursued development policies based increasingly on hyper-extractivism. They have sought economic growth through the ever-increasing and more intensified removal of minerals, oil, wood, and other resources (Burchardt and Dietz, 2014). It is part of a "mindset and pattern of resource procurement based on removing as much material as possible for as much profit as possible" (Willow, 2018, p. 2). While humans have always pulled resources from the Earth, hyper-extractivism indicates a fundamental shift in scale and pace that has resulted in planet-spanning transport and finance networks that are reconfiguring spatial relations and reorganizing capital. Today, innovative technologies and consolidated deployment of capital are enabling unprecedented amounts of oil, aluminum, mercury, iron-ore, lead, copper, natural gas, timber, and wildlife to be removed from the Earth—levels that arguably rival the combined extraction of almost all previous historical periods (Willow, 2018, p. 6). As can be expected, such extractivism is taking an enormous toll on the Earth's ecosystem functionality and, importantly, ruining particular areas and endangering specific communities. Hyper-extractivism as a form of globalized development thus represents a fiercer face of environmental degradation.

Hyper-extractivism is also leading to a new kind of activism. It is pushing the envelope of EJ by focusing specifically on the displacement character of environmental harm and thereby directing activists to focus on unjust elements of the new extractivist intensification. Whether it is mountain-top removal for coal, hydraulic fracturing to capture natural gas, expansion of agricultural land for paper and pulp, or massive clear-cuts for timber, hyper-extractivism almost always involves plundering areas of the politically weak and concentrating the resultant wealth generation into the hands of the few who live, work, and play far from extractivist sites. One result is that people who work within and live near extractive industries are banding together to resist the assault. Dubbed "extra-ACTIVISM" by some scholars to denote a kind of "extraordinary activism" that must confront industrial-level actions on the ground (Kidd, 2016) or simply action against large-scale extractivist industries (Willow, 2018), this kind of collective action fights for sovereignty over people's bodies and over the environmental resources of people's communities (see also Hern and Johal, 2018). The "extraordinary" dimension has to do with the desperate character of such efforts. Unwilling to be continually thrown against the wall—to have their land stolen, bodies poisoned, labor forever exploited, or future blighted—but up against extreme power imbalances, such people are finding ways to resist the vicious excesses of hyper-extractivist practices and, in some instances, the very raison d'être of extractive industries.

It is worth highlighting extrACTIVISM in the context of a politics of post-ponement because of its unique strategic objective. ExtrACTIVISM, as a form of political expression, does not get hung up on global thresholds or planetary apocalypse. This is because, for many extrACTIVISTS, the apocalypse has already happened. It has come as they find themselves desperate simply to stay alive or protect the last ecological sources of their health, livelihood, and cultural identity. They face the apocalypse as they hit bottom in terms landscape despoliation, cultural genocide, financial ruin, chronic illness, and the death of loved ones, community disarray, and abandonment by government and social services. To them, conventional indicators that gauge environmental harm by the so-called "end of the world" miss the "end" of individuals and communities. Such indicators hold out a distant and abstract form of measurement and therewith ignore or belittle the suffering that accompanies significant loss.

As mentioned, extrACTIVISM, while distinct, is a recent manifestation of EJ. It arose out of a heightened concern for the linkages between human rights and environmental protection. Its lineage can be associated with, for instance, the efforts by Bruno Manser and the Penan tribe to stop rampant deforestation in Sarawak, Malaysia (Dauvergne, 2016) and the campaigns by Chico Mendes and fellow rubber tappers to protect particular areas in Brazil from domestic and foreign extractive companies (Bratman, 2020). It can also be seen in the activism that led to a ban of open-pit mining in El Salvador and to pressure for similar bans in Honduras and Guatemala. One can also see elements of it in the Brazilian Landless Movement that halted the expansion of eucalyptus plantations in southern Brazil. Such efforts represent "contentious agency" in that they question orthodox understandings of development and seek alternative strategies (Kröger, 2013). Such questioning has been taking place as extrACTIVISTS resist oil facilities in Ecuador (Fiske, 2017), gold mines of Kyrgyzstan (Wooden, 2017), hydropower in Quebec, Canada (Willow, 2018), hydraulic fracturing and longwall coal mining in Pennsylvania, USA (McCoy et al., 2017), and dams in Southeast Asia and Tibetan parts of China (Eyler, 2019). In each of these instances and many others, extrACTIVISTS fight primarily to block or slow specific extractive projects but end up partly influencing broader resource flows and, at times, even the narrative of development. Importantly, such extrACTIVISM measures success not in parts per million, degrees Celsius, or species disappearance but by their ability to mitigate and ultimately stave off harmful extractivist practices. Put differently, they focus not on the atmosphere, hydrosphere, or lithosphere *per se* but on the lived experience of those at the receiving end of hyper-extractivist exploitation. Their focus is dignity, survival, and justice rather than planetary wellbeing. To them, environmental disaster is not some globalized extinction but the cries of those on the frontlines of justice-displacing extractivism.

For those looking for planetwide political effectiveness—for example, a dramatic reduction in global carbon emissions, deforestation, or globally traded minerals—extrACTIVISM might appear parochial or marginal. However, this would miss its power to contest a politics of postponement. The "wins" of extrACTIVISM are not geographically circumscribed. Like all political action, they reverberate beyond stopping a single mining operation, clear-cut forest, hydropower dam, or hydraulic fracturing facility. As communication networks share extrACTIVISTS' strategies and stories and as others replicate their actions, the effort is beginning to loosen the cultural and practical hold that fossil fuel industries and extractivism more generally have over people's lives and suggest the possibility of a more fair and humane future. They seek local justice or advance place-based criticism against the global reach of extractivist industries and, in doing so, ironically contribute to wider critiques of extractivism as a mindset and practice. They shift, in other words, the balance of legitimacy over extractive industries and the ethics of environmental harm. To be sure, like the steps toward the irrevocable crossing of global ecological thresholds, such shifting comes about not in one dramatic moment but in increments of ethical adjustment. Put differently, extrACTIVISM does not change the state of global affairs through a frontal attack but by altering the lived experience of resistors, the possible horizon of what activists can strive for, and the normative framework within which environmental affairs operate. While certainly not a panacea, extrACTIVISM represents one way that environmentalists are going beyond global threshold thinking and refocusing attention on worldly pain, violence, justice, and reclamation. It signals an evolution of EJ in which justice is not simply a partner to environmental protection but central to the meaning and goals of environmental wellbeing. Moreover, by pulling the environmentalist gaze from the globe to the trenches of environmental conflict, it robs the politics of postponement of its foundational grounding. No longer do planetary boundaries provide the measuring rod for environmental destruction and thus no longer can politicians, corporate executives, and other powerful elites dismiss environmental urgency. The "end" is here. It is etched onto the nervous systems, skin, and muscle of those upon whom environmental harm has been displaced.

Conclusion

The 1960s and 1970s represented a high-water mark of environmentalism. They gave birth to the modern environmental movement. At the time, journalists, activists, scholars, and others recognized the unsustainability of industrial society and launched pleas to change course. Dire warnings sat at the center of their efforts. Environmentalists pointed out the finitude of the Earth. The planet simply could not support indefinite numbers of people with insatiable material appetites. If population, affluence, and technological capability continued to increase and if consumerism remained the world's secular identity, environmentalists warned that eventually the planet itself will buckle under ecological pressure. As Herman Daly put it years ago, "growth is an impossibility theorem" (1993). The planet has only so much regenerative power. Pressed too far, it will eventually weaken and collapse.

Environmentalists of the 1960s and 1970s were able to make such predictions because they subscribed to a neo-Malthusian understanding of biophysical limits and holistic planetary functionality. As a result, they failed to see the elasticity that politics can introduce. Their global focus helped create a visibility problem that obscured how people displace environmental harm onto the lives of others and how political power can both direct such displacement and use it to justify delaying action. Unable to see the "end" as experienced by those on the frontlines of unfolding environmental degradation, the more powerful essentially pushed back the thresholds of cataclysm. They learned to master a politics of postponement.

Environmentalism has evolved since the 1960s and 1970s. Most dramatically, it has spawned the EJ movement wherein activists have come to see environmental harm as an instance and result of broader structural and institutional forms of discrimination, chauvinism, and racist and classist injustice. As a result, they have also come to see that addressing environmental degradation cannot be done without rooting out the deeper animators of social unfairness and oppression. The EJ movement has made significant gains by providing a broader tent for environmentalism by including labor unions, landless workers, human rights advocates, indigenous people, anti-corporate activists, and all who experience or make rejection of injustice their cause. It has also done so by adding an additional ethical dimension to environmentalism's values and purpose by foregrounding the indignities that accompany environmental degradation. At the same time, it must be acknowledged, EJ has also created new risks for environmentalism by exposing activists to the threats and dangers that have long attended social justice advocacy.

As hyper-extractivism has introduced a new, more menacing form of environmental intensification and has increasingly set the pulse of economic practice and reshaped worldwide patterns of collective life, EJ has sprung a new frontier of resistance and contestation—extrACTIVISM, whereby activists are confronting large-scale extractive industries on the ground. They recognize the injustices they continue to endure and the existential dangers such injustice entails and fight for survival and dignity at the local level. Importantly, this means that extrACTIVISTS are responding not only to prudential concerns about dwindling resources, tapped-out pollution sinks, and the despoiling of certain sites, but to the moral foundations that support the entire exploitative, extractivist mindset. They reject a world premised on an economistic accounting of value and animated by capitalist forces that envision life as merely a set of objects to be extracted, transported, manufactured, used, and discarded. It is in this sense that extrACTIVISTS subscribe to a different measure of environmental impact. We must forget planetary thresholds and concentrate on the lived experience of actual communities. Forget the Earth's organic infrastructure and attend to the injustices of ripping land out from under people, polluting people's homes, and pushing workers to extremes in an effort to hyper-extract labor. In short, we must relax concentration on the brittleness of planetary limits and instead concern ourselves with the elasticity of human pain and suffering. In doing so, sensitivity replaces technical prognosis, and observation supplants prediction. Instead of calculating and waiting for the end of the Earth—and encouraging a politics of postponement—a significant slice of environmentalism now notices and resists extractivist assaults wherever they occur, independent of their so-called ultimate, planetary significance. As a result, extrACTIVISTS not only fight to secure the wellbeing of their communities and shift conventional narratives of resource exploitation, but also to undermine justifications for political postponement.

One final word: in highlighting extrACTIVISM and its ethical intervention, I am not arguing that environmentalists should ignore planetary measures. Climate science, conservation biology, toxicology, and so forth must, necessarily, ponder and try to identify tipping points after which cascading, planetary environmental decline happens. Rather, I am posing a warning about the politics of such practice. To the degree that environmentalists focus on global thresholds, they make themselves prone to ignore the dynamics of how some people suffer disproportionately as the world approaches such boundaries and how the world actually breaks down. The world will not disappear in a single evaporation. It will unhinge and *is* unhinging, one landscape and one being at a time. Each instance is an end. Resisting the end is a moral responsibility, and arguably it must lie at the heart of all environmental concern.

References

Adams, V. (2013) *Markets of Sorrow; Labors of Faith: New Orleans in the Wake of Katrina.* Durham, NC: Duke University Press.

Agyeman, J. (2005) *Sustainable Communities and the Challenge of Environmental Justice.* New York, NY: NYU Press.

Assadourian, E. (2015) 'Consequences of consumerism' in Wapner, P. and Nicholson, S. (eds.) *Global Environmental Politics: From Person to Planet.* Abingdon and New York, NY: Routledge, pp. 97–105.

Bailey, R. (2000) 'Earth Day, Then and Now', *Reason Magazine*, May. Available at: https://reason.com/2000/05/01/earth-day-then-and-now-2.

Beech, H. (2018) 'At typhoon's eye in Philippines, whipping debris and fervent prayers', *The New York Times*. Available at: www.nytimes.com/2018/09/16/world/asia/philippines-typhoon.html.

Bluhdorn, I. (2001) *Post-ecologist Politics: Social Theory and the Abdication of the Ecologist Paradigm.* New York, NY: Routledge.

Bluhdorn, I. (2007) 'Sustaining the unsustainable: Symbolic politics and the politics of simulation', *Environmental Politics*, 16 (2).

Bluhdorn, I. and Walsh, I. (2007) 'Eco-politics beyond the paradigm of sustainability: A conceptual framework and research agenda', *Environmental Politics*, 16 (2), pp. 185–205.

Branford, S., Borges, T., and Torres, M. (2018) 'Bolsonaro shapes administration: Amazon, indigenous and landless at risk', *Mongabay*, December. Available at: https://news.mongabay.com/2018/12/bolsonaro-shapes-administration-amazon-indigenous-and-landless-at-risk.

Bratman, E. (2020) *Governing the rainforest: Sustainable development in the Brazilian Amazon.* New York, NY: Oxford University Press.

British Broadcasting Corporation. (2020) 'Australia fires: How do we know how many animals have died?', BBC News, 4 January 2020. Available at: www.bbc.com/news/50986293.

Brown, A. (2020) 'Louisiana Environmental Activists Charged with 'Terrorizing' for Nonviolent Stunt Targeting Plastics Giant', *The Intercept*, 25 June. Available at: https://theintercept.com/2020/06/25/environmental-activists-charged-terrorizing-louisiana-formosa.

Brown, L. (1974) *In the Human Interest: A Strategy to Stabilize World Population.* New York, NY: W.W. Norton & Company.

Bullard, R. (2000) *Dumping in Dixie*. New York, NY: Routledge.

Burchardt, H. and Dietz, K. (2014) '(Neo-)extractivism—A new challenge for development theory from Latin America', *Third World Quarterly*, 35 (3).

Commoner, B. (1971) *The Closing Circle: Nature, Man, and Technology*. New York, NY: Knopf.

Daly, H. (1993) 'Sustainable growth: An impossibility theorem' in Daly, H. and Townsend, K. (eds.) *Valuing the Earth: Economics, Ecology, Ethics*. Cambridge, MA: MIT Press.

Dauvergne, P. (2016) *Environmentalism of the Rich*. Cambridge, MA: MIT Press.

Dauvergne, P. and LeBaron, G. (2014) *Protest Inc.: the Corporatization of Activism*. Cambridge: Polity.

Detraz, N. (2016) *Gender and the Environment*. Cambridge: Polity.

Dyson, M. (2007) *Come Hell or High Water: Hurricane Katrina and the Color of Disaster*. New York, NY: Civitas Books.

Ebell, M. and Milloy, S. (2019) Wrong Again: 50 years of Failed Eco-pocalyptic Predictions, Competitive Enterprise Institute. Available at: https://cei.org/blog/wrong-again-50-years-failed-eco-pocalyptic-predictions.

Ehrlich, P. (1970) *The Population Bomb*. New York, NY: Sierra Club-Ballentine.

Elferink, M. and Schierhorn, F. (2016) 'Global demand for food is rising. Can we meet it?', *Harvard Business Review*, 7 April, 2016.

Eyler, B. (2019) *Last Days of the Mighty Mekong*. London: Zed Books.

Falk, R. (1972) *This Endangered Planet*. New York, NY: Random House.

Falk, R. (1975) 'A new paradigm for international legal studies: prospects and proposals', *Yale Law Journal*, 84 (969).

Faulkner, W. (2012) *Requiem for a Nun*. New York, NY: Vintage.

Federman, A. (2020) 'Revealed: US listed climate activist group as 'extremists' along with mass murderers', *The Guardian*, 12 January, 2020. Available at: www.theguardian.com/environment/2020/jan/13/us-listed-climate-activist-group-extremists.

Fiske, A. (2017) 'Bounded impacts, boundless promise: Environmental impact assessments of oil production in the Ecuadorian Amazon' in Jalbert et al., *ExtrACTION: Impacts, Engagements, and Alternative Futures*. London: Routledge.

Fricker, M. (2009) *Epistemic Injustice: Power and the Ethics of Knowing*. New York, NY: Oxford University Press.

Gaard, G. (2019) *Critical Ecofeminism*. Lanham, MD: Lexington.

Hern, M. and Johal, M. (2018) *Global Warming and the Sweetness of Life: A Tar Sands Tale*. Cambridge, MA: MIT Press. Available at: www.ipcc.ch/2018/10/08/summary-for-policymakers-of-ipcc-special-report-on-global-warming-of-1-5c-approved-by-governments.

Huang, A. (2012) Hurricane Sandy's disproportionate impact on NYC's most vulnerable communities, NRDC. Available at: www.nrdc.org/experts/albert-huang/hurricane-sandys-disproportionate-impact-nycs-most-vulnerable-communities.

IPCC (Intergovernmental Panel on Climate Change). (2018) Summary for policy makers of IPCC special report on global warming of 1.5°C approved by governments.

Islam, S.N. and Winkel, J. (2016) 'UN-DESA policy brief #45: The nexus between climate change and inequalities', UN Department of Economic and Social Affairs.

Israel, B. (2012) 'Coal plants smother communities of color', *Scientific American*. Available at: www.scientificamerican.com/article/coal-plants-smother-communities-of-color.

Kallis, G. (2019) *Limits: Why Malthus was Wrong and Why Environmentalists Should Care*. Stanford, CA: Stanford University Press.

Kasler, D. and Sabalow, R. (2018) 'Trump officials blame 'environmental terrorists' for wildfires. California loggers disagree, Sacramento Bee'. Available at: www.sacbee.com/latest-news/article218559945.html.

Kidd, D. (2016) 'Extra-Activism: Introduction to grassroots response to extraction', *Peace Review*, 20 (1).

Kituyi, M. (2018) '*90% of fish stocks are used up—fisheries subsidies must stop*', UNCTAD. Available at: https://unctad.org/en/pages/newsdetails.aspx?OriginalVersionID=1812.

Kröger, M. (2013) *Contentious Agency and Natural Resources Politics*. New York, NY: Routledge.

Lakhani, N. (2014) 'Surge in deaths of environmental activists over past decade, report finds', *The Guardian*. Available at: www.theguardian.com/environment/2014/apr/15/surge-dea ths-environmental-activists-global-witness-report.

Lerner, S. (2010) *Sacrifice Zones: The Front Lines of Toxic Chemical Exposure in the United States*. Cambridge, MA: MIT Press.

Malthus, T. (1789/2007). *An Essay on the Principle of Population*. Mineola, NY: Dover Publications.

McCoy, C., Coptis, V., and Grenter, P. (2017) 'Harmonizing grassroots organizing and legal advocacy to address coal mining and shale gas drilling issues in southwestern Pennsylvania' in Jalbert, K. et al. (eds.) *ExtrACTION: Impacts, Engagements, and Alternative Futures*. London: Routledge.

McKibben, B. (2019) *Falter: Has the Human Game Begun to Play Itself Out?*New York, NY: Henry Holt.

Meadows, D., Meadows, D., Randers, J., and Behrens, W. (1972) *The Limits to Growth; A Report for the Club of Rome's Project on the Predicament of Mankind*. New York, NY: Universe Books.

Michaels, P. and Maue, R. (2018). 'Thirty years on, how well do global warming predictions stand up?', *The Wall Street Journal*. Available at: www.wsj.com/articles/thirty-years-on-how-well-do-global-warming-predictions-stand-up-1529623442.

Mingle, J. (2020) 'A World Without Ice', *The New York Review of Books*, 14 May, 2020.

Pellow, D. and Brulle, R. (eds.). (2005) *Power, Justice, and the Environment: a Critical Appraisal of the Environmental Justice Movement*. Cambridge, MA: MIT Press.

Roser, M. (2018) Economic growth, OurWorldInData.org. Available at: https://ourworl dindata.org/economic-growth.

Rowe, J. and Carroll, M. (2014) 'Reform or radicalism: Left social movements from the Battle of Seattle to Occupy Wall Street', *New Political Science*, 36 (2).

Rowe, J. and Carroll, M. (2015) 'What the left can learn from Occupy Wall Street', *Studies in Political Economy*, 96 (1).

Scranton, R. (2015) *Learning to Die in the Anthropocene: Reflections on the End of a Civilization*. San Francisco, CA: City Lights Books.

Sellers, C. (2017) 'Storms Hit Poor People Harder, from Superstorm Sandy to Hurricane Maria', *The Conversation*. Available at: https://theconversation.com/storms-hit-poorer-p eople-harder-from-superstorm-sandy-to-hurricane-maria-87658.

Shabecoff, P. (2003) *A Fierce Green Fire: The American Environmental Movement*. Washington, DC: Island Press.

Steffen, W., Richardson, K., Rockström, J., Cornell, S.E., Fetzer, I., Bennett, E.M., Biggs, R., Carpenter, S.R., de Vries, W., de Wit, C.A., Folke, C., Gerten, D., Heinke, J., Mace, G.M., Persson, L.M., Ramanathan, V., Reyers, B., and Sorlin, S. (2009) 'Planetary boundaries: Exploring the safe operating space for humanity', *Ecology and Society*, 14 (2), p. 302.

Steffen, W., Richardson, K., Rockström, J., Cornell, S.E., Fetzer, I., Bennett, E.M., Biggs, R., Carpenter, S.R., de Vries, W., de Wit, C.A., Folke, C., Gerten, D., Heinke, J., Mace, G.M., Persson, L.M., Ramanathan, V., Reyers, B., and Sorlin, S. (2015) 'Planetary boundaries: Guiding human development on a changing planet', *Science*, 347 (6223) 1259855.

Taylor, D. (2014) *Toxic Communities: Environmental Racism, Industrial Pollution, and Residential Mobility*. New York, NY: NYU Press.

United Nations. (2019) 'Nature's Dangerous Decline 'Unprecedented': Species Extinction Rates 'Accelerating'. Available at: www.un.org/sustainabledevelopment/blog/2019/05/nature-decline-unprecedented-report.

USAID. (2012) Nepal climate vulnerability profile. Available at: www.climatelinks.org/sites/default/files/asset/document/nepal_climate_vulnerability_profile_jan2013.pdf.

Waldman, S. (2018) 'Talk of unicorns and 'eco-terrorists' at alternative forum', *E&E News*. Available at: www.eenews.net/stories/1060092985.

White, L. (2018) 'California's wildfires are exposing the rotten core of capitalism', *Vice*. Available at: www.vice.com/en_us/article/8xpgz4/californias-wildfires-are-exposing-the-rotten-core-of-capitalism.

Willow, A. (2018) *Understanding extrACTIVISM: Culture and Power in Natural Resource Disputes*. London: Routledge.

Wooden, A. (2017) 'Images of harm, imagining justice: Gold mining contestation in Kyrgyzstan' in Jalbert, K. et al. (eds.) *ExtrACTION: Impacts, Engagements, and Alternative Futures*. London: Routledge.

Worldometers. (n.d.) Current world population. Available at: www.worldometers.info/world-population.

Zachos, E. (2018) 'Why 2017 was the deadliest year for environmental activists', *National Geographic*. Available at: www.nationalgeographic.com/environment/2018/07/environmental-defenders-death-report.

PART 2

Exacerbated Violence at the Local Level

4

EMPOWERMENT OR IMPOSITION?

Extractive Violence, Indigenous Peoples, and the Paradox of Prior Consultation

Philippe Le Billon and Nicholas Middeldorp

Introduction

The rapid growth of resource sectors over the past two decades has seen many policies seeking to address the harmful effects of extractive activities on environments and communities (Feichtner *et al.*, 2019). Environmental impact assessments (EIAs), corporate social responsibility (CSR) programs, sustainability principles, and the consultation of affected communities have become part of governance tools making land "investable" for extraction (Le Billon and Sommerville, 2017). Many of these instruments—such as EIAs, Human Rights Impact Assessments, and CSR—have come under critique for legitimizing and reproducing extractivist logics and praxis (Brock and Dunlap, 2018).

Here, we focus on the paradox of prior consultation and extractive violence: while Indigenous peoples supposedly benefit from rights to prior consultation over extractive projects, they are still disproportionately facing the various forms of violence associated with extraction (Comisión Interamericana de Derechos Humanos, 2015; Global Witness, 2019). To help to explain this paradox, we examine the role of the *practice* of prior consultation in advancing extractive projects rather than its *envisioned ideal* of enforcing compliance with Indigenous and environmental human rights, which would ensure free, prior, and informed consent (FPIC) by affected communities if obtained before project execution. Many governments, corporations, and international development agencies have now accepted prior consultation of local communities and, in particular, Indigenous ones as a prerequisite for the implementation of "extractive" projects—a term referring to mining and hydrocarbons extraction, but also more broadly to high-impact, land-based projects including conservation (which might require displacement) and renewables as well as large-scale infrastructure projects such as agri-business, hydropower, solar energy, ports, or highways. Project proponents pragmatically see prior consultation as a mechanism to avoid costly conflicts and preempt some of the most violent aspects of land-based development projects (e.g. Food and Agriculture Organization of the United Nations., 2014; GIZ, n.d.).

The prior consultation of local and, in particular, Indigenous communities thus appears as a prominent constituent of progressive forms of extractive governance, while a lack of FPIC is often presented as a root cause of conflict (Global Witness, 2019). Many studies denounce the inadequacies of prior consultation processes (e.g. Flemmer and Schilling-Vacaflor, 2016), the non-recognition of consent as a legal requirement (Miller, 2015; Perreault, 2015), the depoliticization effects of bureaucratic consultation processes (Merino, 2018; Urteaga-Crovetto, 2018), and the often "abysmal disparities in power and resources between the actors involved" (Rodríguez-Garavito, 2011, p. 305). Few studies, however, have attempted to systematically examine how consultation processes contribute to "extractive violence."

By extractive violence, we mean violence associated with extractive logics and projects. Seeing violence as more than an "act" or "consequence," we approach violence as an unfolding process (Springer and Le Billon, 2016), which instils fear, hurts, or lowers the "level of needs satisfaction below what is potentially possible" for both the human and non-human (Galtung and Fischer, 2013, p. 35). From this perspective, the concept of extractive violence allows us to consider the various violent dimensions of prior consultation, including those that could result from the anticipation of future (even if uncertain) project implementation (Groves, 2017). As argued below, prior consultation processes cannot be separated from the violence of dispossession, repression, and pollution, including through their effects on health, livelihoods, wellbeing, culture, and sense of belonging. We therefore argue that prior consultation cannot be counted upon as a panacea for avoiding socio-environmental conflict. *Unless principles of free, prior, and informed consent are more stringently implemented*, prior consultation does little to avoid "extractive violence" at best and could cloud and actually deepen extractive violence at worst.

Our conceptual framework and discussion of the violence of prior consultation is based on a review of 68 studies. These studies were selected through the following process: first, a general identification using the search terms "prior consultation," "consent," "participation," and "FPIC" in three languages (English, Spanish, and Portuguese) and through two search engines (Google Scholar, Web of Science); and second, a selection of studies relevant to the focus of this review (i.e. academic studies of consultation processes examining the wider context and impacts of consultation, rather than publications discussing its technicalities, which are amply found in the gray literature). Of the 68 selected studies, 53 were published in English, 12 in Spanish, and three in Portuguese. Out of this total, 28 discuss one or more case studies, others discuss the application of prior consultation nation-wide, cross-examine or compare practices of prior consultation between countries, or discuss the legal basis and implications of the right to prior consultation in the countries and regions under research. The selected studies represent contributions from a broad range of disciplines, including anthropology, geography, development studies, sociology, political science, and (international) law. (The full literature review chart can be found here: https://bit.ly/34CIvyW.)

All studies except for one were published over the past decade, reflecting the recent and rapid rise in attention for this topic. Two-thirds of the studies empirically focused on Latin America, reflecting the high level of scholarly attention on prior consultation in that region, a high incidence of conflicts over extractive and infrastructural projects, and widespread (nominal) acceptance of the need to consult Indigenous peoples—in addition to slight bias in languages used in our search. The rest include studies on cases in Australia (Walsh *et al.*, 2017), Canada (e.g. Youdelis, 2016; Moore *et al.*, 2017), Germany (Brock and Dunlap, 2018), India (Choudhury and Aga, 2020), the Philippines (Young, 2019), Sub-Saharan Africa (e.g. Ece *et al.*, 2017; Inkman *et al.* 2018; Mitchell and Yuzdepski, 2019), and the United States (Miller, 2015). Although the findings of this review are largely based on studies empirically focused on Latin America, we observed that findings of studies from other regions, regardless of whether consulted populations are Indigenous or not, were largely consistent with the negative experiences documented in Latin America.

In addition to this scholarly literature, we also selected and reviewed gray literature, including policy recommendation reports, prior consultation guidelines, and reports by non governmental organizations (NGO) (e.g. Global Witness) and human rights institutions (e.g. Interamerican Commission of Human Rights). This review is also informed by insights gained through participant observation and group discussions in three subregional forums organized by the Instituto Interamericano de Derechos Humanos (IIDH) on the experiences of prior consultation with a total of 70 Indigenous leaders from 18 Latin American countries in 2015, as well as field research and work visits by Philippe Le Billon across Latin America on socio-environmental conflict and Indigenous rights (both as academic researcher and as NGO consultant). These include visits to Indigenous communities resisting extractive projects seeking to usurp ancestral lands in Honduras; interviews in Guatemala and Honduras with local actors for a proposed environmental justice project by the IIDH; workshops on intercultural justice in Guatemala, Chiapas, and Oaxaca with *operadores de justicia* (judges, Ombudsman staff, public defence lawyers, public attorneys) debating prior consultation; field research in Nicaragua's Caribbean coast on resistance to the proposed Interoceanic Canal project following a fraudulent consultation process; and a field visit to the Colombian Amazon, where a micro-hydropower plant was imposed on the Vaupés River in Indigenous territory, destroying sacred sites. During these visits, the lead author spoke with a range of actors, including Indigenous leaders and activists, private sector representatives, as well as state officials such as Ombudsman staff, Environmental Ministry staff, and *operadores de justicia*.

Following this introduction, we first review debates around extractive violence in relation to consultation and indigeneity and then discuss relations between prior consultation and extractive violence. The paper concludes with suggestions for further research and practical recommendations.

Extractive Violence, Indigenous Peoples, and Consultation Processes

Extractive violence has three main dimensions. First is the violence of *dispossession*, often "compensated" with cash payments, temporary jobs, and "alternative" livelihoods marked by their unfairness, their uneven allocation, and the false "equivalences" they seek to create between incomparable entities across incommensurable ontological and epistemic differences (Leifsen et al., 2017). Thus, dispossession can be both material and ontological, through the delinking of communities from their territories, facilitated by the appeal or imposition of capitalist modernity and the environmental degradation that renders traditional livelihoods increasingly unviable. Here, Indigenous deterritorialization runs parallel with state/company-led (re)territorialization, as a mutually imbricated process (Di Giminiani, 2015).

Second, the violence of *coercion* often is exercised on local communities to impose extractive activities, especially when resistance to projects takes a more organized shape (Navas *et al.*, 2018; Middeldorp and Le Billon, 2019); when clashes occur within or between communities over granting consent (Jaskoski, 2020) such as the case of elite abuse (Vermeulen and Cotula, 2010); or when projects are forcibly imposed despite consent refusal (Steinberg, 2016). If coercion is often understood as direct threats or use of physical violence, we understand it as also taking many other forms, including deception, manipulation, and corruption consolidating the dominance of extractive regimes, affecting decision-making within and by Indigenous communities (Cariño, 2005; Nest, 2017).

Third are the physiological and psychological harms associated with the *pollution and degradation of socio-environmental systems* resulting from extractive activities, including temporally dispersed "slow violence" that is often "invisible," such as insidious health risks for communities exposed to pollution (Nixon, 2011) and "ecological violence" against the non-human within Indigenous territories (Navas *et al.*, 2018).

These three dimensions help to sketch out the outcomes of extractive projects for concerned communities and to understand their relations with prior consultation processes. As discussed below, prior consultation can be interpreted as a soft instrument of dispossession, which involves some forms of coercion and frequently results in exposure to pollution and other socio-environmental impacts as projects often end up being implemented despite a lack of consent.

Indigenous Peoples and Extractive Frontiers

Indigenous peoples are social groups self-identifying as distinct from the settler population and who, "irrespective of their legal status, retain some or all of their own social, economic, cultural and political institutions" (International Labour Organization, 2009, p. 49). At the cores of many Indigenous peoples' cosmovisions and ways of life are their reciprocal relationships with the land and territories both shaped by and confronting historical and contemporary colonial processes (Wildcat *et al.*, 2014). The relational attachment among Indigenous peoples, their territory, and the non-human beings that might inhabit it (Youdelis, 2016) is what gave birth to the right to prior consultation in the first place—Indigenous peoples demand control over their lands and prior consultation was envisioned as a guarantee mechanism (Comisión Interamericana de Derechos Humanos, 2015)—but it is also in part why they are disproportionately affected by extractive violence.

Indigenous peoples are frequently located in extractive frontiers, with their territories being considered by settlers and extractive companies as sparsely populated and underused "resource rich" or "critical infrastructure" areas (Pasternak, 2017). The extractivist logic of exploiting natural resources, implemented in territories inhabited by communities who are predisposed not to exploit land, water, or other livelihood means on an industrial basis, means that Indigenous peoples often face the brunt of environmental degradation and its polluting effects (Moore *et al.*, 2017) and become the first line of opposition against extractive projects (Urkidi, 2011). Here, the claim of rights to consultation and consent is frequently central to the struggles that result from extractive projects. We note that these two terms often are confused or purposively misused by project proponents *replacing* consent (i.e. obtention of a voluntary agreement) with consultation (i.e. providing project information and/or obtention of an opinion).

According to the Environmental Justice Atlas, 48 percent of the conflicts over extraction around the world involved Indigenous groups or traditional communities, with half of these conflicts resulting in the criminalization or other forms of repression, including assassinations (Temper *et al.*, 2015). Between 2014 and 2018 alone, at least 276 Indigenous people were killed while peacefully seeking to protect their land and environment, representing a third of the total number of defenders recorded (Global Witness, 2019). Indigenous peoples are also disproportionately affected considering their supposed right to refuse the forceful imposition of extractive projects. In many countries, and in contrast with other population groups such as non-indigenous agrarian communities, Indigenous peoples officially have stronger land and resource use rights. Not only does it appear that these rights are not well respected, but also, the denial of such rights could result in greater levels of violence against Indigenous peoples, including through prior consultation processes.

A Brief Overview of "Prior Consultation"

ILO Convention 107 (1957), the earliest international attempt to grant rights to Indigenous peoples, described their condition as "less advanced" (art. 1) and was highly assimilationist. In the context of the rise of Indigenous rights movements and acknowledgements in the 1970s and the 1980s, its successor, ILO Convention 169 (1989) moves away from the assimilationist standpoint and acknowledges Indigenous self-determination, including the right to prior consultation. Prior consultation of Indigenous peoples is considered a legal obligation for the twenty countries that ratified ILO Convention 169 and is now widespread within processes around extractive projects involving Indigenous communities. UNDRIP, the United Nations Declaration for the Rights of Indigenous Peoples (United Nations, 2007, art. 32), although not legally enforceable, progressively considers obtaining FPIC through the mechanism of prior consultation as a key requirement for the implementation of extractive projects.

In the Americas, based on its interpretation of Article 21 (the right to property) of the American Convention of Human Rights (1969), the Inter-American Court of Human Rights issued landmark decisions with *pueblo Saramaka vs. Surinam* (2007) and *pueblo Kichwa de Sarayaku vs. Ecuador* (2012), holding the states of Surinam and Ecuador responsible for the violation of the right to free, prior, and informed *consultation* (these decisions remain ambiguous about the meaning and legal requirement of obtaining *consent*) regarding extractive industries operating in Indigenous territories, creating jurisprudence for future cases (Instituto Interamericano de Derechos Humanos, 2016). Prior consultation has become common practice in much of Latin America: with the notable exceptions of El Salvador and Panama, C169 is adopted by all Latin American countries with Indigenous populations, and in some cases, as in Peru, the *consulta previa* is also integrated into national legislation (Urteaga-Crovetto, 2018). Canada also recognizes the duty to consult Aboriginal peoples, although the notion of a veto is rejected (Mills, 2017).

In Sub-Saharan Africa, a debate on prior consultation is emerging (Roesch, 2016), but experiences are scarce. Many African governments were hesitant to accept UNDRIP, fearing that it would encourage tribalism, ethnic violence, and secessionist movements (Mitchell and Yuzdepski, 2019). The concept of "Indigenous" is also disputed in a continent where most ethnic groups lay a claim to autochthony (ibid). However, the African Commission, borrowing heavily from jurisprudence from the Interamerican Court of Human Rights (leading authors to frame FPIC in Africa as a legal transplant (Roesch, 2016) progressively ruled in 2010 that the nomadic Endorois people (Kenya), displaced in the name of conservation, in their condition as Indigenous people have collective rights to territory and to FPIC, setting a legal precedent for future cases. Nonetheless, the ruling remained ambiguous on whether it is the state's duty merely to *seek* or also to *obtain* consent (Inkman et al.,2018). Ten years later, implementation of prior consultation procedures remains the exception rather than the rule in Sub-Saharan Africa, but the concept has entered the vocabulary of communities and NGOs supporting them in the struggle against dispossession induced by mining, conservation, and other industries (Roesch, 2016).

The UN program REDD+ and many multinational companies also pursue prior consultations, in part to address human rights standards and reputational risks concerns. Although some CSR principles are not entirely clear about this, UNDRIP and ILO Convention 169 place the obligation to consult firmly in the hands of the state—not the involved private company (Doyle, 2014). Generally speaking, where prior consultation is not a legal obligation, the procedure appears to be carried out more often by the private company itself, following CSR principles such as the IFC's Performance Standards or the RSPO (Roundtable on Sustainable Palm Oil) prior consultation guidelines, as seen in Canada where consultations are commonly delegated to project proponents as well (Moore et al., 2016).

Mainstream views on public consultation generally recognize three at times overlapping components: notification, which simply informs the public; consultation proper, which seeks to gain the perspectives of members of the public; and participation, which actively involves the public in decision-making over the formulation of the project and its approval (Rodrigo and Amo, 2006). Beyond these general components, prior consultation processes dealing with Indigenous peoples need to be culturally appropriate, inclusive, and integrated with social and environmental impact studies (United Nations, 2013). The notion of consent refers to the permission granted by the public for the project, and thereby constitutes a principle that is more substantive than narrow interpretations of participatory rights (Rodrigo and Amo, 2006). For consent to be valid, several principles need to be followed, with permission being granted through free will (i.e. absence of coercion), in full knowledge of the possible consequences, and prior to the implementation of the project. However, as the next section will show, prior consultation is in practice often wrought with violent dimensions.

Extractive Violence and Prior Consultation

Building on critiques of prior consultation processes, we identify five main ways through which violence can permeate prior consultation processes.

Ongoing Colonialism

The first violent process within prior consultation lies in a wider context of ongoing and often unacknowledged colonialism—symbolic violence in Bourdieu's terms (1979), or ontological violence according to Escobar (2015). The framework of settler colonialism is rarely used outside the Anglo-context (for Latin America, see Speed, 2017), but alternative conceptualizations such as *internal colonialism* exist (Iturralde, 2015). Contemporary colonialism is visible in the lack of acknowledgement of Indigenous rights, lack of basic service provisioning in Indigenous communities, ongoing land conflicts with settlers, ongoing racism and negative stereotyping, and the undermining or outright negation of alternative ways of living (Simpson, 2017). A notable aspect of ongoing colonialism in the context of prior consultation is the imposition of state law and the non-recognition of customary or Indigenous law and jurisdiction, along with the dismissal of the validity of Indigenous cosmovisions (De la Rosa Rondón, 2017) and traditional environmental knowledge (Baker and Westman, 2018).

Denial of Prior Consultation

The second violent process associated with prior consultation is the non-recognition of Indigenous status or territorial rights of affected communities. This symbolic form of violence affects identity rights. The denial of internationally or domestically prescribed consultation rights constitutes a violation *in practice* of the law and a betrayal of the political processes that allowed for these rights to be legally recognized in the first place. In Sub-Saharan Africa, despite the African Commission's conceptualization of Indigenous Peoples as self-identifying as such, having a connection to the land, and being in state of marginalization/dispossession versus other groups, the category of Indigenous remains contested by groups in power (Mitchell and Yuzdepski, 2019). In Canada, Metís communities continue to face difficulties in being recognized as Aboriginal (Mills, 2017). In Colombia, where legislation has recognized collective land rights and consultation rights for Indigenous and Afro-Colombian groups, government authorities and extractive companies have argued that communities and their territories could not be considered "Indigenous" once communities had resettled in new areas, even though displacement was often the result of paramilitary actions; or that there were too few Indigenous or Afro-Colombian people in a community to remain Indigenous, even if in-migration was the primary cause of demographic changes (Rodríguez, 2009). In Peru, the government has hired private consultants with the intention to demonstrate the non-Indigenous status of communities (Urteaga-Crovetto, 2018). Not only does this deny the right to consultation, but also, it actively denies Indigenous identities and the right to self-identification more broadly, with implications for entitlements, territorial claims, and group survival.

Limited Scope of Prior Consultation

A third problem associated with efforts to limit the scope of consultations, and to force Indigenous peoples into the legal formalism and technicalities of a (settler) colonial juridical realm (Rodríguez-Garavito, 2011). Peru is unique in adopting a law on free, prior, and informed consultation (Law 29785, of 2011), but Urteaga-Crovetto (2018, p. 10) points out that the law has "proceduralized" consultation, masking power asymmetries in the absence of state neutrality—while provisions specifying the right to consent (the right to veto a project) were completely taken out of the law (Flemmer and Schilling-Vacaflor, 2016). In some cases, consultations end up being simply a provision of (partial and biased) information on the part of the company and government (Alzate, 2019). The resulting misperception of projects and their impacts by communities can later result in frustration, as well as a sense of exploitation, especially when a lack of corporate disclosure results in greater exposure to harm (Helwege, 2015). These experiences are not exclusive to the Global South nor to Indigenous peoples, as Walsh et al., (2017, p. 167) show in their study of community consultation of a non-indigenous community in Australia regarding mining: "community members felt their livelihoods and landscapes were being destroyed and felt powerlessness to stop or change the project." The scope and technicalities of consultation processes could also result in insufficient support for meaningful participation (Perreault, 2015). This could include translation issues, funds to finance counter-expertise, or time pressures denying communities the time they need to decide. Consultation and consent can also be limited to minor project dimensions, take place at too late a stage in its design, and not be renewed when significant changes—including the extension of activities—are planned (Weitzner, 2002). In turn, these experiences can bring about deeper grievances and motivate an escalation of protests against companies and government authorities (Conde and Le Billon, 2017).

Biased Selection of Consulted Individuals and Communities

A fourth violent process is associated with the biased selection of communities and community members involved in consultation processes. The scope of a project's impact is not determined by the affected people, but by the company and the EIA it has commissioned—possibly leaving entire communities out (Baker and Westman, 2018), while the cumulative impact of different extractive projects is often not considered (Mills, 2017). Local elites frequently take advantage of consultation processes to secure private gains, often facilitated by extractive companies aiming to cultivate mutually profitable relations with local politicians and customary authorities. For example, across Sub-Saharan Africa, cases have been documented of customary authorities abusing their power and circumventing democratic processes for private gain (Roesch, 2016; Ece et al., 2017; Hundsbæk Pedersen and Kweka, 2017). Co-optation of community members and/or local authorities, including payments of per diems, contracts, jobs, and outright bribery not only undermine principles of political representation weakening local democracy (Ece et al., 2017) but also exacerbates intra-community wealth inequalities and tensions. When multiple leadership structures exist, states and companies take advantage to consult those that suit their needs, as currently playing out in Western Canada in the disputes between hereditary chiefs and Band Councils (accountable to the government) over the construction of pipelines through Indigenous lands (Sterrit, 2019). Gender inequality is also a major concern. Sekar (2016, p. 113) studying a case of prior consultation surrounding forestry-induced displacement of a tribal people in India, observed that "the voices of women and some socially marginalized individuals are systematically neglected during discussion." A strong male bias was also found in consultation processes in Latin America (Instituto Interamericano de Derechos Humanos, 2016), while the social impacts of extractive industries are often gendered (The WoMin Collective, 2017). As expressed by a Peruvian female participant in a regional forum held in 2015:

> those who are affected by the disasters [the mining projects], are the women. The men can take their backpacks and leave, looking for work. But the women cannot leave that easily because we have our children, we have our animals to look after. ... But it is a sacrifice to make, to be visible, to be heard, to be consulted. It was not easy to enter into these spaces of dialogue, because they [the men] did not want to give us the microphone.

Participation biases and the associated exacerbation of inequalities can contribute to creating and/or fuelling conflicts within and between communities. The exclusion of some community members or authorities from consultation processes through stigmatization, criminalization, and physical repression constitute major forms of violence. For example, in the context of prior consultation for the Nicaraguan Inter-oceanic Canal, the state directly interfered in Indigenous elections, and took away the legal status of an oppositional Afro-descendant communal government while granting it to a newly founded parallel one headed by Sandinista party members (personal observation, February 2017). As distrust in institutional channels grows, more people engage in protest actions to get their voices heard and interests recognized.

Coercive Imposition of Project Despite Lack of Consent

Finally, there is the violence of imposing a project despite its rejection by affected communities. This dimension includes slow and ecological forms of violence, structural violence associated with inequalities and relative deprivation resulting from extractive activities and revenue (mis)distribution, and direct forms of violence and repression, for example as a result of mine area extension. Although currently published research does not systematically account for the number of cases in which projects are implemented without consent, no single case or country study reviewed in this paper found state/company recognition of binding consent or veto rights, and there are signs that its consequences for increased social protest opposing large scale projects seem well established across many jurisdictions (Hanna *et al.*, 2016a). In Honduras, even the murder of Indigenous activists against the Agua Zarca hydroelectric dam initially did not stop the involved European development banks from backing the project, using an unsubstantiated discourse of FPIC as legitimization (Middeldorp and Le Billon, 2019).

Once a project has started, a lack of appeals process and independent dispute resolution mechanisms exacerbates grievances among communities, which could incite direct action in the form of protests, blockades, and sabotage. Despite the risk of repression, protesters generally see such forms of mobilization as legitimate and necessary to obtain respect for their rights (Hanna *et al.*, 2016b). Opponents of extractive projects not only face the direct physical violence of security forces *during* protests, but also threats, criminalization, and physical abuses, including assassination (Global Witness, 2019). Government authorities often impose a "state of emergency," restricting civil rights and intensifying (sometimes deadly) repression as documented by Young (2019) in the Philippines, further legitimizing these actions through the parallel promotion of restricted "dialogue" processes (Taylor and Bonner, 2017). Across the political spectrum, Indigenous rights defenders are often framed as opponents of "development" or "the public interest," and they are sometimes criminalized for doing so (Birss, 2017; Doran, 2017, Graddy-Lovelace in this volume). Following this effort to map out the potential violence of prior consultation processes, we turn to the main ways in which prior consultation processes can unfold, using possible outcomes based on two sets of variables: whether authorities and extractive companies launch a prior consultation process or not, and whether communities have a strong or weak capacity to organize themselves politically to provide or withhold consent.

Indigenous Peoples and Prior Consultation Scenarios

The relative strength of Indigenous institutions, including their level of internal unity and agreement over decisions, political advocacy capacity, and ability to rally external support are key in demanding prior consultation (Schilling-Vacaflor and Flemmer, 2015), while their willingness to accept extractive projects is key in the negotiation of project benefits (Torres-Wong, 2019). The willingness of communities to accept or oppose extractivism depends on many factors, including the relative success of local livelihood strategies such as agriculture or artisanal mining (Orozco and Veiga, 2018) and the level of (dis)trust in the state and concerns over health and socio-environmental impacts—often informed by prior experiences (Conde and Le Billon, 2017).

Assessing the Record of Prior Consultation in Latin American Countries

As mentioned above, prior consultation is most widely put in practice in Latin America, although the debate around prior consultation and FPIC is emerging in Sub-Saharan and South and South-East Asian contexts. In this section, we specifically provide an overview of experiences with prior consultation in Latin America.

Following an extensive study on prior consultation in Bolivia, Mexico, and Peru, Torres-Wong (2019) found that these states consistently choose not to implement prior consultation regarding mining projects due to expected resistance and that consultation processes are only helpful to Indigenous communities that support extractivism and seek to obtain project benefits. Looking at consultation outcomes in these countries between 2007 and 2017, Zaremberg and Torres-Wong (2018) note that, to their knowledge, prior consultation procedures have not once succeeded in halting undesired extractive projects. However, they contend that prior consultation is "not completely without use in resource-based conflicts" (ibid, p. 44): it lowers the propensity and intensity of state repression, and it allows for pecuniary benefits when Indigenous communities and their institutions are well-organized and willing to negotiate.

While prior consultation processes employed as a "soft tool" to impose extractive projects can reduce the likelihood and intensity of direct forms of violent repression (which more frequently takes place when communities reject both the project and the consultation process from the onset), they can simultaneously promote other forms of violence associated with inter- or intra-community tensions over biased participation and benefit-sharing mechanisms (Peterson St-Laurent and Le Billon, 2015), as well as the "slow violence" of environmental degradation resulting from the legitimization of extractive projects implemented, even without final consent (Nixon, 2011; Holterman, 2014).

There is no equivalent systematic study to that conducted by Zaremberg and Torres-Wong (2018) conducted on Bolivia, Mexico, and Peru. Yet, the literature on prior consultation in Latin America broadly confirms their findings (Comisión

Interamericana de Derechos Humanos, 2015; Wright and Tomaselli, 2019). Latin America is also the world region where the right to consultation is most widely embraced. However, Latin America is likely the world's most conflictive region in terms of extractive violence. Out of 2,584 environmental justice conflicts recorded across the world by the Environmental Justice Atlas, 770 conflicts were located in Latin America. The region also accounted for 72 percent of reported killings of land and environmental defenders worldwide between 2002 and 2018, although—like for EJAtlas—this could in part reflect a higher level of reporting compared to other regions (Global Witness, 2019).

While some Latin American states made progress with the implementation of FPIC (e.g. use of Indigenous languages in administrative decisions, education reforms, and proposed legal reforms), its effective implementation falls behind with regards to "strategic industries" such as infrastructure and extractive projects. The principle of FPIC is consistently violated as governments, across the political spectrum, follow (neo)extractivist development models giving precedence to resource extraction, hydroelectric projects, and large-scale plantations with inadequate or no consultation (Instituto Interamericano de Derechos Humanos,, 2016).

Peru and Colombia are two key cases to consider. Peru adopted a law on prior consultation in 2011, but Urteaga-Crovetto (2018, p. 10) points out that the law has "proceduralized" community consultation, masking power asymmetries in the context of the absence of state neutrality. Provisions on the right to consent were completely taken out of the law; and Doyle (2019) points towards serious implementation flaws favouring the interests of corporations. Despite these serious flaws and Indigenous discontent, this law now stands as a model for law proposals in both Honduras and Guatemala (field visit to Guatemala and Honduras, March 2018), processes facilitated by the German development cooperation GIZ (n.d.). These law proposals do not count on support of the Indigenous peoples themselves.

In Colombia, consultation processes have been consistently applied in the last decade despite the lack of a specific law, as they are rooted in a progressive constitution (Alzate, 2019) and a series of Constitutional Court rulings that have led to a strong jurisprudence on the subject matter (Instituto Interamericano de Derechos Humanos,, 2016) which nonetheless remain ambiguous about binding consent. According to the Ethnic Groups department of the *Defensoría del Pueblo* (Colombia's human rights ombudsman) 8,560 consultations have taken place up to 2017, of which 1,585 in 2017 alone. However, the agency lacks the capacity to supervise the majority of *consultas* meaning that the state, in practice, functions as a mere observer of company-led processes. Furthermore, the right to a veto is not recognized, as expressed by the Ethnic Groups Ombudsman himself: "we are a developing country … is it fair that 3 percent of the population holds 28 percent of the subsoil riches?… we cannot move ahead without exploiting" (personal communication, February 9th, 2018), implying with his answer that the duty to obtain consent before exploiting natural resources would mean an impediment to the development process.

At forums in 2015 Indigenous leaders from Bolivia, Nicaragua, and Venezuela denounced the co-optation of Indigenous institutions by the state by offering private benefits, or the creation of Indigenous governing organizations parallel to the pre-existing ones. Guatemala, Honduras, Peru, Colombia, and Mexico were associated more strongly with threats directed at opposition leaders and the militarization of strategic territories. These contexts of violence were also directly related to prior consultation procedures by Padilla Rubiano (2015) in Guatemala, Llanes-Salazar (2020) in Mexico, and Weitzner (2019) in Colombia. In this context, whether dealing with right-wing or left-wing governments, indigenous rights defenders frequently face stigmatization—being framed as opponents of the public interest of development—as well as criminalization. So far, no study has been conducted that maps the full scale of criminalization of human rights defenders on a regional level. However, it seems to be a widespread phenomenon, with indigenous rights defenders held captive as political prisoners for claiming their rights to territory (Global Witness, 2019). In Ecuador and Chile for example, dozens to hundreds of indigenous rights defenders have been jailed or are facing criminal charges for participating in demonstrations (Doran, 2017). In the years before her murder in 2016, Honduran activist Berta Cáceres was both stigmatized (framed as a murderer in national media) and criminally accused. In April 2018, Guatemalan anti-mining activist Abelardo Curup was sentenced to 150 years in prison for a crime that he claimed he did not commit and died in prison shortly thereafter (Prensa Comunitaria, 2018). In countries otherwise affected by high impunity rates, legal systems function with great efficiency when employed to criminalize rights-claiming populations (Doran, 2017). In addition to the lack of meaningful consultation, female indigenous leaders note that the social impacts of extractive industries are often gendered (The WoMin Collective, 2017) and emphasize that they often have to struggle to have their voices heard as decision-makers and especially as rights defenders, including within their own communities (Instituto Interamericano de Derechos Humanos,, 2016, p. 67).

Conceptualizing Prior Consultation Outcomes

Building on our literature review and field observations, we identify eight major possible prior consultation outcomes based on whether prior consultation is taking place as well as the organizing capacity of communities and their willingness to provide consent (see Table 4.1). We understand the organizational capacity of communities as reflecting a number of variables such as the number of affiliates, the quality of leadership and institutions, the mobilization of supporters and allies, as well as the broader political and repressive contexts—such as the democratic or authoritarian character of the state—in which community responses take place.

Based on the objectives of mainstream prior consultation, Outcome 5 represents a so-called "win-win" for companies and communities, whereby the interactions between communities with a strong Indigenous organizing capacity and consulta-tion procedure followed in good faith by all parties result in consent being

TABLE 4.1 Prior consultation outcomes

Community characteristics	Prior consultation	No prior consultation
Strong Indigenous organizing capacity + opposition to the project	**Outcome 1a: cancellation** or **1b: impasse** The project is cancelled or stalled as no agreement is reached. Possible Indigenous strategies: reconsider the project and its benefits **(Outcome 5)** or opt for social mobilization **(Outcome 2).** The impasse is likely to fuel internal divisions, exploited by the state or private company, who could opt for repression to weaken Indigenous institutions **(Outcome 3)**; co-optation to influence decision-making **(Outcome 5)** or a combination **(Outcome 7).** Examples: Jaskoski, 2020; Young, 2019; Merino, 2018; Rodríguez-Garavito, 2011.	**Outcome 2: confrontation** The project tries to move forward without prior consultation and/or without consent, but social mobilization halts or disrupts the project, the protest often being legitimized through a discourse of the right to FPIC. Could lead to a cycle of protest and repression, which could lead to project cancellation. Unlikely to lead to compensation and project benefits. Examples: Middeldorp and Billon, 2019.
Weak Indigenous organizing capacity + opposition to the project	**Outcome 3: manipulation** The project moves forward despite the lack of consent. The weak position of the Indigenous community vis-à-vis the state and the company is exploited: the consultation procedure is (ab)used to channel and contain discontent and satisfy legal requirements. Unlikely to lead to compensation and project benefits. Examples: Alzate, 2019; Choudhury and Aga, 2020; Brock and Dunlap, 2018; Ece et al., 2017; Marston and Perreault, 2017; Walsh et al., 2017; Padilla Rubiano, 2015; Navarro Smith, et al., 2014; Castillo Meneses, 2012.	**Outcome 4: imposition** The project moves forward without prior consultation. Some repression could be used to quell and discourage social mobilization. Unlikely to lead to compensation and project benefits. Examples: Torres-Wong, 2019.
Strong Indigenous organizing capacity + consent	**Outcome 5: collaboration** The project moves forward after consent is obtained. The consultation procedure is followed in good faith by all parties who negotiate a fair compensation and benefit-sharing agreement. Examples: Torres-Wong, 2019.	**Outcome 6: risking future conflict** The project moves forward without prior consultation and without initial objections, but could face protests once it is operational, as a pressure mechanism to obtain compensation and/or a benefit-sharing agreement, or due to unforeseen negative impacts. Protests could in turn be responded to with repression. Example: Middeldorp et al., 2016.

(Continued)

TABLE 4.1 (Cont.)

Community characteristics	Prior consultation	No prior consultation
Weak Indigenous organizing capacity + consent	**Outcome 7: an unfair deal** The project moves forward after consent is obtained. The consultation process is treated as a mere administrative procedure, and power imbalances, including possible cases of corruption within Indigenous institutions, are exploited by the state/company to avoid (fair) compensation and benefit-sharing. Examples: Vermeulen and Cotula, 2020; Schilling-Vacaflor and Flemmer 2015; Perreault, 2015.	**Outcome 8: exclusion** The project moves forward without prior consultation, regardless of Indigenous attitudes toward the project. Unlikely to lead to project benefits.

Source: Authors.

obtained together with a fair compensation and benefit-sharing agreement between communities and project proponents. However, our literature review suggests that this outcome is often not materialized.

With the exception of Outcome 5 and Outcome 1a, *if* indeed the company and government respect the denial of consent, all other scenarios entail situations that violate the FPIC principle. These could leave communities with uncompensated negative livelihood impacts; including in cases where prior consultation is denied as a result of symbolic violence or structural violence (e.g. denial of Indigenous status or of "significant" impact requiring consultation by biased licencing institutions). As further discussed in the next section, prior consultation processes can thus be considered as being part and parcel of extractive violence.

Rejecting or Reforming Prior Consultation?

As Merino (2018, p. 82) warns in his analysis of prior consultation in Peru, Indigenous demands run the risk of "ending in an institutional vacuum or reproducing social conflicts by providing social actors with no option but to step aside from the participatory game." Merino fears that conflicts will re-emerge, because prior consultation as a participatory mechanism has failed to give Indigenous peoples a relevant voice in environmental governance. Through successful use of the court system, Indigenous people have in some cases managed to temporarily halt extractive projects due to improper or lack of consultation, as documented by Xiloj (2019) in Guatemala But given the lack of court-acknowledged veto power, the right to prior consultation is increasingly rejected by the intended beneficiaries themselves: several Indigenous peoples, across different countries, have declared they no longer wish to be consulted. Examining resistance by Indigenous groups to prior consultation processes around five extractive conflicts within Colombia, Jaskoski (2020, pp. 1–2) found different strategies: trying to obtain environmental protection prior to the arrival of extractive industries, avoiding prior consultation altogether; denouncing the lack of prior consultation when they were excluded from it; and challenging the legitimacy of the prior consultation process when consultation was limited to project proponents; or refusing to engage in the mandatory consultation procedure altogether. A member of Colombia's U'wa declared with strong words during a forum attended in October 2015:

> We do not want a prior consultation, they do not have reason to ask us, because we do not have the right to a veto. If they don't respect the ultimate decision of the people, why negotiate mother earth, why negotiate life, negotiate the territory? So the consultation would be like, how would you, as a people, like to die? By knife? By the bullet? Expropriated? In this moment that is how we think of it. What we want is to live, and to live with dignity. The territory is our life, and that is where we feel we are alive, where we give continuity to our uses and customs. ....while [the government] doesn't respect [the ILO 169] Convention, we will not accept a consultation process.

Some communities have organized alternative consultation processes (Doyle and Cariño, 2013), mostly through self-organized "popular consultations." Such *consultas comunitarias* or *consultas populares* have mostly taken place in Latin America, notably in Guatemala and Peru, often with the objective of demonstrating local opposition to large-scale mining (McNeish, 2017; Walter and Urkidi, 2017) and, increasingly, hydroelectric projects (observation in Honduras and Guatemala, March/April 2018). Originally used by Indigenous peoples claiming the right to FPIC, other rural populations (who lack the legal right to prior consultation) have also started to use this mechanism with great success, as recently shown in the Cajamarca case in Colombia (McNeish, 2017).

One difference between official prior consultations and popular ones is that the latter frequently takes place prior to project development—with, for example, the objective of declaring a region as a "no-go" area for extractives projects—while official prior consultations, ironically, often take place *ex post* project approval. While prior consultation is under state or company control—and thus subject to colonialist logics and manipulation as detailed in this chapter—popular consultation takes place in an arena where involved communities seek to exercise their autonomy and claim the territory. Studies of popular consultations have pointed at outcomes ranging from "changes in project design to political agenda setting, and the opening of spaces for participation and public debate" (Dietz, 2019, p. 145).

States, however, often do not recognize the legal validity of popular consultations and thus seek to delegitimize them as symbolic moves on the part of externally manipulated populations. While popular consultations can be understood as a promising alternative, their political potential needs to be understood as either part of the broader realm of "participatory governance" paradigm, as the strength of popular consultations mostly emerges from contexts of liberal policies of participation, or as part of Indigenous and community autonomy paradigms and law (Xiloj, 2019). In this respect, popular consultations might be a more successful instrument than mainstream prior consultation to prevent or counter extractive violence. Yet, at least two caveats need to be considered. One is that popular consultations also point to the shortcomings of, or increasing discontent with, formal electoral democracy, as it questions the legitimacy of the territoriality of the nation-state (Haesbaert, 2013), with a possible outcome of entrenching sectarian forms of communalism. Second, some popular consultations have themselves been criticized for erasing divergent views within communities (Walter and Urkidi, 2017), hinting at intra-community coercion within these processes.

Conclusion: Taking Violence Out of Prior Consultation

Prior consultation, as enshrined in ILO Convention 169 and UNDRIP, is hailed as a tool of conflict prevention, participation, and benefit sharing. In most cases, consultation processes are considered the responsibility of the state. Nonetheless, this chapter has shown that in practice, states ignore their obligation to respect the human right to consultation, particularly discarding free, prior, and informed consent. As Torres-Wong (2019, p. 144) argues, "in a context of persistent and deep economic inequality and generalized violence, it should come as no surprise that the implementation of the right to prior consultation has failed to protect Indigenous territories in the ways envisioned by its most forthright advocates." Prior consultation is thus often an act of window dressing which has dire consequences for both livelihoods and lives (Wright and Tomaselli, 2019).

National development policies remain guided by the interests of the extractive and infrastructure sectors, which often are in contrast with the needs and cosmovisions of Indigenous communities. As a result, the historical marginalization of Indigenous peoples persists, with distrust often characterizing their relation to states that reproduce a colonial logic well into the twenty-first century. In this context, prior consultation processes are either denied altogether due to lack of recognition of Indigenous status or land title, or are characterized by power asymmetries and top-down "informing of project intentions." All too often, limits are placed on who is consulted through geographical scope, the co-optation of community leaders, as well as gender biases. Furthermore, the abundance of CSR standards and principles has not filled the gap between theory and practice and, if left unchecked, could be used strategically to legitimize projects lacking meaningful consultation.

The disregard for the principle of FPIC contributes to the historical marginalization of Indigenous peoples, who depend on their access to land and natural resources for the continuation of their livelihood and culture. From the Indigenous side, negative experiences with prior consultation have led to an increase of rejection of the entire process and have encouraged the pursuit of alternative self-organized "popular" consultations. Many Indigenous rights defenders have turned to protest actions that are commonly responded to with stigmatization, criminalization, and direct violence. This situation of disrespect for Indigenous human rights, in turn, raises questions about the substantive quality of nominally democratic states.

Our objective here is *not* to reject prior consultation processes altogether, but to point to their commonly biased instrumentalization, as well as their violent dimensions and counterproductive effects. Nor is it our intent to dismiss the political and emancipatory potential of political mobilization *around* consultation processes (see Rodríguez-Garavito, 2011). Indeed, formal participatory instruments can have significant potential when combined with other Indigenous-led strategies and organizational processes (Green, 2014; Machado *et al.*, 2017). The non-implementation of the right to consultation can itself constitute an opportunity for political organization and resistance. Furthermore, as argued by Leifsen et al. (2017), other participatory mechanisms and practices around extractive activities, rather than prior consultation alone, need to be considered in order to capture a more complex picture, which is beyond the scope of this paper. Finally, a more comprehensive analysis would also examine how prior consultation relates to environmental justice. This should not only be in terms of its three main classical dimensions—recognition as a "stakeholder," participation in decision-making and distribution of the burdens of extraction— but also in terms of a decolonial and transformative environmental justice based on Indigenous "pillars of self-governing authority, the undoing of the ontology of land as property, and epistemic justice" (Temper, 2019, p. 108). As a result, the scope of this paper is limited by its object (prior consultation) and its conceptual framework (extractive violence).

There is a need for scholars, in collaboration with communities affected by extractive industries, to envision and put into practice alternatives to state- or company-led prior consultation. Torres-Wong (2019) calls for the exploration of alternative mechanisms to channel anti-extractivist demands. The popular consultation is one such mechanism, albeit community-led and often not legally recognized. It explicitly rejects the state- or company-led process: whereas the latter is used as an instrument of control, popular consultation is a form of resistance and a claim of autonomy. But what can be done to reconstruct the consultation process itself as a fundamental Indigenous right recognizing veto power? Promoting rights awareness, independent monitoring, legal aid, and accompaniment is fundamental to reduce the abuse of power by state institutions, companies and local leaders. Finally, we believe that a strengthening of the (inter)national legal system is paramount, including the ability to hold transnational companies complicit in depriving Indigenous peoples of their rights and make them legally accountable for doing so.

References

Alzate, L. (2019) 'Prior consultation as a scenario for political dispute: A case study among the Sikuani Peoples from Orinoquía, Colombia' in Wright, C. and Tomaselli, A. (eds.). *The Prior Consultation of Indigenous Peoples in Latin America: Inside the Implementation Gap.* London: Routledge, pp. 91–105.

Baker, J.M. and Westman, C.N. (2018) 'Extracting knowledge: Social science, environmental impact assessment, and Indigenous consultation in the oil sands of Alberta, Canada', *The Extractive Industries and Society*, 5, pp. 144–153.

Birss, M. (2017) 'Criminalizing environmental activism: As threats to the environment increase across Latin America, new laws and police practices take aim against the frontline activists defending their land and resources', *NACLA Report on the Americas*, 49 (3), pp. 315–322.

Bourdieu, P. (1979) 'Symbolic power', *Critique of Anthropology*, 4(13–14), pp. 77–85.

Brock, A. and Dunlap, A. (2018) 'Normalising corporate counterinsurgency: Engineering consent, managing resistance and greening destruction around the Hambach coal mine and beyond', *Political Geography*, 62, pp. 33–47.

Cariño, J. (2005) 'Indigenous peoples' right to free, prior, informed consent: Reflections on concepts and practice', *Arizona Journal of International & Comparative Law*, 22, pp. 19–39.

Choudhury, C. and Aga, A. (2020) 'Manufacturing Consent: Mining, Bureaucratic Sabotage and the Forest Rights Act in India', *Capitalism Nature Socialism*, 31 (2), pp. 70–90.

Comisión Interamericana de Derechos Humanos. (2015) 'Pueblos indígenas, comunidades afrodescendientes, industrias extractivas', *CIDH-OEA*. Available at: www.oas.org/es/cidh/informes/pdfs/industriasextractivas2016.pdf.

Conde, M. and Le Billon, P. (2017) 'Why do some communities resist mining projects while others do not?', *The Extractive Industries and Society*, 4 (3), pp. 681–697.

De la Rosa Rondón, M.G. (2017) 'Inobservancia del derecho a la consulta previa', *Revista Jurídica Mario Alario D'Filippo*, 8 (16).

Di Giminiani, P. (2015) 'The becoming of ancestral land: Place and property in Mapuche land claims', *American Ethnologist*, 42 (3), pp. 490–503.

Dietz, K. (2019) 'Direct democracy in mining conflicts in Latin America: Mobilising against the La Colosa project in Colombia', *Canadian Journal of Development Studies/Revue canadienne d'études du développement*, 40 (2), pp. 145–162.

Doran, M. (2017) 'The hidden face of violence in Latin America. Assessing the criminalization of protest in comparative perspective', *Latin American Perspectives*, 44 (5), pp. 183–206.

Doyle, C. and Cariño, J. (2013) 'Making free prior & informed consent a reality: Indigenous Peoples and the extractive sector', Indigenous Peoples Links (PIPLinks), Middlesex University School of Law, and The Ecumenical Council for Corporate Responsibility.

Doyle, C.M. (2014) *Indigenous Peoples, Title to Territory, Rights and Resources: the Transformative Role of Free Prior and Informed Consent*. London: Routledge.

Ece, M., Murombedzi, J., and Ribot, J. (2017) 'Disempowering Democracy: Local Representation in Community and Carbon Forestry in Africa', *Conservation and Society*, 15 (4), pp. 357–370.

Environmental Justice Atlas. (n.d.) World Map. Available at: https://ejatlas.org/.

Escobar, A. (2015) 'Territorios de diferencia: La ontología política de los "derechos al territorio"', *Cuadernos De Antropología Social*, 41, pp. 25–38.

Food and Agriculture Organization of the United Nations. (2014) Respecting free, prior and informed consent. Practical guidance for governments, companies, NGOs, Indigenous peoples and local communities in relation to land acquisition. Rome.

Feichtner, I., Krajewski, M., and Roesch, R. (2019) *Human Rights in the Extractive Industries*. Berlin: Springer.

Flemmer, R. and Schilling-Vacaflor, A. (2016) 'Unfulfilled promises of the consultation approach: the limits to effective Indigenous participation in Bolivia's and Peru's extractive industries', *Third World Quarterly*, 37 (1), pp. 172–188.

Galtung, J. and Fischer, D. (2013) *Johan Galtung*. Berlin, Heidelberg: Springer.

GIZ. (n.d.) 'Implementación del derecho a la consulta previa de los pueblos indígenas como aporte a la prevención de conflictos'. Available at: www.giz.de/en/worldwide/36075.html.

Global Witness. (2019) *Enemies of the State?* London.

Green, M. (2014) *The Development State. Aid, Culture and Civil Society in Tanzania.* Oxford: James Currey.

Groves, C. (2017) 'Emptying the future: On the environmental politics of anticipation', *Futures*, 92, pp. 29–38.

Haesbaert, R. (2013) 'Del mito de la desterritorialización a la multiterritorialidad', *Cultura y Representaciones Sociales*, 8 (15), pp. 9–42.

Hanna, P., Langdon, E.J., and Vanclay, F. (2016b) 'Indigenous rights, performativity and protest', *Land Use Policy*, 50, pp. 490–506.

Hanna, P., Vanclay, F., Langdon, E.J., and Arts, J. (2016a) 'Conceptualizing social protest and the significance of protest actions to large projects', *The Extractive Industries and Society*, 3 (1), pp. 217–239.

Helwege, A. (2015) 'Challenges with resolving mining conflicts in Latin America', *The Extractive Industries and Society*, 2 (1), pp. 73–84.

Holterman, D. (2014) 'Slow violence, extraction and human rights defence in Tanzania: notes from the field', *Resources Policy*, 40, pp. 59–65.

Huizenga, D. (2019) 'Governing territory in conditions of legal pluralism: Living law and free, prior and informed consent (FPIC) in Xolobeni, South Africa', *The Extractive Industries and Society*, 6, pp. 711–721.

Hundsbæk Pedersen, R. and Kweka, O. (2017) 'The political economy of petroleum investments and land acquisition standards in Africa: The case of Tanzania', *Resources Policy*, 52, pp. 217–225.

Instituto Interamericano de Derechos Humanos. (2016) 'La consulta previa, libre e informada. Una mirada crítica desde los pueblos indígenas', San José, Costa Rica. Available at: www.iidh.ed.cr/libroconsulta.

International Labour Organization. (1957). Indigenous and Tribal Populations Convention. Available at: www.ilo.org/dyn/normlex/en/f?p=NORMLEXPUB:12100:0::NO::P12100_ILO_CODE:C107.

International Labour Organization. (2009) Indigenous and tribal people's rights in practice. A guide to ILO Convention NO. 169. Geneva.

International Labour Organization. (n.d.) 'Who are the Indigenous and tribal peoples?'. Available at: www.ilo.org/global/topics/indigenous-tribal/WCMS_503321/lang–en/index.htm.

Inkman, D., Cambou, D., and Smis, S. (2018) 'Evolving Legal Protections for Indigenous Peoples in Africa: some Post-UNDRIP Reflections', *African Journal of International and Comparative Law*, 26 (3).

Iturralde, F. (2015) 'Descolonización y colonialismo interno: Lugar y función de lo colonial', *Bolivian Studies Journal*, 21, pp. 40–60.

Jaskoski, M. (2020) 'Participatory Institutions as a Focal Point for Mobilizing: Prior Consultation and Indigenous Conflict in Colombia's Extractive Industries', *Comparative Politics*.

Le Billon, P. and Sommerville, M. (2017) 'Landing capital and assembling 'investable land' in the extractive and agricultural sectors', *Geoforum*, 82, pp. 212–224.

Leifsen, E., Gustafsson, M.T., Guzmán-Gallegos, M.A., and Schilling-Vacaflor, A. (2020) 'New mechanisms of participation in extractive governance: Between technologies of governance and resistance work', *Third World Quarterly*, 38 (5), pp. 1043–1057.

Llanes Salazar, R. (2020) 'La consulta previa como símbolo dominante: Significados contradictorios en los derechos de los pueblos indígenas en México', *Latin American and Caribbean Ethnic Studies*, 15 (2), pp. 170–194.

Machado, M., López Matta, D., Campo, M.M., Escobar, A., and Weitzner, V. (2017) 'Weaving hope in ancestral black territories in Colombia: The reach and limitations of free, prior, and informed consultation and consent', *Third World Quarterly*, 38 (5), pp. 1075–1091.

McNeish, J.A. (2017) 'A vote to derail extraction: Popular consultation and resource sovereignty in Tolima, Colombia', *Third World Quarterly*, 38 (5), pp. 1128–1145.

Merino, R. (2018) 'Re-politicizing participation or reframing environmental governance? Beyond Indigenous' prior consultation and citizen participation', *World Development*, 111, pp. 75–83.

Middeldorp, N. and Le Billon, P. (2019) 'Deadly environmental governance: Authoritarianism, eco-populism and the repression of environmental and land defenders', *Annals of the Association of American Geographers*, 109 (2), pp. 324–337.

Miller, R.J. (2015) 'Consultation or consent: The United States' duty to confer with American Indian governments', *North Dakota Law Review*, 91, pp. 37–98.

Mills, J. (2017) 'Destabilizing the Consultation Framework in Alberta's Tar Sands', *Journal of Canadian Studies/Revue d'études canadiennes*, 51 (1), pp. 153–185.

Mitchell, M.I. and Yuzdepski, D. (2019) 'Indigenous peoples, UNDRIP and land conflict: An African perspective', *The International Journal of Human Rights*, 23 (8), pp. 1356–1377.

Moore, M., von der Porten, S., and Castleden, H. (2017) 'Consultation is not consent: Hydraulic fracturing and water governance on Indigenous lands in Canada', *WIREs Water*, 4 (1180).

Navas, G., Mingorria, S., and Aguilar-González, B. (2018) 'Violence in environmental conflicts: the need for a multidimensional approach', *Sustainability Science*, 13 (3), pp. 649–660.

Nest, M. (2017) *Preventing Corruption in Community Mineral Beneficiation Schemes*. Bergen: U4/CMI.

Nixon, R. (2011) *Slow Violence and the Environmentalism of the Poor*. Cambridge, MA: Harvard University Press.

Orozco, Z.T. and Veiga, M. (2018) 'Locals' attitudes toward artisanal and large-scale mining—A case study of Tambogrande, Peru', *The Extractive Industries and Society*, 5 (2), pp. 327–334.

Padilla Rubiano, G.A. (2015) 'Los pueblos indígenas y la consulta previa: ¿Normatización o emancipación? Una mirada desde Guatemala', *Revista Colombiana de Sociología*, 39 (1), pp. 193–219.

Pasternak, S. (2017) *Grounded Authority: The Algonquins of Barriere Lake Against the State*. Minnesota, MN: University of Minnesota Press.

Perreault, T. (2015) 'Performing participation: Mining, power, and the limits of public consultation in Bolivia', *The Journal of Latin American and Caribbean Anthropology*, 20 (3), pp. 433–451.

Peterson St-Laurent, G.P. and Le Billon, P. (2015) 'Staking claims and shaking hands: Impact and benefit agreements as a technology of government in the mining sector', *The Extractive Industries and Society*, 2 (3), pp. 590–602.

Prensa Comunitaria. (2017) 'Abelardo Curup, Murió en la Carcel en Resistencia contra una Empresa Cementer a'. Available at: www.prensacomunitaria.org/abelardo-curup-murio-en-la-carcel-en-resistencia-contra-una-empresa-cementera1.

Rodrigo, D. and Amo, P.A. (2006). Background document on public consultation. Paris: OECD. Available at: www.oecd.org/mena/governance/36785341.pdf.

Rodriguez, A. (2009) *El papel de la consulta previa en la pervivencia de los pueblos Indígenas y demás grupos étnicos de Colombia*. Bogota: Universidad del Rosario.

Rodríguez-Garavito, C. (2011) 'Ethnicity.gov: Global governance, indigenous peoples, and the right to prior consultation in social minefields', *Indiana Journal of Global Legal Studies*, 18 (1), pp. 263–305.

Roesch, R. (2016) 'The story of a legal transplant: The right to free, prior and informed consent in sub-Saharan Africa', *African Human Rights Law Journal*, 16, pp. 505–531.

Schilling-Vacaflor, A. and Flemmer, R. (2015) 'Conflict transformation through prior consultation? Lessons from Peru', *Journal of Latin American Studies*, 47 (4), pp. 811–839.

Sekar, N. (2016) 'Tigers, Tribes, and Bureaucrats: The voluntariness and socioeconomic consequences of village relocations from Melghat Tiger Reserve, India', *Regional Environmental Change*, 16 (1), pp. S111–123.

Simpson, L. (2017) *As We Have Always Done: Indigenous Freedom Through Radical Resistance*. Minnesota, MN: University of Minnesota Press.

Speed, S. (2017) 'Structures of settler capitalism in Abya Yala', *American Quarterly*, 69 (4), pp. 782–790.

Springer, S. and Le Billon, P. (2016) 'Violence and space: An introduction to the geographies of violence', *Political Geography*, 52, pp. 1–3.

Steinberg, J. (2016) 'Strategic sovereignty: A model of non-state goods provision and resistance in regions of natural resource extraction', *Journal of Conflict Resolution*, 60 (8), pp. 1503–1528.

Sterrit, A. (2019) 'When pipeline companies want to build on Indigenous lands, with whom do they consult?', CBC. Available at: www.cbc.ca/news/canada/british-columbia/when-a-pipeline-wants-to-build-whose-in-charge-1.4971597.

Taylor, A. and Bonner, M.D. (2017) 'Policing economic growth: Mining, protest, and state discourse in Peru and Argentina', *Latin American Research Review*, 52 (1), pp. 3–17.

Temper, L. (2019) 'Blocking pipelines, unsettling environmental justice: From rights of nature to responsibility to territory', *Local Environment*, 24 (2), pp. 94–112.

Temper, L., Del Bene, D., and Martinez-Alier, J. (2015) 'Mapping the frontiers and front lines of global environmental justice: The EJAtlas', *Journal of Political Ecology*, 22 (1), pp. 255–278.

The WoMin Collective. (2017) 'Extractives vs development sovereignty: Building living consent rights for African women', *Gender & Development*, 25 (3), pp. 421–437.

Torres-Wong, M.T. (2019) *Natural Resources, Extraction and Indigenous Rights in Latin America: Exploring the Boundaries of Environmental and State-corporate Crime in Bolivia, Peru, and Mexico*. Abingdon: Routledge.

United Nations. (2007) *United Nations Declaration on the Rights of Indigenous Peoples*. Geneva: UN.

United Nations. (2013) *UN-REDD Programme. Guidelines on free, prior and informed consent*. Geneva: UN.

Urkidi, L. (2011) 'The defense of community in the anti-mining movement of Guatemala', *Journal of Agrarian Change*, 11 (4), pp. 556–580.

Urteaga-Crovetto, P. (2018) 'Implementation of the right to prior consultation in the Andean countries. A comparative perspective', *The Journal of Legal Pluralism and Unofficial Law*, 50 (1), pp. 7–30.

Vermeulen, S. and Cotula, L. (2010) 'Over the heads of local people: consultation. Consent, and recompense in large-scale land deals for biofuels projects in Africa', *The Journal of Peasant Studies*, 37 (4), pp. 899–916.

Walsh, B., van der Plank, S., and Behrens, P. (2017) 'The effect of community consultation on perceptions of a proposed mine: A case study from southeast Australia', *Resources Policy*, 51, pp. 163–171.

Walter, M. and Urkidi, L. (2017) 'Community mining consultations in Latin America (20022012): The contested emergence of a hybrid institution for participation', *Geoforum*, 84, pp. 265–279.

Weitzner, V. (2002) *Through Indigenous Eyes: Toward Appropriate Decision-making Processes Regarding Mining on or Near Ancestral Lands; Final Synthesis Report*. Ottawa, ON: North-South Institute.

Weitzner, V. (2019) 'Between panic and hope: Indigenous peoples, gold, violence(s) and FPIC in Colombia, through the lens of time', *The Journal of Legal Pluralism and Unofficial Law*, 51 (1), pp. 3–28.

Wildcat, M., McDonald, M., Irlbacher-Fox, S., and Coulthard, G. (2014) 'Learning from the land: Indigenous land based pedagogy and decolonization', *Decolonization: Indigeneity, Education & Society*, 3 (3), pp. 1–15.

Wright, C. and Tomaselli, A. (2019) *The Prior Consultation of Indigenous Peoples in Latin America: Inside the Implementation Gap*. London: Routledge.

Xiloj, L. (2019) 'Implementation of the right to prior consultation of Indigenous Peoples in Guatemala' in Wright, C. and Tomaselli, A. (eds.). *The Prior Consultation of Indigenous Peoples in Latin America: Inside the Implementation Gap*. London: Routledge.

Youdelis, M. (2016) '"They could take you out for coffee and call it consultation!": The colonial antipolitics of Indigenous consultation in Jasper National Park', *Environment and Planning A*, 48 (7), pp. 1374–1392.

Young, S. (2019) 'Indigenous Peoples, Consent and Rights. Troubling Subjects. Chapter 4: FPIC as National legislation' in *The Philippines, the B'laan and the Tampaken Mine*. Abingdon: Routledge.

Zaremberg, G. and Torres-Wong, M. (2018) 'Participation on the edge: Prior consultation and extractivism in Latin America', *Journal of Politics in Latin America*, 10 (3), pp. 29–58.

5

LEVERAGING LAW AND LIFE

Criminalization of Agrarian Movements and the Escazú Agreement

Garrett Graddy-Lovelace

The Crisis of Criminalization

In July 2017, at the Seventh International Conference of La Via Campesina (LVC) in Spain's Basque Country, agrarian leaders gathered to strategize their transnational grassroots coalition, which spans tens of millions of *campesinos*—peasants, indigenous people, landless farmers, fishers, and rural communities across the world. They foregrounded the growing wave of violence against agrarian justice activists, calling it "the criminalization of our movements," with community leaders "being assassinated, jailed, tortured, and threatened." They denounced the 2016 murder of internationally renowned river-protector Berta Cáceras in Honduras, as well as the 2012 Curuguaty Massacre in Paraguay (FIAN and LVC, 2014), the imprisonment of Union of Agricultural Workers Committee leaders in Palestine, the arrest of prominent fisher rights advocates in Pakistan, and the murder of Argentine activists by agribusiness police forces (Motta, 2017). They also condemned the criminalization of social movements in Brazil (Canofre, 2017; Sauer and Mészáros, 2017), particularly those of Afro-Brazilians rural land activists, and the disproportionate policing of Standing Rock indigenous resistance in the USA to pipeline infrastructure in the Dakotas. LVC has issued repeated public demands for rigorous investigations into the hundreds of murders of indigenous and *campesino* activists and movement leaders across continents. In their words:

> We call for no crime to go unpunished, because impunity is one of the main reasons why leaders have been murdered…We call for an end to the repression and criminalisation of those who defend life in Brazil and the rest of the world.
>
> *(LVC, 2018b, no page)*

This chapter presents an analysis from the frontline communities themselves, who argue that, in order to understand and counter the surge of violence against land defenders, it is important to investigate the processes of criminalization that aid and abet, justify, and hide the violence. It traces how criminalization occurs when and where land defenders have mobilized to assert their rights to protect water and defend land. Often, land defenders and water protectors use the language of *dignity, freedom, justice*, or *sovereignty* rather than, or in conjunction with, a language of *human* or *environmental rights*.

International bodies, such as the United Nations (UN), as well as prominent and grassroots civil society organizations have classified these victims of violence and criminalization as environmental human rights activists. Frontline mobilizations are also increasingly striving to articulate their struggles to international forum audiences. This work is done through and within the terrain of law, policy, and state-governance, and against, for instance, transnational mega-dam or agribusiness corporations. In backlash, mega-extractor companies have deployed their considerable political-economic power in these same terrains of law, policy, and state-governance. Not only does this inflict great harm on the low-resource frontline communities, from bankruptcy to social fissures, from slander all the way to murder, but it also co-opts and commandeers the realm of law and governance itself.

Legal justice system resources are observed to be used both to help land defenders to protect their lands and themselves and to mobilize against them. When the legal system is tied up with the criminalization of those on the frontline of extraction and displacement, it directs attention away from the environmental crimes of mass extraction and displacement. This maintains a colonial orientation of legal and governance systems that protects those with capital from those who might disrupt cycles of further capital accumulation. Moreover, it maintains the racialized and gendered exclusion and expropriation of this colonialist orientation.

To address these harm-laden developments, grassroots agrarian organizations and movements are forging wider and deeper coalitions and communications strategies. This entails, among other tactics, reasserting and reclaiming the terrain of law, policy, and state-governance with even more transnational force and leverage. A key example of this is the 2018 Regional Agreement on Access to Information, Public Participation and Justice in Environmental Matters in Latin America and the Caribbean, adopted in Escazú, Costa Rica (the Escazú Agreement). This is the first international agreement to address, denounce, and strive to prevent the criminalization of environmental and human rights defenders.

Research is needed on the negotiations that are currently underway to advance signing, ratification, and implementation of this treaty. As an initial contribution, the chapter makes three interrelated theoretical arguments. First, criminalization is observed as a backlash against agrarian and indigenous land defenders' reclamation of legal realms. It shows how powerful and contested the realms of law and its legitimacy are, even as it proves how threatening land defenders' use of the law is to agro-extractors. Second, the Escazú Agreement is recognized as a powerful counterweight to this backlash. It has the potential to strengthen land defenders' resources, protections, capacities, and alliances, as they work to reclaim the contested terrain of law to defend themselves and their lands—as they in turn criminalize the extraction itself and concurrent violence associated with it. Finally, the chapter locates the current struggle over law and legitimacy, over land and life, as an agricultural contestation. Land defenders are essentially defending their agrarian ways of land-based life: foodways, foraging, agroforestry, fishing, community nourishment and food sovereignty. Conversely, extractors are often agribusinesses, large-scale planters, and their paid enforcers and deputies. Accordingly, the chapter ends arguing that the follow-up to the Escazú Agreement needs to focus even more on the extractive and violent nature of agroindustrial production policies and paradigms.

Say Their Names: Violence Against Land and Water Defenders

On March 2, 2016 renowned Honduran activist **Berta Caceres** was brutally murdered in her home. A mother of four from the Lenca indigenous community, she was targeted and killed owing to her bold, public, and increasingly effective opposition to the Agua Zarva hydropower dam on her people's ancestral lands and the Gualcarque River, which holds crucial food, ecological, and spiritual significance for them. She and **Lesbia Yaneth Urquia**, from the same organization, COPINH, were two of 14 prominent land defenders killed in Honduras in 2016 alone. In the previous year, the dam company that Caceres opposed brought false charges against her for "usurpation, coercion and continued damage" (Global Witness, 2017, p. 16) of its property. It also called on the Honduran government to "act with all resources at its disposal to persecute, punish and neutralise" (Global Witness, 2017, p. 16). After a year-long probe, the investigative panel International Advisory Group of Experts found that Honduran government officials and senior executives of the dam company had colluded in the "planning, execution, and cover-up" of Cáceras' assassination (International Advisory Group of Experts, 2017, p. 46).

In September 2009 the Morona Santiago indigenous communities of Ecuador protested against the government for ecological and hydrological destruction from mining projects and for the murder of indigenous leader **Bosco Wisuma** (LADB Staff, 2009). In 2010 **José (Pepe) Acacho González**, president of the Shuar Federation. and six other Ecuadorian indigenous leaders were officially classified as perpetrators of terrorism and, in 2013, were sentenced to 12 years in prison (Human Rights Watch, 2018). In 2012 **Andrés Francisco Miguel** from Barillas, a leader of the Maya Peoples' Council, was murdered in Huehuetenango in the Guatemalan highlands (Geglia, 2012), while community leaders **Esteban Bernabé** and **Pablo Antonio Pablo** were also seriously injured during a peaceful demonstration against mining and agroindustry expansions. Later, a Guatemalan court brought a bevy of charges, including terrorism, against Esteban and Pablo despite a lack of evidence (Guatemala Human Rights Commission, n.d.). In January 2016 **Nilce "Nicinha" de Souza Magalhães** was declared missing. A prominent leader of the longstanding Movement of People Affected by Dams, she publicly denounced the human and environmental violations of the Sustainable Energy of Brazil consortium in their construction of the Usina Hidrelétrica in Porto Velho. Six months later, her body surfaced in the river next to the dam site; her arms and feet had been tied with rope and weighted with rocks. According to the Brazilian Human Rights NGO Comitê Brasileiro, de Souza was one of 66 environmental activists murdered in 2016 in Brazil alone (HRD Memorial, n.d.; Front Line Defenders, 2016). In January 2018 **Márcio Matos**, leader of the Brazilian Landless Workers Movement was shot to death outside his home in Bahia (Friends of the MST, 2018). In northwest Mexico in 2019 indigenous Tarahumara elder **Otilia Martinez Cruz** and her 20-year-old son, **Gregorio Chaparro Cruz**, were shot dead at their home in El Chapote for their leadership fighting illegal deforestation in their ancestral Sierra Madre lands (Global Witness, 2020).

Overall, Global Witness recorded over 200 assassinations of land defenders worldwide in 2017—more than any previous year (Global Witness, 2018). That year also tallied the most massacres (with seven cases of more than four people murdered at the same time), and agrobusiness overtook mining as the industrial sector most responsible for such violence and murder (Global Witness, 2018). In 2018 Global Witness documented another 164 killings of land and environmental defenders, with many more attacked and imprisoned; over half of the murders took place in Latin America (Global Witness, 2019). Global Witness's 2019 report counted a record 212 murders of land defenders, outpacing 2017 records (Global Witness, 2020). Despite the seemingly high numbers, it is also important to note that national level observations by academics and the media demonstrate them to be vast underestimations of the number of land defenders targeted for assassination. Furthermore, between 2017 and 2019 Colombia went from being third to first in the list of countries plagued by this violence (after a brief decline between 2016 and 2017 following the signing of the Peace Accords in 2016) (Global Witness, 2018; Global Witness, 2019; Global Witness, 2020). Additional reports by international organizations, such as Amnesty International, indicate that threats and assassinations of human rights and land defenders have increased during the COVID-19 pandemic quarantine measures (Estupiñán, 2020).

Naming the victims is a hallmark of the resistance effort, helping to put human faces to the violence and ensure that the victims are not forgotten. Echoing the commitment of the Black Lives Matter movement against police murders of Black Americans, the LVC press releases and manifestos (as well as communiqués from other frontline grassroots organizations and coalitions) emphasize the stories of those murdered and the importance of continuing to *say their names*. Criminalization works to "disappear" the social mobilization's messages, even as the concurrent violence literally shoves them into carceral invisibility, abduction, or death itself. The oppositional politics of memory renders visible what criminalization aims to hide: the humanity of those under threat.

Calling It Criminalization

The journalists covering this phenomenon increasingly follow the lead of the activists themselves by focusing not only on the violence but also on the insidious parallel phenomenon of criminalization. Beginning with the 2018 report, Global Witness has begun to track, investigate, and record instances of criminalization as well as violence—and to note how ubiquitous the parallel phenomenon has become (Global Witness, 2019).

What is the purpose of calling state-sanctioned justifications for the violence against indigenous and agrarian movement leaders "criminalization"? Focusing on criminalization at the instigation of the victims partakes of the same processes and dynamics noted by Martinez-Alier *et al.* (2014), wherein political ecology scholars learn and analyze as informed by terms and concepts coined by activist environmental justice organizations: As they write, this "activist-led and co-produced social sustainability science" advances both academic scholarship and activism (Martinez-Alier et al., 2014, p. 19).

Francisco Morales, of the Maya Poptí settlement in western Huehuetenango, Guatemala, is from the Maya People's Council, an indigenous political movement. He chronicles that after a half millennia of fighting colonization and violent extraction, Mayan communities have developed clarity around how to organize and defend themselves:

> The State's response to this democratic and legitimate participation of the people has been violence, the criminalization of the exercise of rights. This also shows the colonial structure of the State, because when we indigenous people claim rights, the State consider us delinquents, terrorists.
>
> *(LVC, 2017, no page)*

LVC has put forth its own analysis of why focusing on criminalization is so important to understanding the broader processes of violence and dispossession and imposed trajectories and conceptions of "development."

In many places, the people who defend themselves against and resist this "development" model face being demonized and criminalized, which in turn leads to prosecutions, imprisonment, violence at the hands of state or private security forces, and even murders. These are not random "incidents", they are occurrences reported by almost every organization. In this respect, States are not only failing in their duty to protect the people from these outrages but are in fact important actors in advancing this model. (LVC, 2020, no page).

Within a few months, LVC sharpened its analysis, stating that the criminalization of "the peasant movements and social struggles" leads to widespread violence against them. Such dynamics are:

> part of a violent and repressive policy, which aims to contain the movement for an agrarian and popular reform that can put agriculture at the service of the

people instead of turning it into a tool to generate profits for a handful of corporations.

(LVC, 2018a, no page)

Other frontline groups are articulating sharp *in situ* political analysis. The Honduran National Front of Popular Resistance issues communiques which locate the Honduran oligarchy as a driver of such violent criminalization under the post-coup dictatorship. The armed forces and state-backed repressions serve as tools while oligarchic control over media seals the situation. Cacerás herself led workshops in order to clarify the messy webs of graft among landed elite, corrupt government, military and controlled media—the very colluding forces that would ultimately kill her (Méndez, 2018). The Brazilian Landless Workers Movement (MST) issues regular communiques, as well, and asserts that the struggle for land is an exercise in citizenship (Carter, 2010; Housing & Land Rights Network, 2016). MST was categorized as criminal for the first time in Law No. 12,850/2013, which defines criminal organizations. In response the organization stated:

> Framing the MST as a criminal organization is the most inconsequential way to combat social movements. There is already extensive jurisprudence of the Supreme Court of Justice, the Federal Supreme Court and Courts of Justice stating that the struggle of the landless is an exercise of citizenship and therefore is not to be confused with crime.

(Housing & Land Rights Network, 2016, no page)

Another famous example of criminalization of indigenous protest is Standing Rock. By 2016, the Standing Rock camp had become the largest gathering of Native Americans and indigenous communities in more than 100 years. In February 2017, after a year of police surveillance and harassment, the National Guard and police officers arrived heavily armed with military equipment and riot gear and evicted protesters from the camp. A leak of over 100 documents to news website *The Intercept* (from an employee of the private security firm contracted by Energy Transfer Partners) evidenced "intrusive military-style surveillance and a counterintelligence campaign" against the Standing Rock Water protesters and their allies, whom they branded "jihadists" (Juhasz, 2017, no page). Following the Standing Rock protests, at least 18 of the 50 U.S. states have proposed criminalizing protests (Cagle, 2019). In addition, in 2017, the Trump Administration reinstated the controversial "Presidential Executive Order on Restoring State, Tribal, and Local Law Enforcement's Access to Life-Saving Equipment and Resources" program that transfers surplus military equipment and weapons of war to police departments across the country to use against "rioters," as in protestors and land defenders (International Center for Not-for-Profit Law, 2020).

Meanwhile, in January 2019 the Royal Canadian Mounted Police threatened to deploy police forces on behalf of TransCanada mining to forcibly remove Wet'-suwet'en from their indigenous lands—sovereign, unceded territory in British Columbia. A year later, talks between Wet'suwet'en leaders and Canadian officials collapsed, the former now fearing police repression and violent backlash from the latter (McIntosh, 2020). In short, the processes of criminalization within the context of hyper-extraction and violence continue to unfold across both the Global North and the Global South (Amnesty International, 2018; Global Witness, 2019; Global Witness, 2020).

Focusing on and calling out the criminalization of agrarian protest recalibrates the view of these instances as isolated conflicts to seeing them as a part of broader state-sanctioned, industry-oriented modes of repression, vilification, and violence against agrarian movements. It foregrounds how industry and state collusion operates, on-the-ground and on the bodies of the most marginalized frontline communities and the most committed frontline natural resource defenders. It draws attention to the juridical, political, and political-economic aspects of policing and police brutality, even as it foregrounds the racialized tactics and tendencies embedded within the psychosocial aspects of criminalization.

Temporally and spatially, criminalizing social movements works to quash movements by hiding them. Hence the tactic, particularly common in Latin America, of "disappearing" victims, of creating off-site torture and imprisonment centers, hiding indigenous victims' bodies, expunging records, and controlling media and journalist investigations. Writing in a deliberately academic venue, LVC leader F. Torrez (2011) contended:

> This criminalization and repression of the struggle for agrarian reform, involving the police, the army, and the private security of corporations, translates into assassinations, judicial persecution. There is a high level of impunity because nothing is investigated nor are the responsible parties punished despite evidence.
>
> *(p. 54)*

In this context, the criminalization of agrarian activists is an attempt to hide the violence of policing—hide it logistically, legally, physically, politically, and ideologically. It switches the focus. It secures impunity in the name of security. It frames the offenses of state-oppression as defensive. And it frames the defending of land, water, and local communities as inimical to the public good.

The clarifying power of the "criminalization" lens also helps disclose racialization in this policing. It elucidates the racialized and class-based privilege of *not* experiencing this violence—and the luxury of not seeing it. Violence has long been ubiquitous within the political economies of colonialism, and now within agro-industrialization and ongoing coloniality (Graddy-Lovelace, 2017). Yet criminalization works to legitimate it, as well as to tuck it away as a legal issue to be resolved between the beneficent state and the individualized criminal.

Calling It Agrarian

Recognizing the iterative criminalization of agrarian movements helps illuminate the *longue durée* of this industry-state collusion in police repression and brutality. In the context of the Americas, for instance, it conjures the land-based resistance to colonialist assault, imperialist capture, and neoliberal extraction and appropriation—all of which have deep agro-industrial roots. The violence is as ancient as the earliest colonization, with its genocidal attempts to expunge and extract—to capture and control indigenous people. Criminalization tactics were used throughout agrarian revolutions, the land occupations, the Dirty Wars, and now continue through a resurgence in agrarian resistance to extractivism. As Grandia (2019) explains, "a surprising number of 'contemporary' environmental justice conflicts are actually recurrent threats to indigenous peoples and their territories from decades past" (p. 7); he and others in LVC member groups track "how many places in corporate crosshairs today were previous targets of state repression during the Guatemalan civil war" (p. 10).

The *longue durée* lens helps to trace the colonial roots of this violence. It shows how the current wave of extractivism is a logical extension of colonialist modes of thinking and doing, and that agrarian resistance has long received the brunt of its strong-armed apparatus of dispossession-by-vilification. It also shows the fault lines and the inherent dysfunctions of agro-industrial extractive production. The 2017 Global Witness Report grimly tallied the number of people murdered in 2017 while protesting against large-scale agriculture had more than doubled compared to the previous year: "Ultimately, attacks against land and environmental defenders stem from our voracious appetite for agricultural goods like palm oil and coffee, and for fossil fuels, minerals and timber" (Global Witness, 2018, p. 7). The 2018 report tallied the agribusiness sector as second only to mining and extraction in violence perpetration; the 2019 report documented 34 killings linked directly to large-scale agriculture, an increase of more than 60 percent in one year (Global Witness, 2020).

A *longue durée* agrarian lens, thus, clarifies how violence against land and water defenders undermines their agricultural livelihoods even as it enables agro-industrial extraction. Moreover, it recalls how such agricultural dispossession has long fallen along race, gender, and class lines, so that the agrarian realm has been demeaned and made illegitimate. Accordingly, LVC has long prioritized a historical, anti-colonial framework in its protests against contemporary agroindustrial extraction.

The Terrain of Law: Negotiating Legality and Legitimacy

Criminalization attempts to deploy the law against communities defending natural resources. It depends upon impunity, extra-judicial imprisonment, and executions even as it nominally defers to and upholds legality and public good. This disjuncture, however, reveals a potential crack in the logic that shows a grasping at legitimacy. Agrarian movements are growing around the world, and they are increasingly and proactively working through legal and political channels—for instance, to seek justice and punishment of guilty parties at the UN level and in Truth and Reconciliation reports. They are leveraging it to call out the real criminality at work in agro-industrial extraction and violence against agrarian movements and communities.

Extractors collude with the state to occupy and wield the terrain of law's legitimacy. Glazebrook and Opoku (2018) describe how this unfolds in Central America:

> Given the embeddedness of capital in Honduran governance, it seems that the fate of the dam and of accountability in Cáceres' murder is a negotiation (i.e. a transactional hedging of bets) within capital based on public relations, cost-benefit analyses, and hollow corporate responsibility commitments.
>
> *(p. 88)*

Ultimately, "Governments seem to be at best the tool of capital, at worst, its weapon" (p. 90).

Here, an analysis of criminalization divulges the bitter battle of the very terrain of law and legal juridical systems themselves. On one side, frontline groups work assiduously, against historic odds, to make the legal system work for them in their fight for survival. For example, López Rodríguez and Excell (2017) document how the *Ríos Vivos Antioquia* movement in Colombia worked closely with the *Corporación Jurídica Libertad* (a lawyer collective) to research and file a claim before the Council of State to petition that they revoke the environmental license for the Hidroituango dam. The movement has built an extensive multi-sector coalition to work with the Ministry of Interior to develop a protection plan for land and river defenders in this struggle to maintain control over these legal terrains and their legitimizing force. For example, in May 2017, the "ruralista front" representing Brazilian agribusiness and large-scale landholders reinitiated a Commission of Parliamentary Inquiry investigating possible fraud and abuse in the demarcation process by indigenous and Afro-Brazilian groups. They listed more than a hundred people—from federal attorneys to social scientists to indigenous leaders to activists—to be prosecuted for land fraud. Such broad criminalization attempts to stop the demarcation process of returning indigenous and Afro-Brazilians to traditional lands (Da Cunha *et al.*, 2017). All the while, the recent rise in criminalization of rural and agrarian justice movements, such as the Landless Workers Movement, denies the extensive jurisprudence and constitutional reforms in Brazil, which legitimize the struggle for land reform as a key exercise of citizenship—not a crime. Bolsonaro's election has raised the lethal stakes of this battle for constitutional leverage even higher.

Civil society organizations are learning from each other how to co-ordinate consultations, prepare lawsuits, study legal history, and wield juridical openings for support; they are launching louder challenges to hyper-extraction in their lands using legal avenues. However, their successes fuel retaliation and more aggressive persecution—through and beyond legal channels. UN Special Rapporteurs and other transnational human rights organizations have found rampant misuse of criminal law with judges accepting false testimony and misinterpretations of the law to incriminate indigenous defenders (Comisión Interamericana de Derechos Humanos, 2015). Scholarly findings concur. As Rasch writes, "Trust in political institutions and democracy is further hampered by using penal law and anti-terrorist legislation to obstruct social mobilization, and by declaring a state of emergency to justify the detention of activists" (Rasch 2017, p. 134).

The Escazú Agreement: Reclaiming the Legal Terrain via Treaty

Amidst this convoluted contestation for the terrain of law—and thus land-based life itself—a watershed moment occurred in 2018. On March 4, the Regional Agreement on Access to Information, Public Participation and Justice in Environmental Matters in Latin America and the Caribbean (the Escazú Agreement) was adopted by regional leaders in Escazú, Costa Rica and opened for signature at the UN in September 2018. By mid-2020 a total of 22 of 33 countries had signed, eight had ratified, and only 11 more were needed for the Agreement to go into force.

The Escazú Agreement situates itself as the logical next step from prior international environmental governance, namely the Rio Declaration's Principle 10 (on access rights) and the UNECE Convention on Access to Information, Public Participation in Decision-Making and Access to Justice in Environmental Matters (Aarhus Convention). The Agreement asserts the principles of "equality" and "non-discrimination" followed by "intergenerational equity" and "*pro persona.*" In a November 2018 interview, Roberto Avendaño, the Costa Rican diplomat who led the negotiation and drafting of the Agreement, emphasized the groundbreaking assertion of these foundational principles. Article 4 (entitled "General provisions" but actually meaning "obligations" according to both Avendaño and Maritza Chan, the other lead diplomat in implementing the Agreement, also interviewed in November 2018) goes even further: "Each Party shall guarantee an enabling environment for the work of persons, associations, organizations, or groups that promote environmental protection, by recognizing and protecting them." The interview with Chan and Avendaño, conducted at the Costa Rican embassy in Washington, DC, delved into the Agreement's promotion of public participation in environmental decision-making processes, as well as "free technical and legal assistance" (5). This entailed supplying translators and interpreters of dozens of indigenous languages and, as Chan stressed, more culturally grounded, multimedia translation services than have ever been employed in multilateral agreement negotiations and implementations—far beyond the standard, major colonial tongues.

Yet it is Article 9, "Human rights defenders in environmental matters," that addresses the crisis of criminalization head-on:

> Each Party shall take adequate and effective measures to recognize, protect and promote all the rights of human rights defenders in environmental matters, including their right to life, personal integrity, freedom of opinion and expression, peaceful assembly and association, and free movement, as well as their ability to exercise their access rights, taking into account its international obligations in the field of human rights, its constitutional principles and the basic concepts of its legal system.
>
> *(United Nations ECLAC, 2018, p. 29)*

Even more explicitly, it demands that: "Each Party shall also take appropriate, effective, and timely measures to prevent, investigate and punish attacks, threats or intimidation that human rights defenders in environmental matters might suffer while exercising the rights set out in the present Agreement" (p. 3). This final self-reference serves to solidify the Agreement's potency in countering criminalization and violence mechanisms. Avendaño explained that Costa Rican negotiators proposed this key article in Argentina in August 2017. Intersessional meetings spent hours, days, weeks, and months packing their most pressing concerns into Article 9.3, before a December 2017 meeting in Chile. Once consensus was reached that the Agreement must include this demand, more virtual meetings with civil society organizations had to be called in advance of the March 2018 meeting to come to consensus on phrasing.

Article 10, "Capacity-building," continues with reclaiming legal and governance terrains to defend land and land defenders. Each Party needs to "develop and strengthen environmental law and access rights awareness-raising and capacity-building" for "the public, judicial and administrative officials, national human rights institutions and jurists" (United Nations ECLAC, 2018, pp. 29–30). Sriskandarajah (2018) notes the significance: "In times when civic space and opportunities for citizens to participate seem to be shrinking, Escazú [Agreement] sets an important example on the importance of *reimagining democracy* to face the environmental challenges of our world" (no page).

There were precursors, such as the European Union's 1998 Aarhus Convention on Access to Information, Public Participation and Access to Justice in Environmental Matters Convention, supplemented by the 2005 Almaty Guidelines. The Escazú Agreement calls for the election of Representatives of the Public, people with "integrity" to coordinate the input of diverse publics. In reflecting on the Aarhus Convention's attempt to incorporate key civil society organizations and coalitions, the Escazú Agreement learned that two key and elusive capacities are needed for effective civil society collaboration in negotiations: Coordination and expertise. This aligns with calls by López Rodríguez and Excell (2017) in Colombia for "providing legal support for judicial processes and technical support to make scientific studies," (p. 3) as well as communication strategies to counter defamation, cross-community coordination and learning. Along these lines, in 2015, the Paris Conference of the Parties (COP) launched the Local Communities and Indigenous Peoples Platform (LCIPP) as a forum for integrating traditional, Indigenous, and local knowledge systems into the United Nations Framework Convention on Climate Change (UNFCCC) processes (Reidel and Bodle, 2018); in 2018 at COP 24, the Parties established the LCIPP Working Group composed of State and indigenous representatives, allowing them to negotiate at equal levels (United Nations Framework Convention on Climate Change, n.d.).

Another precedent was the UN Environment Programme's (UNEP) "Promoting Greater Protection for Environmental Defenders" Policy, which aimed to reclaim and "Promote the critical role of the rule of law in environmental matters" (n.d.) by establishing an "internal accountability mechanism" through an email hotline. The

UNEP launched a new Law Division to develop a communications template to guide its public responses to violence against land and water defenders, and now it aims to expand and strengthen its legal assistance team with an Environmental Rights Initiative and Campaign (UN Environment Programme, 2018). Meanwhile, expanding on a precedent in Cambodia (Embree, 2015), the International Criminal Court expanded its remit in September 2016 to include politicians and businesspeople, who can now be charged and indicted under international law for crimes such as land grabbing and environmental destruction. "International law has, however, very little bite to hold accountable the actual criminals who embezzle, profit off resource theft, and commit murder and other atrocities against defenders" (Glazebrook and Opoku, 2018, p. 93); nonetheless, civil society organizations have not given up hope that this terrain might do more.

Principle 10 of the 1992 Rio Declaration states famously that environmental decisions should require full participation of those concerned and affected. While the Aarhus Convention strives to implement this in Europe, the Escazú Agreement goes further by articulating and implementing the conditions under which such participation can happen—particularly with the majority of impacted communities being indigenous, remote, and illiterate in official languages. As a "bespoke Principle 10 for Latin America and the Caribbean," the Escazú Agreement "allows for different legal cultures to elaborate on core values of international environmental law" (Barritt, 2019, p. 1).

The regional scale of governance carries risks. In last-minute negotiations on the Agreement's final text, powerful negotiators from large countries changed the definition of "public" to "one or more natural or legal persons and the associations, organizations or groups established by those persons, that are nationals or that are subject to the national jurisdiction of the State Party" (Stec and Jendroska, 2019, p. 544). Citizenship risks trumping personhood here, with clear political loss for migrants, undocumented people, and indigenous communities at odds with their respective states. Likewise, the preliminary text of the Escazú Agreement Preamble recognized that: "Everyone has the right to a healthy environment in harmony with nature, which is essential for the full development of human beings and for the achievement of sustainable development, poverty eradication, equality, and the preservation and stewardship of the environment for the benefit of present and future generations." Guerra and Parola (2019) document how concluding debates deleted this bold vision, making it a goal and not a right. Nonetheless, the Escazú Agreement solidifies an emphasis on indigenous rights, historical violence and extractivism, and the importance of traditional knowledge whilst drawing on the deep and long histories of agrarian and indigenous resistance in South and Central America and the Caribbean.

In June 2018, Amnesty International and the Access Initiative created a "Campaign Strategy for the Signing and Ratification of the Escazú Agreement" to ensure coherence between the goals outlined and their implementation. Civil society networks now leverage the success of the Agreement for further gender, racial, ability, class, age, and ontological diversity and equity. Other international governance realms are taking notice. The Escazú Agreement was highlighted at two events at UNFCCC COP 24 meeting in Poland in December 2018 (CEPAL, 2018a); at a United Nations Economic Commission for Latin America and the Caribbean (ECLAC) briefing on promoting public participation in climate policies (CIEL, 2018); and a side event the Office of the UN High Commissioner for Human Rights, led with many co-organizations. This side event focused on the urgency of including vulnerable frontline communities in climate negotiations and governance to uphold human rights (CEPAL, 2018b). In May 2020 the ECLAC issued a press release centering the Escazú Agreement as core to climate change survival in the Americas (CEPAL, 2020).

Following the Lead of Indigenous and Agrarian Defenders

Under neoliberalism, extractivism has spread like cancer across the Americas and worldwide, and now it festers under contemporary neo-authoritarianism, indicating a new iteration of a well-established dynamic. This chapter has built upon recent activism and scholarship on extractivism to focus on two aspects of its perpetual violence: the mechanism of criminalization, and the context of agriculture and agrarianism. Activists and scholars show how deep extractive modes go, beyond literal mining and plantations into new realms of extraction of data, logistics, finance, and biocapital, moving from the "Washington consensus" to the "commodities consensus" (Mezzadra and Neilson, 2017, p. 186).

Agribusiness looms large in these nodes of power, and in the sprawling sector of data accumulation and surveillance, which serves as a powerful and elusive site of dispossession-by-vilification. The genetic resources of plants for food and agriculture are often framed as a commons, but extractivism scholars point to allegedly "open access" commons-discourse as a slippery trope, a means to unbundle value for capture (Dahlin and Fredriksson, 2017). Violence against land defenders grabs (some) headlines, but a quieter domination ensues as extractive industries deploy exclusion-inclusion modalities and extraction-based subject formation. Frederiksen and Himley (2020) lay out three steps to this process: consolidating exclusion (extracting), compounding exclusion (criminalizing those who protest, through discursive exclusion overlaid upon material exclusion), and counteracting exclusion (constructing the "beneficiary" by employing some of the affected community in the actual extraction). Within the realm of agriculture, this entails hiring dispossessed indigenous and agrarian communities as farmworkers in new palm oil or soybean plantations.

This chapter has shown that the process of criminalization works in iterative layers. Indigenous and *campesino* agrarian defenders seek redress through the law; agro-extractors counter with criminalization tactics; agrarian defenders work with lawyers and international political forums to try to protect their rights further; extractive industries collude with (and help elect) authoritarian rulers to repress agrarian defenders even more. Dunlap (2019), for example, demonstrates the "whole-of-government" counterinsurgency apparatus at work in Peru and elsewhere.

Meanwhile, Chile embodies the pivotal potential—and paradoxes—of the Escazú Agreement and its need to address agroindustry extractivism more directly. The country has grappled with criminalization of agrarian and indigenous defenders for generations, particularly after the 1973 U.S.-backed coup toppling President Salvador Allende. All the while, a wide and strong coalition has worked to counter this phenomenon (Figueroa Hernandez and Herrera, 1998). Thereafter, Chile became a poster child for export-oriented agribusiness and neoliberal agro-developmentalism, with hydro-extractivism, high-input-productivism, and displacement of indigenous lands and foodways (Panez *et al.*, 2020).

Francisca Linconao, a renowned Mapuche elder and traditional healer, was arrested and labelled a terrorist for her alleged role in protests against Chilean corporate elite, largely from the agribusiness sector. Indefinite pre-trial detention ensued, along with a hunger strike and public outrage. She was found not guilty in 2017. According to Bernauer *et al.* (2018), her "case is not unique, but rather the latest in a long history of Chile's use of its criminal justice system to repress Mapuche resistance to the dispossession of Wallmapu, the Mapuche homeland" (p. 34). The seeds for the Escazú Agreement were sown in 2012 in Chile, with coalitions of indigenous and civil society leaders and scholar activists thinking expansively about how to prevent criminalization, control corruption, ensure transparency, and advance land and human rights such as food and land sovereignty. Though the Agreement's instigation and leadership of "an unprecedented example of 'deliberative democracy', which allowed all concerned to contribute to the process using their knowledge and experiences in open sessions" (Valencia and Nagalech, 2019, no page) began in Chile alongside Costa Rica, Chile has not yet signed much less ratified, the Agreement. Nonetheless, the Strategy for Civil Society Engagement in the Escazú Agreement proposed the registration of a nonprofit dedicated solely to implementation, to be headquartered in Chile and Jamaica, with a dedicated Secretariat.

In short, widening the lens beyond the site of overt violence expands the realm of culpability. Suddenly, seemingly benign developmentalism becomes implicated in extractivism and its apparatus of criminalization. Andreucci and Kallis (2017) challenge scholars to "to explain how the tension between resource-based development and the violence that sustains it is recomposed through a discursive 'othering', targeting those who oppose extraction" (p. 95). Focusing on Peru, they show how discourses of "idle lands" and commodity crop productivism enable neoliberal dispossession-by-vilification, wherein agrarian and indigenous defenders suffer charges of "crimes against the public order", rebellion, sedition, conspiracy against state and constitution.

What would counter such powerful forces? The 2019 UN State of the World's Indigenous Peoples calls for "solidarity and support for indigenous rights defenders and for coordinated, high-level prevention and defense mechanisms to guarantee their safety and security and the freedom to lawfully defend indigenous rights" (United Nations, 2019b, p. 57). Yet, activists and scholars warn of simplistic, romanticized compartmentalization of peasant or indigenous difference. The "radical difference of these [indigenous] ontologies cannot serve as the endpoint of analysis," advise Neale and Vincent (2017, p. 432). Indigenous communities have suffered extreme physical and ontological violence and appropriation for centuries, as their articulations of opposition have been vilified on one hand, and demeaned as quaint on the other.

Conclusion

At least four people working to defend land and water and agrarian dignity are murdered each week. The latest 2019 Global Witness report points out that land defenders have been on the front line of defense against climate crisis causes and impacts for years. Journalists, scholars, and UN policymakers have recognized that land defenders are not merely worth defending—but worth lauding and following. "If we want to end climate breakdown, then it is in the footsteps of land and environmental defenders we must follow. We must listen to their demands and amplify them" (Global Witness, 2020, p. 8). In particular, this entails following and learning from indigenous and Afro-Diaspora women agrarian justice leaders who, like Angelica Ortiz and Francia Marquez, are facing regular death threats as they defend ancestral lands in Colombia.

Yet, they act with impunity. The hitmen are rarely charged, while the high-level agents and broader structure remain unpunished. The LVC and other front-line agrarian organizations and coalitions around the world are mobilizing their grief at the murder of their community leaders into a deeper analysis of the role of criminalization in the violence at hand. Following this *in situ* lead, journalists are investigating in order to count the murdered and help to bring injustice to account. Scholars follow suit, working across fields and disciplines to contextualize the *longue durée* of the tactic of racialized criminalization within the broader colonialist strategy of dispossession. These journalistic and scholarly analyses work to connect seemingly isolated instances into a broader understanding of the systemic nature of criminalization, and how it arises as a backlash against mobilizations to defend land and protect water and life—mobilizations that are themselves aiming to work in and through the terrains of law, policy, and state-governance.

The backlash merely engenders more mobilization: amidst heightened repression and criminalization, "we also see solidarity and internationalism as a potent strategy of peoples' resistance against extractive capital" (LVC, 2017, no page). This transnational solidarity increasingly works to wield legal and juridical means for defending land and land defenders. All the while, these processes of protests, then criminalization, then more resistance "shape new subject positions" (Rasch, 2017, p. 131). As more call out and recall the names of those killed defending water and land for their communities, mobilizations grow and multiply. Criminalization aims to depoliticize; calling out criminalization roundly re-politicizes. Grant and Le Billon (2019) document the layered psychological and political impact of these processes: "Rather than simply repressing and disciplining forest dwelling populations, violence against defenders shapes their subjectivity and re-politicizes their lives" (p. 768).

As we have seen in this chapter, the Escazú Agreement stands as the culmination of unprecedented civil society engagement and leadership in international negotiations. Though as-of-yet waiting to come into force, it carries considerable potential to re-leverage law and scales of reference and reckoning. Who gets to call whom a criminal? And who brings whom to justice and how? The Escazú Agreement—both as product and process—serves as a key reclamation of the terrain of law and formal governance.

This discussion of criminalization and counter-criminalization leaves many questions unanswered. In particular, can a formal treaty address and redress deep political grievances against the colonial roots and legacies underlying current conflicts? After all, defending the defenders might necessitate a:

> radical transition from patriarchal capital, i.e. from patriarchy as inherent in international and national governance, and from capital as an economic system that values human existence as the individual accumulation of private wealth, to non-gender-privileging governance aimed at the flourishing of life.
>
> *(Glazebrook and Opoku, 2018, pp. 102–103)*

It remains to be seen whether civil society engagement parameters will allow for decolonial plurality of ecological and agrarian ontologies, and whether a formal UN Agreement mechanism can encompass such transformative directions.

References

Amnesty International. (2018) *A Recipe for Criminalization: Defenders of the Environment, Territory and Land in Peru and Paraguay*. London: Amnesty International.

Andreucci, D. and Kallis, G. (2017) 'Governmentality, Development, and the Violence of National Resource Extraction in Peru', *Ecological Economics*, 134, pp. 95–103.

Barritt, E. (2019) 'Global values, Transnational Expression: From Aarhus to Escazú' in Heyvaert, V. and Duvic-Paoli, L.A. (eds.) *Research Handbook on Transnational Environmental Law*. Northampton, MA: Edward Elgar.

Bernauer, W., Heller, H., and Kulchyski, P. (2018) 'From Wallmapu to Nunatsiavut: The Criminalization of Indigenous Resistance', *Monthly Review*, 69 (8), pp. 33–40.

Cagle, S. (2019) '"Protesters as terrorists": Growing number of states turn anti-pipeline activism into a crime ', *The Guardian*. Available at: www.theguardian.com/environment/2019/jul/08/wave-of-new-laws-aim-to-stifle-anti-pipeline-protests-activists-say.

Canofre, F. (2017) 'Criminalizing Indigenous Rights: The Battle for Land in Brazil', *World Policy Journal*, 34 (3), pp. 64–68.

Carter, M. (2010) 'The Landless Rural Workers Movement in Brazil', *Latin American Research Review*, 45, pp. 186–217.

CEPAL. (2018a) The Escazú Agreement at COP24 on climate change. Available at: www.cepal.org/en/notes/escazu-agreement-cop24-climate-change#:~:text=jpg-,The%20recently%20adopted%20Regional%20Agreement%20on%20Access%20to%20Information%2C%20Public,Convention%20on%20Climate%20Change%20.

CEPAL. (2018b) UNFCCC COP24 Side Event "Voices from the Climate Frontlines: Protecting the rights of the most vulnerable (and furthest behind)". Available at: www.

cepal.org/en/events/unfccc-cop24-side-event-voices-climate-frontlines-protecting-rights-most-vulnerable-and.

CEPAL. (2020) ECLAC and the OECS Establish an Enhanced Programme of Action on the Escazú Agreement in the Eastern Caribbean. Available at: www.cepal.org/en/pressreleases/eclac-and-oecs-establish-enhanced-programme-action-escazu-agreement-eastern-caribbean.

CIEL. (2018) 'Public Participation in Climate and Environment-related Frameworks'. Available at: www.ciel.org/wp-content/uploads/2018/12/PromotingParticipation_Entry Points_COP24-draft-FINAL.pdf.

Comisión Interamericana de Derechos Humanos. (2015) Criminalization of Human Rights Defenders. Available at: www.oas.org/en/iachr/reports/pdfs/criminalization2016.pdf.

Da Cunha, M.C.*et al.* (2017) 'Indigenous peoples boxed in by Brazil's political crisis', *HAU: Journal of Ethnographic Theory*, 7 (2), pp. 403–426.

Dahlin, J. and Fredriksson, M. (2017) 'Extracting the commons', *Cultural Studies*, 31(2–3), pp. 253–276.

Dunlap, A. (2019) '"Agro sí, mina NO?' The Tia Maria copper mine, state terrorism, and social war by every means in the tambo valley, Peru', *Political Geography*, 71, pp. 10–25.

Embree, J. (2015) 'Criminalizing Land-Grabbing: Arguing for ICC Involvement in the Cambodian Land Concession Crisis', *Florida Journal of International Law*, 27 (3), pp. 399–420.

Estupiñán, D. (2020) Colombia's social leaders are still being killed during the quarantine. Available at: www.amnesty.org/en/latest/news/2020/06/lideres-sociales-nos-siguen-matando-durante-cuarentena.

FIAN and LVC. (2014) 'Land Conflicts and the Criminalization of Peasant Movements in Paraguay: the Case of Marina Kue and the "Curuguaty Massacre"', *Land & Sovereignty Brief*, 6, Food First, Oakland, CA.

Figueroa Hernandez, D. and Herrera, J.R. (1998) 'Notas para la criminalización de algunas ofensas ambientales en Chile', *Revista De Derecho*, 5 (1), pp. 111–127.

Frederiksen, T. and Himley, M. (2020) 'Tactics of dispossession: Access, power, and subjectivity at the extractive frontier', *Transactions of the Institute of British Geographers*, 45 (1), pp. 50–64.

Friends of the MST. (2018) 'MST grieves murder of a leader: "The shedding of workers' blood is one more reason to fight"'. Available at: www.mstbrazil.org/news/mst-grieves-murder-leader-%E2%80%9C-shedding-workers%E2%80%99-blood-one-more-reason-fight%E2%80%9D.

Front Line Defenders. (2016) Case History: Nilce de Souza Magalhães. Available at: www.frontlinedefenders.org/en/case/Case-History-Nilce-de-Souza-Magalhaes.

Geglia, B. (2012) '"We Are All Barillas": A new moment in Guatemala's anti-extraction movement'. Available at: wagingnonviolence.org/2012/06/we-are-all-barillas-a-new-moment-in-guatemalas-anti-extraction-movement.

Glazebrook, T. and Opoku, E. (2018). 'Defending the Defenders: Environmental Protectors, Climate Change and Human Rights', *Ethics and the Environment*, 23 (2), pp. 83–109.

Global Witness. (2017) Honduras: The Deadliest Place to Defend the Planet. Available at: www.globalwitness.org/en/campaigns/environmental-activists/honduras-deadliest-country-world-environmental-activism.

Global Witness. (2018) 'At What Cost? Irresponsible business and the murder of land and environmental defenders in 2017'. Available at: www.globalwitness.org/en/campaigns/environmental-activists/at-what-cost.

Global Witness. (2019) Enemies of the State? How governments and businesses silence land and environmental defenders. Available at: www.globalwitness.org/en/campaigns/environmental-activists/enemies-state.

Global Witness. (2020) Defending Tomorrow: The climate crisis and threats against land and environmental defenders. Available at: www.globalwitness.org/en/campaigns/environmental-activists/defending-tomorrow.

Graddy-Lovelace, G. (2017) 'The Coloniality of US Agricultural Policy: Articulating Agrarian (In)Justice', *Journal of Peasant Studies*, 44 (1), pp. 78–99.

Grandia, L. (2017) 'Ecocide in the Americas: Continuities and Connections', *Brújula*, 11, pp. 1–26.

Grant, H. and Le Billon, P. (2019) 'Growing Political: Violence, Community Forestry, and Environmental Defender Subjectivity', *Society & Natural Resources*, 32 (7), pp. 768–789.

Guatemala Human Rights Commission. (n.d.) Santa Cruz Barillas. Available at: www.ghrc-usa.org/our-work/current-cases/santa-cruz-barillas.

Guerra, S. and Parola, G. (2019) 'Implementing Principle 10 of the 1992 Rio Declaration: A Comparative Study of the Aarhus Convention 1998 and the Escazú Agreement 2018', *Revista Jurídica*, 1 (55), pp. 1–33.

Housing & Land Rights Network. (2016) Brazil: Goiás Land Struggle, Exercise in Citizenship. Available at: https://hlrn.org/news.php?id=pnBmYw==.

HRD Memorial. (n.d.) Nilce de Souza Magalhães. Available at: https://hrdmemorial.org/hrdrecord/nilce-de-souza-magalhaes.

Human Rights Watch. (2018) Amazonians on Trial: Judicial Harassment of Indigenous Leaders and Environmentalists in Ecuador. Available at: www.hrw.org/report/2018/03/26/amazonians-trial/judicial-harassment-indigenous-leaders-and-environmentalists.

International Advisory Group of Experts. (2017) Dam Violence: The Plan that Killed Berta Cáceras. Available at: https://gaipe.net/wp-content/uploads/2017/10/GAIPE-Report-English.pdf.

International Center for Not-for-Profit Law. (2020) US Protest Law Tracker. Available at: www.icnl.org/usprotestlawtracker.

Juhasz, A.(2017) 'Paramilitary security tracked and targeted DAPL opponents as "jihadists," docs show', Grist. Available at: https://grist.org/justice/paramilitary-security-tracked-and-targeted-nodapl-activists-as-jihadists-docs-show.

LADB Staff. (2009) 'Ecuadoran [sic]President Correa Faces Discontent from Indigenous Communities'. Available at: https://digitalrepository.unm.edu/cgi/viewcontent.cgi?article=14820&context=notisur.

López Rodríguez, A. and Excell, C. (2017) 'Violence Against Land and Water Defenders in Colombia', Open Society Foundation, Washington, DC.

LVC. (2017) 'Affected of the World, Unite!'. Available at: https://viacampesina.org/en/affected-world-unite.

LVC. (2018a) 'La Via Campesina International condemns the Marcinho's murder and demands that the culprits be brought to justice!'. Available at: https://viacampesina.org/en/la-via-campesina-international-condemns-marcinhos-murder-demands-culprits-brought-justice.

LVC. (2018b) 'La Via Campesina: "We speak out on behalf of those who defend lives, Marielle Franco is present with us"'. Available at: https://viacampesina.org/en/la-via-campesina-we-speak-out-on-behalf-of-those-who-defend-lives-marielle-franco.

LVC. (2020) '#TimetoTransform: Why do we regard an Integral and Popular Agrarian Reform as a matter of urgency?'. Available at: https://viacampesina.org/en/timetotransform-why-do-we-regard-an-integral-and-popular-agrarian-reform-as-a-matter-of-urgency.

Martinez-Alier, J., Anguelovski, I., Bond, P., Del Bene, D., Demaria, F., Gerber, J.F., Greyl, L., Haas, W., Healy, H., Marín-Burgos, V., Ojo, G., Porto, M., Rijnhout, L., Rodríguez-Labajos, B., Spangenberg, J., Temper, L., Warlenius, R., and Yánez, I. (2014) 'Between activism and science: grassroots concepts for sustainability coined by Environmental Justice Organizations', *Journal of Political of Ecology*, 21, pp. 19–60.

McIntosh, E. (2020) 'Wet'suwet'en await 'imminent' RCMP action as Coastal GasLink negotiations break down', *Canada's National Observer*. Available at: www.nationalobserver. com/2020/02/04/news/wetsuweten-await-imminent-rcmp-action-coastal-gaslink-nego tiations-break-down.

Méndez, M.J. (2018) '"The River Told Me": Rethinking Intersectionality from the World of Berta Cáceres', *Capitalism Nature Socialism*, 29 (1), pp. 7–24.

Mezzadra, S. and Neilson, B. (2017) 'On multiple frontiers of extraction: Excavating con temporary capitalism', *Cultural Studies*, 31(2–3), pp. 185–204.

Motta, R. (2017) 'Peasant Movements in Argentina and Brazil' in B. Engelsand and K. Dietz (eds.) *Contested Extractivism, Society and the State*. London: Palgrave Macmillan.

Neale, T. and Vincent, E. (2017) 'Mining, Indigeneity, Alterity: Or mining indigenous alterity?', *Cultural Studies*, 31, pp. 2–3, 417–439.

Panez, A., Roose, I., and Faúndez, R. (2020) 'Agribusiness Facing its Limits: The Re-Design of Neoliberalization Strategies in the Exporting Agriculture Sector of Chile', *Land*, 9 (66), pp. 1–26.

Rasch, E.D. (2017) 'Citizens, Criminalization and Violence in Natural Resource Conflicts in Latin America', *European Review of Latin America and Caribbean Studies*, 103, pp. 131–142.

Reidel, A. and Bodle, R. (2018) 'Local Communities and Indigenous Peoples Platform'. Available at: www.ecologic.eu/sites/files/publication/2018/2139-local-communities-a nd-indigenous-people-platform_0.pdf.

Sauer, S. and Mészáros, G.(2017) 'The political economy of land struggle in Brazil under Workers' Party governments', *Journal of Agrarian Change*, 17 (2), pp.397–414.

Sriskandarajah, D. (2018) 'Treaty pushes for environmental justice in Latin America and the Caribbean'. Available at: www.openglobalrights.org/treaty-pushes-for-environmental-jus tice-in-latin-america-and-the-caribbean.

Stec, S. and Jendroska, J. (2019) 'The Escazu Agreement and the regional Approach to Principle 10: Process, Innovation, and Shortcomings', *Journal of Environmental Law*, 31, pp. 533–545.

The Guardian. (2018) 'The defenders'. Available at: www.theguardian.com/environment/ series/the-defenders.

Torrez, F. (2011) 'La Via Campesina: Peasant-led agrarian reform and food sovereignty', *Development*, 54 (1), pp. 49–54.

United Nations. (2019a) Human Rights Council resolution 40/L.22. "Recognizing the contribution of environmental human rights defenders to the enjoyment of human rights, environmental protection and sustainable development". Available at: https://digitallibra ry.un.org/record/3804641?ln=en.

United Nations. (2019b) 'State of World's Indigenous Peoples'.

United Nations Framework Convention on Climate Change. (n.d.) Facilitative Working Group of the Local Communities and Indigenous Peoples Platform. Available at: https:// unfccc.int/LCIPP-FWG.

United Nations ECLAC. (2018) 'Regional Agreement on Access to Information, Public Parti cipation and Justice in Environmental Matters in Latin America and the Caribbean'. Available at: https://repositorio.cepal.org/bitstream/handle/11362/43583/1/S1800428_en.pdf.

United Nations Environment Programme. (2018) 'What Is the Environmental Rights Initiative?'. Available at: www.unenvironment.org/explore-topics/environmental-rights-a nd-governance/what-we-do/advancing-environmental-rights/what-1.

United Nations Environment Programme. (n.d.) 'Promoting Greater Protection for Environ mental Defenders (Policy)'. Available at: https://wedocs.unep.org/bitstream/handle/20.500. 11822/22769/UN%20Environment%20Policy%20on%20Environmental%20Defenders_08. 02.18Clean.pdf?sequence=1&isAllowed=y.

United Nations Human Rights, Office of the High Commissioner. (2016) 'Berta Cáceres Murder: UN Experts Renew Call to Honduras to End Impunity'. Available at: www.ohchr.org/EN/NewsEvents/Pages/DisplayNews.aspx?NewsID=19805.

United Nations Human Rights, Office of the High Commissioner. (2018) 'End of Mission Statement by Michael Forst, United Nations Special Rapporteur on the Situation of Human Rights Defenders on his Visit to Honduras, 29 April to May 12'. Available at: www.ohchr.org/EN/NewsEvents/Pages/DisplayNews.aspx?NewsID=23063&LangID=E.

Valencia, J. and Nagalech, C. (2019) 'CoP25 Host Chile Must Sign the Escazú Agreement. Dialogo Chino'. Available at: https://dialogochino.net/20448-cop25-host-chile-must-sign-the-Escazú-agreement.

6

EXTRACTION AND THE BUILT ENVIRONMENT

Violence and Other Social Consequences of Construction

Victoria Kiechel

Introduction

A domed indoor forest enclosing some 3,000 trees, Terminal 3 at Singapore's Changi Airport houses the world's largest indoor waterfall, the 40-meter-tall Rain Vortex. Finished in 2019 as the capstone of the nearly $1 billion Jewel development, the technological audacity, innovative construction techniques, and biophilic beauty within this constructed biodome makes the airport itself a destination, providing, according to Changi's Chief Executive Officer, "a unique proposition of world-class shopping and dining, seamlessly integrated with lush greenery" to fulfill "the needs of increasingly discerning travelers for a meaningful and experiential journey" (Morris, 2019). The online sustainability journal *Treehugger* dubbed the Rain Vortex a "show-stopping centerpiece" featuring rainwater harvesting in an "exceptionally clean 'n' green Singapore" (Hickman, 2019). *The Dirt*, the online journal of the American Society of Landscape Architects, praised the airport terminal's restorative powers, citing how "immersion in nature can reduce stress, restore cognitive ability, and improve mood" (Green, 2019). *Engineering News-Record* quotes members of the design team describing it as a "new typology," "powerful," and "amazing" (Post, 2019) as a technical tour-de-force of construction.

But a consideration of the broader social and environmental context of Terminal 3 reveals a global reality: many of our most "clean," "green," and spectacular buildings and infrastructure owe their existence to extractive violence, the harmful effects of which could be remote or only slowly revealed.

As the celebratory products of human imagination, buildings and infrastructure embody social status, wealth, and power. They confer prestige upon their sponsors and their sites while, as in the case of Changi's Terminal 3, they supercharge futuristic civic identities, a move deemed essential for competitive urban advancement. Less visible is the trail of extractive impacts and social and ecological violence contained within some of the most renowned buildings and infrastructure of recent generations. Beautiful projects, even ones that are certified or considered to be "green," might come at a tremendous human and environmental cost. Singapore, like other globally competitive cities, relies on labor extracted from nations in the Global South (Hirschmann, 2020), with workers living in conditions tantamount to indentured servitude; builds on reclaimed land built on sand extracted from Indonesia, Malaysia, and Cambodia, nations that have begun to refuse sand export as they see the ecological and human consequences of this depletion (Subramanian, 2017); and depends on extraction of material components (aggregate for concrete, silica for glass, and more) used in construction—a process causing human health and environmental harm in the geographies where these components are sourced, remote from building sites. Worldwide, spectacular and iconic objects–high towers, sports stadia, bridge and road infrastructure, recreational spaces, landmark office buildings and residential developments—that through their technological brilliance, superior utility, and/or the beauty of their sinuous material form excite and attract us, could indeed be among the more lastingly violent of human achievements. Consider that the construction of the U.S. highway system in the twentieth century is estimated to have displaced 500,000 U.S. urban households, most of which comprised low-income people of color (Schmitt, 2016; Halsey, 2016). Destruction of urban neighborhoods and their social fabric has been a feature of sports venue development for recent Olympic Games (Donahue, 2020). Even after global attention to the issue (Amnesty International, 2016), estimates of migrant worker deaths related to the construction of the Qatar stadia for the 2022 World Cup remain in the thousands (DAMfirm, n.d.).

In a seeming paradox, construction involves destruction—not only of raw material stocks, but often of local economies excluded from benefit, of the socio-spatial fabric of neighborhoods, and of construction workers themselves. The built environment, whether beneficial or oppressive in its social and ecological effects, owes its very existence to extractive actions. If in choosing to build we cannot eliminate extraction, we can deconstruct the violence of extractive impacts by making their harm, and their influence over time, publicly visible, and by giving the power of decision-making to the communities most vulnerable to these impacts. In doing so we must consider the status of the architectural object and the role and power of domestic and international boundary-crossers, both globalized and globalizing, in the extractivist world of construction—real estate investors and their financial capital; architects and designers and their knowledge capital; construction firms and their human capital (construction workers); the materials and methods of design and construction—to forge a path towards the goal of a less extractive, less discordant, and more just and equitable built environment.

This essay identifies three forms of extractivism inherent in the design and construction of the contemporary built environment: (1) global real estate investment, which invests capital in the built environment for the future extraction of profit for investors; (2) the trend towards "hyperbuilding" as cities and regions compete for global recognition and investment—a trend in which global design and material supply chains feed a delirium of resource extraction and the mismatch between global expectations and local realities, with the potential for increased social conflict and vulnerability to environmental degradation and climate change; (3) the consequent displacement of humans and other species, whether due to necessity or perceived opportunity, experienced as self-extraction and self-imposed labor migration.

This chapter considers why current green building rating schemes do not include these forms of extraction as a focus, and in doing so it addresses the question of how to reformulate the frame and assessment methods of built environment projects better to reveal extractivism. Here, a redefinition of the status of the constructed object is in order, to transition our conception of it from solely a resource-consuming material thing towards an expanded alternative: that built environment projects are, and should be designed and regulated as, a connective web of social and ecological relationships that include literal and figurative territory well beyond their specific sites and material embodiments. The conclusion puts forward suggested preconditions for ground-up new construction so conceived, including the need for an outcomes-orientated process with planning and assessment methods to bring together speculators, their designers, and their affected communities in order to localize extractive effects, project their place-based influence over time, and thus deter both the slow and sudden violence arising from the consequences of development.

Hyper-Extraction and Global Real Estate Investment

While many factors contribute to making this era a hyper-extractive age in terms of the built environment, the driving impetus is the nature, origin, and intensification of the global flows of capital seeking investment in real estate and infrastructure. According to a 2019 OECD report, "investment needs" drive the construction sector, with 90 percent of global construction used for investment purposes (OECD, 2019, p. 94). Indeed, as a percentage of composition of 2017 investment expenditure by commodity, in most global geographies construction outweighed agriculture, equipment, and services—and, often, outweighed these three combined (OECD, 2019, p. 114). The commercial U.S. real estate services company Jones Lang Lasalle reported that in 2019 global commercial real estate investment in facilities such as offices, retail, industrial structures, multifamily housing, and hotels reached an all-time high of $800 billion (Jones Lang Lasalle, 2020). Global investment in infrastructure projects is estimated at many times more, with one database tracking a mere selection of projects representing nearly $15 trillion (Global Infrastructure Outlook 2019–2023, 2019).

So how might we define what the OECD terms the "investment needs" that drive the construction sector? These are a primary consideration in "accumulation planning" on the part of investors who seek diversified investment portfolios in order to reduce investment risk and create reliable returns. Investors seek the architectural or infrastructural object as a way to invest their capital as an alternative to global markets. The problem hinges less on the practice, *per se*, of investment; rather, problems arise from investor motivation and priorities, physical distance, limited awareness of impacts, and lack of accountability. In our financially globalized world, holders of fortunes, whether individuals, families, businesses, non-profits, sovereign wealth funds, pension funds, or states and/or corrupt states, are capable of speculative investment from great distances, detached from location and thus from the direct social effects of development impacts. There are many who aim to do social and ecological good through the medium of environmental, social, and governance (ESG) investing. Indeed, there exists the potential for good through investment in projects intended to counter extractive traditions such as racial segregation and urban ecological fragmentation and exclusion. Yet the guidelines for ESG investment are vague enough to be open to many concerns; examples include carbon-intense, violence-inducing, or socially destructive and exclusionary projects in energy and transportation infrastructure, as there is no common protocol for assessing the impacts of these projects.[1] Even "green" renewable energy projects, both wind and solar, face criticism for their potential extractive impacts in their capacity for destroying habitats and reducing biodiversity. And as a leading ESG investment firm representative points out, "infrastructure debt is mostly private" (Fiastre, 2019), without publicly available information or required reporting of any kind.

Absent public input, information, and required reporting, the goal of global infrastructure and built environment investment prioritizes investor benefits in the form of return on investment: the extraction of capital rather than the construction of community. The demands for near-term return on investment could leave out considerations of longer-term economic health inherent in the production of social or ecological "capital" for enhanced future well-being and conflict reduction. Returns extracted from built environment investments rival or exceed those of the stock market, averaging nearly ten percent per year for the commercial real estate sector (Maverick, 2020) and more than ten percent per year for infrastructure (PricewaterhouseCoopers, 2017, p. 6). Consequently, global investment for the extraction of profit is widespread practice. For example, in a five-mile radius from the Washington, DC home of this author, there is a housing and retail development funded largely by the government of Qatar (CityCenter: Fisher, 2013), a waterfront mixed-use development constructed with significant foreign investment from Chinese investors (the Wharf: Clabaugh, 2015), and a toll-road between the suburbs and the city managed, for profit, by an Australian infrastructure group (Dulles Greenway: Atlas Arteria, 2020).

The Violence and Delirium of Extraction

One of the factors driving contemporary hyper-extractivism is the dynamic of global cities competing with one another for business, investment, and prestige, as they seek to evince the scale, spectacle, and technological innovation of their built environments, with grandiose new projects often designed by the same global design firms. Enter the concept of *hyperbuilding*, a term linked to the way governments, corporations, and investors can demonstrate the magnitude of their political, economic, and social power through built form. The renowned global architectural practitioner Rem Koolhaus[2] promulgated the term beginning in at least 1996, but others quickly broadened its application. Aihwa Ong writes about it with respect to the emergence of Asian cities:

> Hyperbuilding as a physical landmark stages sovereign power in the great city, or in cities aspiring, through these edifices, to greatness. The interactions between exception, spectacle, and speculation create conditions for hyperbuilding as both the practice and the product of world-aspiring urban innovations.
>
> *(Ong, 2011, p. 207)*

The concept of "hyperbuilding" is the manifestation of hyper-extraction, where hyper-extraction is expressed either directly or indirectly: directly, as in "look at the sheer scale of this development and the amount of extracted materials it uses!" and indirectly, as in, "I have surplus wealth from my extractive economy, and it must be invested." Hyperbuilding, as embodied capital, results in building and infrastructure grand in scale, whether in height, in gross square footage and land area covered, in design or technological audacity, or in all of these together. Building thus audaciously requires considerable extraction in human, ecological, and material terms. Among the increasingly profligate, even violent, forms of hyper-extraction are the following: (1) the human extraction involved in labor migration, with worker conditions tantamount to enslavement; (2) the human extraction arising from the slow violence of community displacement; (3) the extraction of urban spatial territory for unproductive or unneeded use, leading to further displacement; (4) the project-specific extraction of ground, displacement of soil, and disruption of soil ecology; and (5) supply chain corruption for extraction of components of the material most essential to hyperbuilding: concrete.

The Pull and Push of Human Extraction

Demand for construction labor, particularly less-skilled labor, represents economic opportunity for individual workers and their families and for nations whose economies depend on remittances, which in 2018 totaled $689 billion worldwide with $529 billion going to developing countries (KNOMAD, 2020). The coalition *Who Builds Your Architecture? (WBYA?)* documented, for a 2016–17 exhibition at the Art Institute of Chicago (Who Builds Your Architecture?, n.d.), the trail of the global architectural and construction labor force for projects in four cities, Doha, Istanbul, New York, and Chicago:

> Whether majestic skyscrapers, eye-catching museums, or sprawling residential complexes, buildings emerge from intricate, lengthy processes of design and construction that involve a host of different actors, from architects and engineers to clients and banks to contractors and construction workers. These relationships operate within a global network of knowledge transfer,

manufacturing, and labor—people and materials moving around the world, often in uneven and unequal ways.

(Who Builds Your Architecture?, n.d.)

The WBYA? coalition's 2017 *Critical Field Guide* (Who Builds Your Architecture?, 2017) is in part a mapping of extractive flows, and in part a call to action to architects, beginning with a pledge for fair labor in countries where "the task of construction is designated to migrant workers who are indentured, exploited, and all but stripped of their rights" (2017, p. 9). The *Guide* describes the labor recruitment and migrant transport process in detail and visually maps the geographies of origin of the 685,000 migrant construction workers in Qatar, the 45,000,000 internal migrant construction workers in China, and the 500,000 migrant construction workers in the United Arab Emirates (UAE) (2017, p. 19). Even with reforms, labor abuse of construction workers in the UAE under the *kafala* system—including withholding of paychecks, inadequate living conditions, and long working shifts of twelve hours or more—is well-documented (Jacobs, 2018). U.S. construction also relies on authorized and unauthorized immigrant labor, a workforce willing to take on low-paying and often more dangerous jobs while receiving disparate treatment and lower compensation than their U.S.-born peers with annual salaries in metro areas equaling 60 percent of U.S.-born construction workers (Martin, 2016). The Brookings Institution cites that in 2010 U.S. immigrants represented about 22 percent of construction employment, making up over 60 percent of the workforce in the low-skill occupation of reinforcing iron and rebar workers and 30 percent of the carpenter, pipelayer and plumber, and mason and marble-setter workforce (Brookings Partnership for a New American Economy, n.d.). A Pew Research Center 2018 study determined that unauthorized immigrants comprise fifteen percent of the U.S. construction workforce (Passel and Cohn, 2018). The impacts of migrant extraction range from the individual and familial scale—assumption of debt to labor brokers and moneylenders, possible erosion of wages essential to familial support, ill-treatment and isolation, no recourse for injuries, absence of skills-based training—to the societal scale, with impacts on household gender roles, increased national food insecurity in the loss of agricultural labor, and the use of familial remittances to fund uneven development in peri-urban communities (Who Builds Your Architecture?, 2017, p. 17).[3]

In parallel with its documentation of construction labor, the *WBYA? Guide* visually represents selective flows of design work and capital earned, citing $1,700,000,000 as 2013 gross billings for international projects for U.S.-based architecture firms. It also describes the distance within the process of transference of architectural knowledge, another form of the global flows of extraction, from design studios to remote project sites (Who Builds Your Architecture, 2017, p. 46); and maps material networks by major corporations, such as Saint-Gobain (the 2019 materials market leader with $49.3 billion in sales; see Wang, 2019), LaFarge-Holcoim, and Georgia Pacific, suppliers of curtain wall, cementitious products, and gypsum board, respectively (Who Builds Your Architecture, 2017, pp. 48–9).

As a pull factor, migrant construction labor is a type of human extraction experienced early in the life-cycle of built environment projects. Harder to quantify are the delayed effects of this pull and of the push of human extraction that happens over time in response to hyperbuilding and other forms of construction. Scholarly studies describe the scope of more immediate displacement wrought by urban development projects.[4] But gradual community displacement due to declining affordability, gentrification, or shifts in land and building use types, such changes in zoning from housing to office use or agricultural to industrial use, is slower violence, and its extractive effects can and should be similarly highlighted. A 2019 National Community Reinvestment Coalition report based on U.S. Census Bureau and other data, found that:

> … many major American cities showed signs of gentrification and some racialized displacement between 2000 and 2013. Gentrification was centered on vibrant downtown business districts, and in about a quarter of the cases it was accompanied by racialized displacement. Displacement disproportionately impacted black and Hispanic residents who were pushed away before they could benefit from increased property values and opportunities in revitalized neighborhoods. This intensified the affordability crisis in the core of our largest cities.
>
> *(Richardson* et al.*, 2019)*

In circumstances favorable to community activism, local mobilization and resistance have recently emerged to counter this form of extractivism. A combination of tactics, including direct action, lobbying, and litigation, is succeeding in making visible instances where communities have experienced displacement owing to increasing unaffordability as a result of governmental changes in allowed built environment use type and/or density. In New York City in 2019 and 2020, three neighborhoods[5] have "thwarted or stalled both private and public efforts to develop thousands of new apartments," the majority of which, including the units offered at below-market rents, would be too expensive for longtime local residents:

> With a glut of empty luxury apartments and the industry's waning influence … momentum is building for neighborhood groups that are pushing back against new building projects because they believe such plans offer little community benefit … Tall towers that critics say exceed height limits are being held up in litigation. Zoning loopholes that enabled skyscrapers on mid-rise blocks are being scrutinized, and could even result in the shortening of some towers. And in neighborhood rezoning battles, mostly in lower-income communities of color, opponents are fighting efforts to spur new and largely market-rate construction that they say would displace longtime residents
>
> *(Chen, 2020)*

Legal injunctions and temporary restraining orders have provided delays intended to provide more thorough reviews of impacts: salutary tactics, but only in the short term. The development and formalization of an assessment and citizen review process remain to be achieved.

Extracting Urban Spatial Territory for Unproductive or Unneeded Use

Leaving built environment decision-making to investors and to local and national governments eager to expand their tax base and/or prestige could result in a form of real estate waste: property vacancy. Vacancy is the result of a mismatch between development desires and human needs: between developer projections that happen in a too-narrow sphere of analysis, without considering social context or longer time frames. For example, in 2019 as employment in the neighborhood fell by 0.7 percent Washington, DC's downtown business district office vacancy rates rose to 15.5 percent, their highest level since the start of market tracking in 1993 (DowntownDC, 2020). Office vacancy rates in 2018 for global cities included in ESRI's "Top 5 Most Homeless Cities Around the World" were 15.2 percent for Los Angeles, nine percent for Moscow, 14 percent for Mumbai, and 7.7 percent for New York (ESRI ArcGIS Story Map, n.d.). Sometimes owner speculation and projections for future use keep recently built vacant properties vacant, since costs are minimal to hold onto an unoccupied asset (although some local governments are imposing punitive real estate taxes, or "vacancy taxes," on vacant space, especially street-level space; Loh and Rodriguez, 2018)). The social costs of vacancies include diminution of community life and a rise in demand and property rents that could displace people by making neighborhoods unaffordable or inaccessible. A process that results in new construction less wasteful of space, society, and site environment—a process that would demand reconceptualization of the status of the architectural and infrastructural object beyond its material form and individual site—would alleviate some of the violence which real estate waste and unaffordability exact upon communities.

Soil Extraction

Waste is inherent in the act of construction. By some estimates, construction waste accounts for 30 percent of the total weight of building materials delivered to a building site (Osmani, 2011). Materials are cut and fitted with their remnants discarded, or they are damaged in construction or over-ordered, circumstances increasingly scrutinized in search of efficiency gains to achieve a circular economy. Yet material flows studies of construction and demolition (C&D) waste typically exclude the waste of soil excavation from construction sites, making the practice of soil excavation another form of less-visible if potent extractivism in the construction of the built environment.

Various regional estimates suggest its extent. One study of European Union waste put the average generation of C&D waste in 2011 in the EU at 700 million tons without excavated soil, estimating that if excavated soil was included, this value would double or even quadruple to between 1,350 to 2,900 million tons/year (Biointelligence Service, 2011, cited in Córdoba *et al.*, 2019). Soil excavation occurs in part because of the need for safe and stable foundations for buildings and infrastructure, where ground is extracted to a depth determined by structural engineering concerns, and it happens as a consequence of site grading (cutting and filling) undertaken to level sloping ground for human use. The question becomes, to what extent is this disruption of soil and site ecology required to service the demands of competitive hyperbuilding and the quest for the iconic architectural object sustainable? Are super-tall buildings really that necessary in locations where geological conditions do not favor their construction?

In the era of hyperbuilding, super-tall structures, defined as buildings in excess of 300 meters (about 1,000 feet) in height, have increased in global number by more than ten times in the last 25 years (Poulos, 2016), to a current total of 170 (*Architect Magazine*, 2019). Foundational depth allows super-tall buildings to meet their site-specific challenges, whether wind loads or site geology and soil strength or seismic factors, usually by creating an extensive below-ground support system for stability; for example, the 632-meter (2,073 feet) high Shanghai Tower required 980 foundation pilings of 86 meters (or 282 feet) depth and a foundation mat 6 meters (20 feet) in depth (Risen, 2013). The essential material of contemporary foundations is reinforced concrete, the production of which advances the consideration of another less-visible source of modern hyper-extraction: sand mining.

Sand Mining and the Extractive Burden of Concrete

Concrete is not a primary material. The most widely-used man-made material on earth, it is a mix of components (binders, aggregates, admixtures and other additions, and water) whose manufacture makes it one of the most carbon-intensive building materials, the source of about 8 percent of the world's carbon emissions (Rodgers, 2018). In addition to its use in foundations and infrastructure (currently intensifying in global hydroelectric dam construction) as prized for its solidity, concrete has become the material of choice for barrier methods (like flood walls) for climate change adaptation even as it, paradoxically, contributes to climate change through its emissions. The British-based newspaper *The Guardian*'s 2019 "Concrete Week" reporting describes the environmental and human health impacts of concrete, and also its capacity as an instrument of political corruption and construction kickbacks (Watts, 2019). The chain of corruption and extractive impacts is particularly vivid in the mining of concrete aggregate, most commonly sand and gravel.

With the global use of materials by the construction sector estimated to more than double between 2017 and 2060, to almost 84 gigatons (GT) use, the OECD (OECD, 2019, p. 90) projects strong increases in the use of non-metallic minerals which represent the largest share of total materials use, projected to grow from 44 to 86 GT between 2017 and 2060, with the largest growth in tons for sand, gravel, and crushed rock (2019, p. 118):

> Sand, gravel and crushed rock for construction alone represent almost 24 percent of materials extraction … [in terms of] materials extraction across regions and development levels … non-metallic minerals are the largest group, given that these consist of relatively low-value bulk commodities (like sand and gravel) that are expensive to import and thus normally sourced domestically.
>
> *(OECD, 2019, p. 120)*

But it cannot always be sourced domestically. The world mines and uses 50 billion tons of aggregate per year (Beiser, 2019). Desert sand, with edges eroded from wind, is ineligible for use as aggregate; concrete demands angular sand granules dredged from such sources as sea floors, lake beds, floodplains, and shorelines. Dubai, for example—a city bordering the Arabian Desert—imports its sand from Australia (Beiser, 2019). A 2019 commentary in the journal *Nature* describes sand extraction as a landscape of "unsustainable exploitation:"

> This extraction of sand and gravel has far-reaching impacts on ecology, infra-structure and the livelihoods of the 3 billion people who live along rivers … For example, sand mining on the Pearl River (Zhujiang) in China has lowered water tables, made it harder to extract drinking water and hastened river-bed scour, damaging bridges and embankments.
>
> Most of the trade in sand is undocumented. For example, between 2006 and 2016, less than 4 percent of the 80 million tonnes of sediment that Singapore reported having imported from Cambodia was confirmed as exported by the latter. Illegal sand mining is rife in around 70 countries, and hundreds of people have reportedly been killed in battles over sand in the past decade in countries including India and Kenya, among them local citizens, police officers and government officials … In many countries, sand mining is unregulated and might involve local 'sand mafias.' Methods of extraction range from dredging boats and suction pumping to digging with shovels and bare hands, both in daylight and during the night.
>
> Extraction of sand and gravel from active sources can cause great environ-mental, social and economic harm … the Vietnamese government estimates that nearly 500,000 people will need to be moved away from river banks that are collapsing as a result of sand mining in the channel.
>
> *(Bendixen, 2019)*

To add to its list of harms, in various geographies concrete is by weight the most wasted building material at the end of its life, since recycling is much rarer for non-metallic minerals (although concrete waste may be used as road filler; OECD, 2019, pp. 144–145). The cumulative tally of construction minerals extraction impacts includes further examples of slow harm, such as the loss of biodiversity, habitat alteration, soil compaction, and the interruption of site hydrology through the sealing of land area (2019, p. 184).

Green Building Development, Rating Schemes, and Extractivism

These forms of extractive violence—harmful labor practices, displacement, vacancy, and hyper-building and -extraction—are not a particular focus of green building rating schemes. What, then, is a "green" project? If development constructed with exploitative labor practices, corrupt supply chains, and displacement of humans and other species can receive green certification, does this mean the green building movement masks these forms of violence?

Not, perhaps, intentionally. But this is an era for green building certifications, which most commonly apply at the scale of the single building or piece of infrastructure, to redefine the built environment object in other than material or resource-consuming terms, or as it prioritizes its occupants in terms of indoor environmental quality and proximity to services and recreation. We have travelled far from the early 1990s context and origins of the two pioneering new construction-orientated certifications: Leadership in Energy and Environmental Design (which in addition to the United States counts China, Singapore, the United Arab Emirates, Brazil, and India as among its primary geographies for market uptake; Gregor, n.d.) and the Building Research Establishment Environmental Assessment Method (used in 86 countries, but most intensively in Europe; see BREEAM, n.d.). Still urgently necessary are the quantitative measurements these certifications demand for the reduction of carbon and embodied carbon: after all, in the service of their credibility and market uptake, green building rating schemes have been built on what they can transparently and objectively measure. The vision which they embrace, that of voluntary "market transformation," hinges on the data-driven willingness of the leaders of our real estate economy— manufacturers, developer/investors, suppliers, building owners and tenants, and others—to adopt resource efficiency and carbon reduction measures: first, because they result in economic savings, and secondly because of all other reasons, whether regulatory or values-driven (the need to adhere to local carbon limits and/or corporate sustainability plans). Market transformation wrought by the green building movement has led to the production and acceptance (as a consumer standard) of lower-carbon, resource-efficient products and systems, and the subsequent adoption of efficiency standards by local and regional governments. Resource efficiency and governmental policy for low-carbon outcomes have bolstered economic health while contributing to a decline in carbon emissions in geographies like California, where, according to a 2019 National Resources Defense Council report, between 1975 and 2016 fossil fuel consumption relative to GDP output fell 70 percent (Komanoff et al., 2019). But even if the revenue gathered from state-imposed carbon taxes or cap-and-trade programs is redistributed to communities or ecologies judged to need them most, these benefits do little to address slow, systemic violence of the kind discussed here.

Because (by design) they reward time-limited predicted or actual performance, green building rating schemes do not measure the longer-term consequences of the interaction of buildings and infrastructure with human communities and local and regional ecologies, and the fact that over time, building can destroy as much or more than it creates. It is time for green building rating schemes to reframe the architectural or infrastructural object as a socio-ecological force first, and a material thing second, and to widen its subject territory to include communities and ecologies well beyond its specific site, and over the span of its entire life-cycle and beyond.

Reframing the Constructed Object in Space and Time

Sociologist Saskia Sassen observes that urban areas are at risk of becoming "conflictive spaces," "overwhelmed by inequality and injustice" (Sassen, 2017). How can we incorporate an awareness of the impacts of an extractive built environment into the design process to help forestall such outcomes? How can we better understand, across time, the social and ecological complexity of the constructed object and its impacts?

We can begin by conceiving of built environment objects as forces in a web of socio-ecological relationships and querying them accordingly. Buildings and infrastructure are socio-spatial actors in social and ecological systems in a range of nested scales, from the local, to the regional, to the global. We must learn, as citizens and non-experts, to ask tough questions of existing and (especially) of proposed objects in the built environment, since these will outlive us, with impacts of correspondingly long duration. Who will benefit from this construction, and how? Who or what will occupy the constructed object? Who will manage or police it? Is it accessible to "outsiders"? Whom does it privilege and exclude, in terms of race, gender, age, or affluence? How does it relate to its place in scale and orientation? What social, racial, and economic fault lines will it exacerbate in its neighborhood and regional surroundings in the short, medium, and long terms? How will its construction alter the site and regional ecologies? How does it address current and future needs in terms of social equity and climate resilience outcomes? Who built or will build this, and how? What are the primary materials of construction, what social or power-wielding status do their choices imply, and what extractive impacts do they embody? Is the constructed object adaptable for other uses as community demographics, needs, and the very climate itself change?

A citizenry accustomed to asking these and other questions impels the need for new processes to precede, accompany, and assess the impacts of new construction over time: in other words, a longer design and review life-cycle than at present, one that parallels the life-cycle of the constructed object. This reimagined process would include the following sequential actions:

1. Transparent reporting, in the public sphere, of investor sources and amounts of capital investment in proposed commercial real estate and infrastructure projects.

2. Prior to design, a socio-spatial assessment of neighborhood and regional conditions, demographics, and needs, to be funded and led by design and investor teams and ground-truthed by citizens.
3. As part of the design process, an investor-funded analysis of how the proposed development will increase community capacity for climate change adaptation and resilience.
4. An accounting, by the design team, of the extractive impacts of proposed materials choices.
5. A projected plan for socially-equitable project access and management over time, to note neighborhood and regional connections, and outcomes capable of violence reduction.
6. A plan, developed by project designers, for adaptive reuse of the constructed object and for eventual end-of-life demolition, as or if necessary.
7. As a precondition of construction, a time-limited citizen-led project review process, through compensated citizen service or expectations of public service akin to jury duty.
8. During construction, transparent reporting of contractor and subcontractor labor practices, including hiring and workforce training.
9. Using a small percentage of subsequent investor profit, the creation of a public escrow fund to document future development effects over time, at intervals to be determined, for incorporation as part of the public record of development effects.

The end goal is an *outcomes-orientated* design and monitoring process accounting for the trajectory of projects over their life spans and beyond. Through a holistic, life-cycle approach to embedding social and climate resilience considerations within the design and continuing assessment of buildings and infrastructure, this process would recognize and highlight the variety of social and climate justice impacts tangibly manifest in development schemes, and incorporate disclosure and transparency as the first steps towards action and eventual social transformation.

Getting Beyond the Extractive Status Quo

One of the aims of this reframing and its accompanying process is to shrink the distance between the actors in development—investors, designers, and community—to arrive at a common understanding of the potential for social and ecological violence from development impacts, and to forestall these. In some measure, this process seeks to parallel the fundamental goals of the environmental impact statement (EIS) required for many infrastructure projects[6] as mandated in 1969 by the National Environmental Policy Act, which:

"… does not prohibit harm to the environment, but rather requires advanced identification and disclosure of harm … An EIS outlines the status of the environment in the affected area, provides a baseline for understanding the potential consequences of the proposed project, identifies positive and negative effects for the environment, and offers alternative actions, including inaction, in relation to the proposed project."

(Middleton, 2018)

Conceived as a process to influence outcomes and alter the norms of decision-making, over time the EIS process has tended towards *pro forma* production by EIS expert "shops," open to underestimation of, or political influence in, its depiction of environmental outcomes (Cashmore, 2004). The alternative process of socio-spatial and climate impact assessment proposed here would differ from the EIS and typical green building certification processes in significant ways—differences in phasing, authorship, scope, methods, proof of adaptable use and climate change resilience, and an inventory of extractive concerns for material and labor choices—*differences intended to reduce or avoid the violent consequences of extractivism* in the built environment—specifically:

- *A required pre-design assessment.* Unlike traditional environmental impact assessment, which as a screening tool occurs and is used after the selection of a preferred design, the first steps of assessment would occur before the initial design phase in order to inform and form the design path.
- *At pre-design, an enlarged contextual scope of analysis.* The socio-spatial assessment aims to make, as the foundation of design, context, community, and the potential for change itself. It would document historical patterns, reaching backward and then forward in time to analyze and project the social, spatial, and ecological impacts of the new construction project over a longer life-cycle than traditional impact assessments. It would account for traditions of use and social segregation in order to posit future, less violent social, ecological, and economic scenarios.
- *Authorship by the design and investment team in consultation with the community, rather than by independent authorities.* Traditional environmental impact assessments or statements are usually created at significant cost by independent authorities, who as third-party experts are not necessarily connected to the geographical location of the project and certainly not connected to the project design team. This new process would turn the assumed benefit of independent authority on its head, instead requiring that design and development teams embed in the proposed project location for purposes of research. The aim is to shrink the distance between local community realities and needs, and the assumptions of increasingly globally-based design firms and investors. For this reason, in order to enhance awareness and accountability the assessments at pre-design and design stages are emphatically NOT to be carried out, as are environmental impact and social impact statements, by experts, or "independent" third party authorities, but rather by the sponsors themselves.
- *Different and varied assessment methods.* The project team should consider interacting with the community in which it proposes to develop in a manner

different from the usual: one in which investors and designers, with the involvement of citizens and local governments, act as historians, ethnographers, and documentarians of the context of their proposed developments.[7] Rather than relying on self-selecting participation in community input meetings or charrettes— participation which often favors participation by the least vulnerable—design and investment teams should aim to meet community members where they live and work, through street-intercept surveys as well as targeted interviews with a representative cross-section of members of the affected community in a predetermined radius around the proposed project, and conducting time-noted, place-based observational studies of spatial use by humans and other species of the project's physical context. This differs from standard "best" practices in participatory design: community meetings which tend to involve those with the leisure and motivation to participate, rather than a true cross-section of community members. Project teams should consider going beyond more typical social impact assessments by using a variety of research and documentation methods: surveys, interviews, short films, observational studies of public space and ecological conditions, and oral histories, in addition to the more quantitative data-driven economic and demographic trend summaries and projections.

- *Demonstrated resilience of the proposed project to climate change, and the adaptability of the design to alternative future socio-spatial uses.* Projecting the adaptability of a project as part of the design process would help defer obsolescence of the constructed object, and adding to its longer life—and thus to fewer extractive new construction cycles requiring capital and materials expense.
- *Making extractive impacts visible through the documentation of flagged concerns for site development, labor practices, and materials choice and intensity.*

Disadvantages would ensue in thus disrupting the current status quo, although some might consider these disadvantages to be benefits. These include a lengthier design process—although less lengthy than the time added by potential lawsuits and injunctions; less profit—though perhaps more secure returns—for investors, who would fund this process, and added administrative record-keeping and long-term process management on the part of local governments, although this ongoing oversight could have the beneficial effect of improving outcomes. The possible advantages in implementing this alternative assessment process are many, including (1) awareness towards remediating the harm of persistent legacies of capitalism, globalization, and colonialism in design and materials and labor choices, through the engagement and accountability of informed citizenry and investors; (2) a resulting increase in longer-term community political capacity, and social, ecological, and economic health and conflict reduction, and thus a potential increase in longer-term profits for investors; (3) increased adaptive reuse of existing buildings as a lower-carbon alternative to new construction (the renovation of which would not be subject to this process), in which ecological and social costs are already embedded; (4) increased human and climate resilience; (5) opportunities to learn from the historical record of development and decision-making thus created; and (6) the more widespread establishment of the habit of mind of systems thinking, as applied to the life-cycle of development projects across local, regional, and global scales.

A timely next step rests with green building rating and certification schemes. The challenge for these is to reframe their essential definitions and inform concepts towards a new idea of what it means to be green. Defining an architectural or infrastructural object as a socio-ecological force first, and a material thing second, means that we cannot continue to ignore the impacts of construction which last far beyond their material lifespan. Enlarging a certification project's subject territory to include communities and ecologies well beyond its specific site brings awareness of, and accountability for, a wider scale of harm. And by considering the impacts of the whole project life-cycle and beyond, green building certification schemes would be better able to lead those with the power to commission such projects towards a future of less violent consequences.

Notes

1 The Climate Bonds initiative has evolved a taxonomy (Climate Bonds Taxonomy, 2020) of green investment based on comparison with a conventional baseline of energy production, water and transport infrastructure, land use, and more, which only minimally addresses extractivism, and that mostly in the waste category.
2 Designer of projects such as the CCTV Tower in Beijing, with a construction budget of over $1 billion and a floor area of over 4,000,000 square feet.
3 There is, in addition, a rich literature on "remittance houses" as a push factor in migration and their influence on the culture of Latin American peri-urban towns (Blitzer, 2019; Janetsky and Stunt, 2020) and as chronicled in the work of Sarah Lynn Lopez on "house envy" and the remittance landscape (Lopez, 2015). There is a perverse circular effect in the construction of remittance houses in towns in the Global South: a phenomenon enabled by the Global North's employment of remittance senders as construction workers, who thereby fund a construction workforce back home and engender a locally competitive desire for the prestige which comes with large house ownership.
4 For an example, see a Brookings Institution tally (Robinson, 2003, p. 19).
5 The resisting neighborhoods have learned from the experience of two rapidly gentrifying New York City neighborhoods, Williamsburg and Greenpoint, which, as the result of zoning policies favoring market rate developments, experienced "the highest median rent increase of the decade, from $1,207 in 2010 to $1,854 in 2018—a 54 percent jump, according to the NYU. Furman Center" (Chen, 2020).
6 As regards environmental impact assessments, the American Bar Association says that in addition to the USA, "over 100 countries, including Australia, China, India, Nepal, and Ukraine, have adopted similar environmental assessment protocols" (Middleton, 2018).
7 A relevant and interesting set of essays on "expanding modes of [design] practice" is to be found in the journal Log (Davidson, 2020). Byrony Roberts' introductory essay observes that traditional modes of architectural design are ill-equipped to deal with issues of social complexity, and that the tools of urban planning and analysis provide a better foundation. The volume chronicles actually existing design practices which rely on social and empathetic design practices including community collaboration, ethnographic studies, use of temporary installations, and more.

References

Amnesty International. (2016) 'The Ugly Side of the Beautiful Game: Exploitation of Migrant Workers on a Qatar 2020 World Cup Site'. Available at: www.amnesty.org/download/Documents/MDE2235482016ENGLISH.PDF.

Architect Magazine. (2019) 'Record Number of Supertall Buildings Completed in 2019', 13 December. Available at: www.architectmagazine.com/design/record-number-of-supertall-buildings-completed-in-2019_o.

Atlas Arteria. (n.d.) *About Dulles Greenway.* Available at: www.atlasarteria.com/portfolio/dulles-greenway.

Beiser, V. (2019) Why the world is running out of sand, BBC Future. Available at: www.bbc.com/future/article/20191108-why-the-world-is-running-out-of-sand.

Bendixen, M., Best, J., Hackney, C., and Iversen, L.L. (2019) 'Time is running out for sand', *Nature,* 571, pp. 29–31. Available at: www.nature.com/articles/d41586-019-02042-4.

BioIntelligence Service. (2011) '*Evolution Of (Bio-) Waste Generation/Prevention And (Bio-) Waste Prevention Indicators Final Report*', in Córdoba, R.*et al.* (2019) 'Alternative construction and demolition (C&D) waste characterization method proposal', *Eng. Sanit. Ambient,* 24 (1). Available at: www.scielo.br/scielo.php?script=sci_arttext&pid=S1413-41522019000100199.

Blitzer, J. (2019) 'The Dream Homes of Guatemalan Migrants', *The New Yorker Magazine,* 5 April, 2019. Available at: www.newyorker.com/news/dispatch/the-dream-homes-of-guatemalan-migrants.

BREEAM Worldwide. (n.d.) Available at: www.breeam.com/worldwide.

Brookings Partnership for a New American Economy. (n.d.). Immigrant Workers in the U. S. Labor Workforce. Available at: www.newamericaneconomy.org/sites/all/themes/pnae/img/Immigrant_Workers_Brookings.pdf.

Cashmore, M., Gwilliam, R., Morgan, R., Cobb, D., and Bond, A. (2004) 'The Effectiveness of EIA', *Impact Assessment and Project Appraisal* 22 (4), pp. 295–310.

Chen, S. (2020) 'The People Vs. Big Development', *The New York Times,* 7 February. Available at: https://www.nytimes.com/2020/02/07/realestate/the-people-vs-big-development.html.

Clabaugh, J. (2015) 'D.C. development projects attract EB-5 Chinese investors', *Washington Business Journal,* 11 November. Available at: www.bizjournals.com/washington/breaking_ground/2015/11/d-c-development-projects-attract-eb-5chinese.html.

Climate Bonds Taxonomy. (2020). Available at: www.climatebonds.net/files/files/CBI_Taxonomy_Tables_January_20.pdf.

DAMfirm. (n.d.) The human cost of construction. Available at: www.damfirm.com/human-cost-construction/.

Davidson, C. and Roberts, B. (2020) *Log 48: Expanding Modes of Practice.*

Donahue, B. (2020). 'The Price of Gold', *The Washington Post Magazine* (Weekend edition), 6 July. Available at: www.washingtonpost.com/magazine/2020/07/06/inside-troubling-legacy-displacing-poor-communities-olympic-games-one-villages-resistance-brazil/?arc404=true.

DowntownDC. (2020). *State of Downtown 2019 Report,* Downtown DC Business Improvement District, Washington, DC. Available at: www.downtowndc.org/report/state-of-downtown-2019.

ESRI ArcGIS Story Map. (n.d.). Top Five Homeless Cities Around the World. Available at: www.arcgis.com/apps/MapJournal/index.html?appid=e56c3fcc502442ca8f86d89809fbf287.

Fiastre, P. (2019) 'ESG: The foundation of responsible infrastructure investment', *IPE Real Assets Magazine,* January/February. Available at: https://realassets.ipe.com/infrastructure/esg-the-foundation-of-responsible-infrastructure-investment/10029601.

Fisher, M. (2013) 'Qatar is suddenly investing heavily in the U.S., bankrolling D.C.'s City Center, other projects', *The Washington Post*, 17 December. Available at: www.washing tonpost.com/local/qatar-is-suddenly-investing-heavily-in-the-us-bankrolling-dcs-city-cen ter-other-projects/2013/12/17/1ffaceca-5c6a-11e3-95c2-13623eb2b0e1_story.html.

Global Infrastructure Outlook 2019–2023. (2019) Globe Newswire, 17 October. Available at: www.globenewswire.com/news-release/2019/10/17/1931425/0/en/Global-Infrastructure-Outlook-Report-2019-2023-South-South-East-Asia-has-the-Highest-Number-of-Infrastru cture-Projects-Valued-at-US-3-2-Trillion.html.

Green, J. (2019) 'Singapore's New Garden Airport', *The Dirt*. Available at: https://dirt.asla. org/2019/04/08/singapores-new-garden-airport.

Gregor, A. (n.d.) Global LEED: Two Polar Opposites Cities Take the Lead in Environ-mental Design, USGBC+. Available at: http://plus.usgbc.org/global-leed/.

Halsey, A. (2016) 'A crusade to defeat the legacy of highways rammed through poor neigh-borhoods', *The Washington Post*, 29 March. Available at: www.washingtonpost.com/local/tra fficandcommuting/defeating-the-legacy-of-highways-rammed-through-poor-neighborhoo ds/2016/03/28/ffcfb5ae-f2a1-11e5-a61f-e9c95c06edca_story.html.

Hickman, M. (2019) 'The World's Tallest Indoor Waterfall Will Be Located in Already-Spectacular Airport', Treehugger. Available at: www.treehugger.com/worlds-talles t-indoor-waterfall-will-be-located-in-spectacular-airport-4868431.

Hirschmann, R. (2020) Number of Construction Workers in Singapore 2013–2019, Statista. Available at: www.statista.com/statistics/1054354/singapore-foreign-construction-workers-e mployed/.

Jacobs, H. (2018) 'Dubai's glittering, futuristic metropolis came at the cost of hundreds of thousands of workers, and recommending it as a tourist destination feels wrong', *Business Insider*. Available at: www.businessinsider.com/dubai-development-tourism-workers-p roblem-2018-12?r=US&IR=T.

Janetsky, M. and Stunt, V. (2020) Working thousands of miles from home—to build a new one, Christian Science Monitor. Available at: www.csmonitor.com/World/Americas/ 2020/0507/Working-thousands-of-miles-from-home-to-build-a-new-one.

Jones Lang Lasalle. (2020) 'Global real estate investment hits record high in 2019', The Investor. Available at: www.theinvestor.jll/news/world/others/global-real-estate-investm ent-hits-record-high-in-2019.

KNOMAD. (n.d.). *Remittances Data*. Available at: www.knomad.org/data/remittances.

Komanoff, C, Cavanagh, R., and Miller, P. (2019). California Stars: Lighting the Way to a Clean Energy Future, National Resources Defense Council Report. Available at: www. nrdc.org/sites/default/files/california-stars-clean-energy-future-report.pdf.

Loh, H.D and Rodriguez, M. (2018) 'Why is that house or storefront vacant?', Greater Greater Washington. Available at: https://ggwash.org/view/68318/why-is-that-house-or-storefront-vacant.

Lopez, S. (2015) *The Remittance Landscape: Spaces of Migration in Rural Mexico and Urban USA*. Chicago, IL: University of Chicago Press.

Martin, C. (2016) 'Building America: The immigrant construction workforce', Urban Wire, The Urban Institute. Available at: www.urban.org/urban-wire/building-america-immigra nt-construction-workforce-0.

Maverick, J. (2020) 'Average Annual Returns for Long-Term Investments in Real Estate', Investopedia. Available at: www.investopedia.com/ask/answers/060415/what-average-a nnual-return-typical-long-term-investment-real-estate-sector.asp.

Middleton, T. (2018) 'What is an Environmental Impact Statement?', American Bar Asso-ciation. Available at: www.americanbar.org/groups/public_education/publications/tea ching-legal-docs/teaching-legal-docs--what-is-an-environmental-impact-statement-.

Morris, H. (2019) 'The world's tallest indoor waterfall has been unveiled at the world's best airport', *The Telegraph*. Available at: www.telegraph.co.uk/travel/news/jewel-singapore-changi-opening.

OECD. (2019) *Global Material Resources Outlook to 2060: Economic Drives and Environmental Consequences*. Paris: OECD Publishing. Available at: https://read.oecd-ilibrary.org/environment/global-material-resources-outlook-to-2060_9789264307452-en.

Ong, A. (2011) 'Hyperbuilding: Spectacle, Speculation, and the Hyperspace of Sovereignty', in *Worlding Cities: Asian Experiments and the Art of Being Global*, pp. 205–226.

Osmani, M. (2011) *'Construction Waste' in Waste* [Online].

Passel, J. and Cohn, D. (2018) Unauthorized immigrant workforce is smaller, but with more women, Pew Research Center Hispanic Trends. Available at www.pewresearch.org/hispanic/2018/11/27/unauthorized-immigrant-workforce-is-smaller-but-with-more-women/.

Post, N. (2019) 'Singapore's Jewel Mall Project Was No Walk in the Park', *Engineering News-Record*. Available at: www.enr.com/articles/47117-singapores-jewel-mall-project-was-no-walk-in-the-park.

Poulos, H.G. (2016) 'Tall building foundations: Design methods and applications', *Innovative Infrastructure Solutions*, 1 (10). Available at: https://link.springer.com/article/10.1007/s41062-016-0010-2.

PricewaterhouseCoopers. (2017) 'Global infrastructure investment: The role of private capital in the delivery of essential assets and services'. Available at: www.pwc.com/gx/en/industries/assets/pwc-giia-global-infrastructure-investment-2017-web.pdf.

Richardson, J., Mitchell, B., and Franco, J. (2019) 'Shifting neighborhoods: Gentrification and cultural displacement in American cities', National Community Reinvestment Coalition, Washington, DC. Available at: https://ncrc.org/gentrification.

Risen, C. (2013) 'How to Build A 2,073-Foot Skyscraper', *Popular Science*. Available at: www.popsci.com/technology/article/2013-02/how-build-2073-foot-skyscraper.

Robinson, W. (2003) 'Risks and Rights: Causes, Consequence, and Challenges of Development-Induced Displacement', The Brookings Institution, Washington, DC. Available at: www.brookings.edu/wp-content/uploads/2016/06/didreport.pdf.

Rodgers, L. (2018) 'Climate change: The massive CO2 emitter you may not know about', BBC News. Available at: www.bbc.com/news/science-environment-46455844.

Sassen, S. (2017) 'Beyond Differences of Race, Religion, Class: Making Urban Subjects' in Mostafavi, M. (ed.) *Ethics of the Urban: The City and the Spaces of the Political*. Harvard University Graduate School of Design. Zurich: Lars Müller Publishers.

Schmitt, A. (2016) 'Anthony Foxx Wants to Repair the Damage Done By Urban Highways', StreetsblogUSA. Available at: https://usa.streetsblog.org/2016/03/30/anthony-foxx-wants-to-repair-the-damage-done-by-urban-highways/.

Subramanian, S. (2017) 'How Singapore is creating more land for itself', *The New York Times Magazine*, 20 April. Available at: www.nytimes.com/2017/04/20/magazine/how-singapore-is-creating-more-land-for-itself.html.

Wang, T. (2019) Leading manufacturers of construction materials worldwide as of April 18, 2019, based on sales, Statista. Available at: www.statista.com/statistics/314988/leading-buildinc-material-manufacturers-worldwide.

Watts, J. (2019) 'Concrete: The most destructive material on earth', *The Guardian*. Available at: www.theguardian.com/cities/2019/feb/25/concrete-the-most-destructive-material-on-earth.

Who Builds Your Architecture? (2017) A Critical Field Guide. Available at: http://whobuilds.org/wp-content/uploads/2017/02/WBYA_Guidebook_spreads.pdf.

Who Builds Your Architecture? (n.d.) 'Projects from Doha, Chicago, New York, and Istanbul'. Available at: https://graphcommons.com/graphs/15aab298-4e75-45eb-9383-3ffd98bb67bd.

PART 3

New Ways of Thinking about Extraction

7

RETHINKING EXTRACTIVISM ON CHINA'S BELT AND ROAD

Food, Tourism, and Talent

Yifei Li and Judith Shapiro

Introduction

China's Belt and Road Initiative (BRI) is an umbrella term for a mind-boggling range of extractive activities. The BRI is so vaguely conceptualized and understood, even within China, that the phrase has become a metaphorical catch-all for nearly all of China's international investments and projects. As we will see in this chapter, there is a Dairy Belt and Road, a Tourism Belt and Road, and an Educational Belt and Road. The BRI even includes a "Polar Silk Road" and outer space and cyberspace Belts and Roads. The Middle Kingdom's BRI ambitions and projects range from the purchases of African grainfields and Brazilian soybeans to an expansive new milk supply chain sourced in New Zealand. BRI plans even go so far as missions to exploit minerals on the far side of the moon.

In this chapter, we examine select examples of the Chinese state's initiatives on the Belt and Road, focusing on less well-understood aspects of extractivism such as securing food supply chains, commodifying cultural heritage for mass tourism, and appropriating human talent and intellectual property. We thus adopt an expanded definition of "extraction" to encompass not only the well-known, often conflict-plagued harvesting of raw materials and natural resources such as minerals, fossil fuels, and timber, about which there is a rich scholarly and activist literature, but also a range of other activities that are less widely conceptualized as part of an extractive agenda. Our primary concern is to show how the pattern of social and ecological destruction of the traditional extractive economy—the transfer of key resources to benefit the people and economy of the destination at the cost of the people and environment of the origin—has not only spilled over to a wider range of economic sectors, but has also taken a seemingly more benign form that appears not to be extractive at first glance. The three dynamics we explore are: first, the pivot of global food systems toward China through controlling and integrating supply chains for dairy and other foodstuffs; second, the export of mass tourism and the commodification and extraction of landscapes; and finally, the extraction, appropriation, and monopolization of knowledge. This last dynamic ranges from President Xi Jinping's courtship of a "brain drain" of talent through state-backed recruitment programs to China's entry into the arena of transnational intellectual property rights through controlling stakes in international businesses.

To analyze these three facets of China's present-day hyper-extractivism, we draw on public affairs literature, trusted journalistic investigations, and scholarly work on the BRI; on our own decades of experience living in and observing China; and on our recent research and writing on China's "green" authoritarianism (Li and Shapiro, 2020). We hope that this chapter will add a significant dimension to the effort to understand novel forms of extraction and violence that characterize our ever-globalizing age and highlight the particular role that China has been playing in catalyzing and intensifying those dynamics.

As at October 2019 the Belt and Road Initiative had expanded its partnerships to more than 137 countries and 30 international organizations. Since its inception in 2013 the Belt and Road Initiative has met with skepticism and pushback from multiple constituencies in member countries, but such pockets of resistance seem insignificant when compared to the many less-developed countries who have had little choice but to welcome China's investments as injections of much-needed cash into their national economies. China's ready provision of expertise, equipment, and even labor power for the rapid construction of roads, railroads, power plants, stadiums, and deep-water ports has drawn considerable admiration and compares favorably to the often slow pace of funding and cumbersome social and environmental screenings required by Western donors and lending institutions. Even developed European countries—most notably Italy—have embraced Chinese financing as a lifeline to revitalize a stagnant economy. High-profile endorsements of the Belt and Road from the leaders of multilateral institutions and from government officials have often overwhelmed dissenting voices. This is despite the well-publicized negative experience of Sri Lanka, which was forced to agree to give China a 99-year lease of the Hambantota Port when it could not repay its loans, and other situations giving rise to questions about Chinese influence over local sovereignty from Cambodia to the Central Asian republics to Ecuador. Even as some governments—particularly those in Malaysia, Myanmar, Sierra Leone, Zambia, Tanzania, and Ghana—are having second thoughts about BRI risks of dependency, debt, and poor social and environmental protections, it remains difficult for them to challenge China's ready cash and avoid falling into hyper-extractive logic.

Double Extraction and Double Violence on the Belt and Road

The extractive frontier along the Belt and Road finds expression in a series of *hyper*-extractive undertakings. Here, we highlight two qualities that characterize the extractive activities on the Belt and Road, while fully recognizing that colonialism by many nations has over the past centuries decisively transformed the developing world by locking it into relationships of dependency and exploitation.

First, extraction on the Belt and Road often amounts to a full-fledged two-way movement in which raw materials, foodstuffs, scenery, talents, and other benefits move toward the Chinese market, while coal-fired power plants, polluting manufacturing plants, land degradation, intensification of production, cultural appropriation, and many other forms of social and environmental harms flow from the Middle Kingdom. In this sense, China's BRI partners are being doubly extracted, both by the exodus of resources and talents, and by the influx of ecological stress and other externalities of China's global expansion. This double extraction acts as what environmental sociologists liken to a treadmill that keeps "running in place without moving forward" (Gould *et al.*, 2004).

Second, Belt and Road hyper-extractivism is a kind of double violence, since the material activities are wrapped in the overpowering discourses of win–win development and mutual benefits which are formalized in official state-to-state documents between China and its partner countries. The BRI not only inflicts material violence on the people of Eurasia, Africa, Latin America and beyond, but also subjects them to discursive violence, depriving their state officials of the language to account for the injustices they suffer or to articulate resistance to projects that are framed as patently beneficial. Even when grassroots citizen groups have attempted to challenge BRI projects on the basis of their mismatch with ostensible promises, state officials in these countries have until recently been unsympathetic, if not antagonistic, toward dissenting voices. Our discussion of discursive violence draws from international development literature that extends the Foucauldian conception of discourses as practical manifestations of power (Escobar, 1984). This double violence is driven by the unprecedented speed with which China has arrived as the major player on the international investment and development scene, coupled with the depth of its financial pockets. It has meant that impacts on local commodities and ways of living have been sharply transformative of livelihoods and traditions. Examples range from the decimation of East African donkey populations—fundamental for local transport and now slaughtered for export of meat and hide—to distortions of highland Peru's agricultural cycles for planting maca tubers—rare endemic rhizomes that are prized for their aphrodisiac qualities, which are being unsustainably overproduced to meet the demands of the Chinese market.

Considered together, the extraction and violence of the Belt and Road produce a profoundly intensive experience of exploitation and appropriation on the receiving end of the BRI, which warrants the label *hyper*-extractivism.

Such hyper-extractive logic is by no means unique to China; similar logic has marked the history of colonialism and the concomitant co-optation of local cultures, power structures, and voices. In describing it as a "logic," we do not pretend to have knowledge of the mental calculus of China's top decision-makers. Instead, our analysis is based on the manifest qualities of China's dealings on the Belt and Road. These manifestations, as we show below, reveal a pattern that resembles the extractive past of human history but is also far more intensive and consequential within a shorter frame of time. In the current globalized era, the Chinese state exemplifies the hyper-extractive logic in its rawest and most visible form. The speed and thoroughness of China's impact are historically unprecedented. China's footprints can be found from the most highly developed to most remote corners of the globe, from the scale of the village to that of the planet.

Rethinking extractivism on the Belt and Road has broad implications for understanding the Anthropocene, our human-created epoch. The empirical manifestation of modern-day hyper-extractivism is predicated on two enabling conditions: high levels of interdependence among countries, economies, and industries on the one hand, and unprecedented levels of global inequality on the other. Characteristics of this intense globalization include increases in the flow of goods, people, ideas, and capital, and a *de facto* shrinkage of time and space. Supply chains have shortened and become vertically integrated, as has the time between the extraction of raw materials to their consumption in the form of finished product and "disposal" in global waste dumps. The gap between winners and losers in these transactions, however, has never been greater. These are key features of the Anthropocene, in which the "great acceleration" has never meant the speed-up of betterment of the human condition in every corner of the Earth, but rather suggests the accelerated transfer of resources, products, and talents away from disadvantaged locations and of waste and pollutants towards marginalized ecosystems. Seen in this light, the Anthropocene presents itself as both a condition and an outcome of hyper-extractivism. Our unique epoch is marked by continuous positive and negative feedback loops of extraction and consolidation, impoverishment and enrichment, dissolution of physical space, and acceleration of change.

The sectors that we have chosen to investigate—food, tourism, and education—offer a grounded analysis of the BRI's hyper-extractivism as found in unexpected places. Our selection of these three cases is intended to maximize the theoretical value of this chapter. We follow the methodological conventions of analyzing "least likely cases" to advance the cutting edge of social science work on hyper-extractivism (George and Bennett, 2005). As we show below, these cases provide ample fodder for thinking about the depth of the hyper-extractive reality of today's world.

Securing the Food Supply

China's anxiety about feeding itself is deeply embedded in the national psyche. Owing to the country's unfavorable arable-land-to-population ratio, with most of the arable land limited to the more developed Eastern parts of the country, famines have marked dynastic history. The Mao-era's "three hard years" of 1959–61, which caused as many as 30 million unnatural deaths, was only the most recent and most devastating. The effort to turn "wasteland" to grainfields during the 1966–76 Cultural Revolution has roots in this well-founded fear of famine, as described in the Maoist slogan, "Take Grain as the Key Link" or *yi liang wei gang* 以粮为纲. With the current population of 1.4 billion, food-based anxieties have come to a head.

In 1995 American environmentalist Lester Brown published a book, *Who Will Feed China? Wake-up Call for a Small Planet*, outlining the myriad challenges of China's rising demand for global food imports (Brown, 1995). Chinese policymakers, for whom self-reliance is a deeply cherished virtue, regarded the book as an insult and published a white paper in rebuttal, touting China's ability to feed itself (State Council, 1996). But Brown's book drew attention to the rapid loss of arable land to development and urban sprawl, and the Communist Party and state declared an agricultural "red line" to protect no fewer than 129 million hectares as farmland, most of it for grain. At the same time, a rapidly globalizing China was quickly learning that many natural resources and raw materials can simply be purchased on the international open market. Even better, the satisfaction of those needs can be guaranteed through the legal purchase or long term lease of foreign agricultural land and fisheries, as China has done with Peru's rich offshore anchovy stocks. Similarly, mines for ores and oil and gas fields for fossil fuels can be secured through investments and loans—in the Sino-Ecuador relationship, for example, critics have called this "debt for oil." In other words, Chinese Communist Party policy makers have learned that the discipline of self-reliance can play out beyond national borders, as well as within them. The quest for security in foodstuffs, energy, and other resources has thus become one of the underlying drivers for China's wave of hyper-extractivism along the Belt and Road.

Unfortunately, what is "legal" in some contexts amounts to land-grabbing from another perspective, particularly when governments are unaccountable and local people disenfranchised and unconsulted. In this section, we focus on China's efforts to secure food through the vertical integration of the stages of production, capturing supplies at their source, purchasing and running production facilities, and ensuring transportation routes. To do so, we highlight dairy production as a novel form of extraction in the current globalized age.

One of the catchphrases in the age of global China is President Xi Jinping's notion of building a global "community with a shared human future" or *renlei mingyun gongtongti* 人类命运共同体. Sparing no effort to pledge his allegiance to the vision, Lu Minfang, the CEO of a Chinese dairy conglomerate, Mengniu, promised a "global community with a shared dairy future" or *quanqiu ruye gongtongti* 全球乳业共同体 (Mengniu, 2020). The Chinese dairy giant has established a truly planetary footprint under the auspices of the BRI, owning dairy processing plants in Oceania, operating research centers in Europe, and marketing an assortment of ice creams and yogurts across Southeast Asia and beyond.

The company's active role in promoting the so-called Dairy Belt and Road or *yidai yilu naiye lianmeng* 一带一路奶业联盟 might seem like a convergence of Chinese political and private business interests, except that the Chinese government effectively owns and controls the supposedly private company. Specifically, the Chinese state owns nearly a third (31.35 percent) of Mengniu. The 7.05 percent shares of the second-largest owner—a British investment company, Schroders—pale in comparison (Mengniu, 2020). It is, therefore, no coincidence that the company's global undertakings align closely with the ambitions of the Chinese state. Other Chinese dairy conglomerates such as Yili and Bright are hardly different.

Chinese-led dairy globalization must be understood in two waves. The first was driven by the Chinese market's rising demand for global dairy products in the 2000s. As a result of high-profile food safety scandals with domestic dairy producers in which melamine, a toxic chemical added to diluted milk to give the appearance of high protein levels, was repeatedly found to have been deliberately added to powdered milk and infant formula, China's consumer confidence in domestic products tanked. The demand for foreign-made alternatives skyrocketed, so much that Hong Kong had to limit the amount of infant formula that people could carry or ship back to the Mainland. Consequently, global dairy prices were pushed up, most notably in 2008 and 2013 (Salois, 2016). In an effort to regulate dairy prices, the Chinese government intervened by purchasing and stockpiling nearly a year's worth of milk powder during the first three months of 2014 (Leightner, 2017). Having secured a robust national stockpile, Chinese state buyers then sharply reduced their purchase levels, sending global dairy prices into a plunge (Howard, 2016). As a result, small dairy producers around the world struggled to avoid bankruptcy. Some farmers had to cull their herds to stay afloat, whereas others faced pressure from banks over loan repayment issues (Browne, 2015). From the perspective of ordinary dairy farmers in places like New Zealand, China's voracious appetite for global dairy turned from a blessing to a curse. Being able to access the Chinese market must have seemed promising, but the experience of state-sponsored price manipulation from a foreign country was an entirely different story.

In this context, beginning in the mid-2010s, the second wave of Chinese-led dairy globalization was set in motion. Chinese dairy conglomerates were no longer content with buying from suppliers in faraway countries. Instead, they embarked on a global shopping spree to buy up entire farms and even supply chains. At a time when dairy farms struggled to turn a profit, an offer—any offer—from an overseas buyer was received like a lifeline. The second wave pressed ahead with ease. In New Zealand alone, Yili acquired South Canterbury's Oceania Dairy for NZ $214 million, Shanghai Pengxin took over Crafar farms for NZ $200 million, and Mengniu invested in a NZ $220 million infant formula factory in the village of Pokeno in Waikato. In each one of these Chinese takeovers, the new owners expanded their production facilities, purportedly implementing state-of-the-art production technologies. In Mengniu's plant in Pokeno village, which has a population of approximately 400, the idyllic rural landscape has been forever changed. Chinese state media quotes the plant's manager in boasting that "when we came here in 2013, it was all pasture here. Within just six years, the population of Pokeno has doubled, and the employment and infrastructure construction in town have also improved a lot" (Xu, 2019; official translation). The manager's account is in stark contrast with the perspective of local residents, who call the plant "an absolute abomination" for shaking up the placid way of life in rural New Zealand (Browne, 2015).

Perhaps most disquieting is the fact that even though the firms that participate in the Dairy Belt and Road are presented as private corporations, they are anything but. They all benefit substantially from Chinese state ownership, subsidies, and regulatory support. Thus, their allegiance to the BRI is not a matter of political formality but an orchestrated developmental plan that brings these dairy firms into perfect synchrony with the Chinese state's global ambitions. Since the 2010s these substantially state-owned dairy companies from China have continued their push for the Dairy Belt and Road, filing requests to acquire larger and larger farms outside China. The BRI's continued penetration into the dairy heartland of New Zealand has seen some measure of backlash. Shanghai Pengxin's plan to buy the iconic Lochinver Station farm collapsed when regulators were not convinced of the alleged benefits for New Zealanders (Naidu-Ghelani, 2015). More recently, Yili's offer of NZ \$588 million for Westland, a century-old dairy co-operative with more than four hundred farmer shareholders, generated intense debates and protests in New Zealand (O'Sullivan, 2019). The deal went through in the end, as the financially struggling co-op was unable to resist the cash offer.

Unlike copper or timber, dairy might not seem like a commodity that conjures up the conventional image of extraction, much less violence. Yet through China's rapid deployment of the Dairy Belt and Road, a subtle form of resource extraction has taken shape. Here, big-box dairy plants function much like bulldozers and excavators at a conventional copper mine, turning one country's natural endowment into tradable commodities for another. The main difference, however, is that dairy plants feature modern-looking, hyper-sterilized production lines attended to by workers in hazmat suits—an image that is the polar opposite of how one imagines an extractive industry characterized by open pits, chemical spills, dusty roads, clearcut forests, and human misery. Underneath this difference of appearance, the economic activities are strikingly similar. Dairy farms, like mines and timber operations, extract from nature raw supplies of a valuable resource. They then supply large quantities of said resource to dairy plants for processing and refinement, transforming the resource into specific outputs that are tradeable on the global market. In fact, the extractive qualities of dairy production are troubled by additional ethical problems because of the involvement of live animals. In other words, if a copper mine or timber operation subjects people (the miners and loggers) and the Earth to its extractive logic, a dairy supply chain subjects people, the Earth, and animals to the extractive cycle. Therefore, even though the Dairy Belt and Road might seem like a comparatively benign form of extraction, its material and ethical consequences are no less disastrous than those of the traditional extractive economy.

By the same token, from the perspective of people on the receiving end of the Dairy Belt and Road investments, the "slow violence" (Nixon, 2011) of Chinese dairy globalization disrupts their ways of life, displaces them from farms they have called home for generations, and destroys the bucolic landscape that is at the heart of their rural identity. Of course, none of these losses has monetary value in modern economic terms, which makes it all the more challenging for small farm owners in New Zealand to defend themselves against the deep pockets of the Dairy Belt and Road. Their experience of hyper-extractivist violence is therefore both material and discursive.

Moreover, the dairy frontier results in not only slow violence but also the "quick" violence of global market risk, price turbulence, and financial unpredictability. As the Dairy Belt and Road continues to consolidate the global supply chain, more and more farms are being sucked into the orbit of the global market. As a result, the life of a Waikato dairy farmer can be suddenly upended by a Beijing official's decisions. When dairy prices can be halved or doubled in the matter of weeks if not days, farming households are sometimes overjoyed by the thrill of a windfall, but other times overwhelmed by the risk of bankruptcy. Yet, ordinary farmers have no escape from the "quick" violence of global financialization, when the Dairy Belt and Road is framed as the future, or, to be exact, the "global community with a shared dairy future."

It must be noted that food is by no means the only sector in which Chinese state capital has gone full-on expansionist. As we show below, in Belt and Road mass tourism development, China's state-owned business, media, and scientific establishments join forces to reinterpret world history in China's favor and remake world heritage in China's image.

Appropriating History, Culture, and Heritage

China is no stranger to revisionist history or the commodification of culture. The dominant Han Chinese have long sought to manage, categorize, and stereotype ethnic minorities within China's borders. The leadership has crafted an official version of the relationship among the various groups as that of one happy family sharing an ancient, unbroken history within a unified country. In fact, China's borders have been fluid and contested; non-Han groups dominated during the Yuan and Qing Dynasties; and relations today are often hostile, particularly between the Han and the Uyghurs, who have been herded into massive reeducation camps, and the Tibetans, who see themselves as targets of what the Dalai Lama has called cultural genocide. Minority cultures have been pacified in the name of national unity and ethnic harmony even as they have been commodified for the consumption of a middle class only recently empowered to enjoy mass tourism. The romanticization of Tibet in the minds of young Han Chinese is just one prominent example. Han tourists flood monasteries that are increasingly emptied of monks other than those co-opted into service provision. In Southwest China's Kunming, at the Yunnan Nationalities Museum and Ethnic Village, tourists can "experience" more than twenty regional cultures in a sort of living human zoo.

Even as Chinese tourists seek to experience different cultures within their borders, foreign cultures are being brought to China for those who do not have the time or means to travel. In Shenzhen (and in smaller versions in Beijing and Changsha), tourists can visit 130 great sights in a theme park called Window of the World, where the Great Pyramids, Mt. Fuji, the Eiffel Tower, Mount Rushmore, and the Taj Mahal sit cheek-by-jowl within about 100 acres. Wealthy Chinese can purchase homes in gated communities built in European planning and architectural styles imagined and recreated by Chinese builders. For those tourists who can afford foreign tours, the traveling experience is carefully managed: As Evan Osnos describes in *Age of Ambition*, Chinese food is provided throughout the trip to avoid upsetting inexperienced stomachs, tourists are told how to think about what they see, and the goal of most whistle-stops is a photo op (Osnos, 2015). While to many observers, such behavior seems harmless, if a bit odd, there is a more destructive dynamic at play. Chinese-style mass tourism might seem an unlikely candidate for consideration as a form of hyper-extractivism, but as we show below, it has implications that reach beyond a simple lack of cross-cultural experience and sensitivity.

"Tourism is the least controversial and most promising area of collaboration," declared Chinese Premier Li Keqiang to his counterparts from member states of the Association of Southeast Asian Nations (Li, 2017). On the face of it, tourism could seem indeed to embody China's promise of win-win development on the Belt and Road. Since the BRI's inception, the number of Chinese tourists to BRI countries has grown from 15.49 million in 2013 to 27.41 million in 2017, suggesting an impressive average annual growth rate of 15.3 percent. Under the rubric of the BRI, China has signed 76 bilateral agreements to promote tourism. Chinese authorities claim to have brought a whopping US $118.7 billion of tourism revenue to BRI partner countries in 2017 alone (Zhao, 2019). In this context, the benefits of the tourism boom for the livelihoods of the people along the Belt and Road seem undeniable, especially when the influx of visitors from China brings with it economic opportunities and infrastructural improvement.

Tourism has been an integral part of the Belt and Road since the very beginning. Three days after his official unveiling of the BRI in a speech in Kazakhstan, President Xi Jinping found himself in the historic city of Samarkand in Uzbekistan on September 10, 2013. Official historiography has it that, after enjoying a panoramic view of the town center from the ancient Ulugh Beg Observatory, the paramount leader felt so poetically inspired as to remark that the Belt and Road "gives us a special feeling. We are far away in distance, but we are also so near to each other in our soul. It is just like time travel" (Wu, 2013; official translation). Indeed, China now arrives in Samarkand at warp speed, with the full force of state-backed capital, development, construction, media, and archeology.

The arrival of Chinese capital has forever altered the legendary Central Asian town. With financial support from China's Silk Road Fund, the state-owned CSCEC Design Group (CSCEC stands for China State Construction Engineering Corporation) is behind Samarkand's latest master plan, which features a new tourism complex known as the "Samarkand City project" in the historic heart of the UNESCO-listed town. With the facelift, local authorities expect to double the number of foreign visitors from the 2015 figure of some 142,000. The slick new Samarkand City is complete with hotels, villas, an amphitheater, bars, restaurants, and 24/7 entertainment and shopping centers (Yeniseyev, 2017). Despite its apparent newness, this modern-looking tourism complex boasts no shortage of traditional design elements, except that these elements are not derived from Uzbek culture, but from the culture of Xi'an, which is the Chinese terminus of the ancient Silk Road. The main structure of the tourism zone is a bulky gray-brick fortification-style hotel modeled after the city wall of Xi'an, crowned by a pagoda-like upper structure in the shape of Xi'an's iconic bell tower. The shopping street leading from the hotel adopts the same gray-brick look, accentuated by traditional Chinese stone carvings and double eaves.

Xi'an has not only been transposed to Samarkand in construction and physical appearance; Chinese state media present the cultural history of Samarkand in a wholly Sinicized light. The locally crafted paper on which artists make hand-drawn paintings is said to have been produced using papermaking techniques from ancient China's Han Dynasty. A mural in the local history museum is said to feature the Tang Dynasty Empress Wu Zetian. The city is even described as the possible ancestral origin of Chinese people with the Kang family name (Zhang, 2016). In fact, through its expanding consortium of state-controlled media outlets that target a global audience, the Chinese state is eager to construct a cultural narrative of historical links between China and Uzbekistan. Several Chinese archaeological missions have been deployed to the country. One group has drawn top-level praise from China's paramount leader for unearthing evidence of the migratory path of the Yuezhi people, pastoral nomads who lived in China's present-day Gansu Province during the first millennium BC (Zhang and Yang, 2016).

The experience of Samarkand typifies an increasingly salient strategy of Chinese foreign policy. Geopolitical and state capital interests are wrapped in the seemingly neutral narratives of archeological science and tourism promotion (Storozum and Li, 2020). For instance, as part of a bilateral agreement between China and Saudi Arabia, Chinese archeologists have made multiple underwater excavations in the Al-Serrian site near the west coast of Saudi Arabia. Even though the scientists retrieved a wide range of relics, from Persian pottery to Arabian stone tools, Chinese state media is uniquely keen on reporting the discovery of a singular Chinese porcelain piece, which is speculated—with a good stretch of imagination—to be evidence of Chinese explorer Zheng He's voyage across the Indian Ocean in the fifteenth century (Wang, 2018). In fact, narratives of Zheng He, many of which blur the line between history and fiction, are similarly peddled in countries like Indonesia, Malaysia, and Singapore to appeal to Chinese tourists (Lim, 2016). In this way, swayed by the profits of tourism, Belt and Road countries indulge the Middle Kingdom as it perpetuates a Sinicized discourse of culture and history.

For some developing countries, the economic benefits of Belt and Road tourism seem impossible to resist. Cambodia, for example, even launched a "China Ready" policy in 2016, in partnership with the Chinese government, the Confucius Institutes, and the state-owned China International Travel Service. Under "China Ready," the tourism-dependent Southeast Asian country pledges to retrofit public signage to include Mandarin translations, expand the number of commercial establishments that accept payments in Chinese renminbi currency, and increase the share of Mandarin-speaking employees in the hospitality industry. A local dancing show, "Smile of Angkor," even received the Chinese central government's "National Cultural Export" award, recognizing the show's adoption of Chinese cultural preferences, technologies, and support. The state's recognition is hardly surprising, given that the show was developed and operated by the state-owned Yunnan Cultural Industry Investment Company, which reaped annual revenue of US $6.76 million from the show in 2013, the only year for which data is available (Li, 2015). To be sure, these tourism initiatives bring employment opportunities and economic benefits to the local people, but they also draw their lives into China's sphere of influence and rendition of culture and history and essentially "extract" and replace indigenous power over narrative and identity.

In recent years, as an increasing number of countries become dependent on Belt and Road tourism, the Chinese state has begun to extract geopolitical and diplomatic leverage from its outbound tourism economy. For example, in June 2020, after Australia called for an independent investigation into the origin of the coronavirus pandemic, China took offence and retaliated by warning its citizens against traveling to Australia. A month later, after Australia suspended its extradition treaty with Hong Kong in light of Beijing's new national security law governing the former British colony, China's response was an upgrade of the travel warning. China's wrath against similar moves in Canada and the United States has also taken the form of travel warnings. The BRI's promise of shared prosperity can be abrogated as casually as it was made.

The Belt and Road Initiative represents a conscious effort on the part of the Xi presidency to tap into the cultural and historical legacies of Chinese civilization's overland and maritime connections to the outside world. As such, the BRI has been seen as an example of "historical statecraft" (Mayer, 2018) or "cultural diplomacy" (Winter, 2020). By invoking the imagination of a shared past on the ancient Silk Road, the BRI finds a compelling cultural justification for China to regain geopolitical dominance and economic leadership. While the many Chinas of past eras have played undeniably significant roles in the history of the Silk Road, the Chinese government's official narrative of the Belt and Road is at best a selective reading of the Silk Road's complex history. The weight of this history is reduced to a one-dimensional narrative of China's glorious civilizational superiority, thus reverting to an imagined world order where the Middle Kingdom lies at the center of everything (French, 2017). Using this parochial frame of reference, Belt and Road tourism development fails to do justice to world history or cultural diversity. Rather, it deprives Chinese tourists of the authentic experience of the outside world and dispossesses local people of epistemic authority over their own past. What remains is cultural appropriation, historical half-truths, and developmental hegemony.

Enveloped in the rosy discourse of a shared past and the hopeful outlook of a common future, the hyper-extractive logic that drives BRI tourism development can easily go unnoticed. The transnational history of the ancient Silk Road rightfully evokes larger-than-life feelings of awe, wonder, and even belonging among the people of Eurasia. However, when this history is filtered through the lens of Chinese geopolitical and state capital interests, it becomes instrumentalized and even weaponized to serve the growing ambitions of the Chinese state. In this context, even though the influx of tourists brings economic benefits to the local people, Chinese state-owned enterprises have secured the lion's share of revenues from projects such as the Samarkand City and the Smile of Angkor. The material violence of Belt and Road tourism thus manifests in both physical and monetary terms, which mutually reinforce each other. The physical "upgrade" of cultural heritage into mass tourism destinations enables Chinese state-owned businesses to commodify other countries' national treasures into sources of profit for China. The profit is then reinvested in ever more "upgrades" further along the Belt and Road. The more subtle experience of violence occurs when China's state-sanctioned version of Silk Road history is foisted upon developing countries in the name of archeological science and tourism development. Such subservience to Chinese-proffered revisionist history is inextricably linked to the BRI's systematic extraction of ideas, knowledge, and talent, to which we now turn.

Extracting Ideas, Knowledge, and Talent

Our third novel form of hyper-extractivism expressed on the Belt and Road is the extraction of ideas, knowledge, and talent. Developed countries have long understood the value of knowledge and ideas and have prospected for them in less developed countries and repurposed them for commercial purposes. Pharmaceutical companies and agribusinesses have patented and sold local knowledge about plants, particularly those with medicinal applications. International researchers have even sought to patent the human genetic material of indigenous tribes for the controversial Human Genome Project. The Convention on Biological Diversity has been critiqued for excessive protection of "intellectual property" and insufficient protections of the rights of indigenous people and traditional knowledge (Andersen, 2013). Hyper-extraction flows in two directions, as we noted earlier in the chapter, with benefits captured by extractors while harms are forced onto vulnerable recipients. In the case of knowledge, for example, agribusinesses have marketed high-yield genetically modified seeds to developing countries, making it impossible for poor farmers to continue the practice of seed-saving from season to season and trapping them into planting cycles that require the purchase and application of commercial pesticides and fertilizers. Such coerced dependency has been debated and resisted, with Indian activists like Vandana Shiva leading the way (Shiva, 2016).

China, inwardly focused for so long with language and culture that take outsiders many years to grasp, is a latecomer to the arena of hyper-extraction of knowledge in the contemporary globalized context. Nonetheless, reflecting the two-way traffic in ideas that is characteristic of our era, China has started to extract and inject knowledge from and into its partners. China has purchased a commanding stake in major agribusinesses like the Swiss biotech giant Syngenta, which is known for selling its patented genetically modified seeds to developing countries. At the same time, there is a widespread perception that China is exporting unemployed farmer laborers to work in grainfields and biofuels plantations that it has purchased or leased in Africa and elsewhere, although there is debate about such accusations (Brautigam, 2015). In addition to extracting material goods and intellectual know-how from overseas, China has also made it a top priority to export its own technical standards, manufacturing expertise, and economic theories to BRI countries. This bi-directional flow of knowledge and personnel, enveloped in high-minded discourses of educational opportunities, leaves developing countries on the Belt and Road with few options but to join the orbital rings of development fitted around the "great rejuvenation" or *weida fuxing* 伟大复兴 of the Middle Kingdom.

After the initial brain drain that followed China's reform and opening in the 1980s, China sought to entice Chinese citizens who had earned PhDs and achieved professional success abroad to return to China. In addition, the Thousand Talents Program, which was founded in 2008, has brought scholars from prominent laboratories and research universities into its friendly fold, including those from the United States and other countries that are not part of the Belt and Road. The program is explicitly designed to recruit experts with skills important to China's scientific competitiveness, providing them with grants and short-term appointments at Chinese universities and research institutions. Recruited scholars are lavished with tax rebates, housing, discretionary funds, and other emoluments. The program recently came under fire when a prominent Harvard chemist was charged with breaking U.S. law for failing to disclose generous payments to set up a research lab in China (Subbaraman, 2020). As with the controversial Confucius Institutes (the Chinese-subsidized soft-power educational institutions that have increasingly been shut down in developed countries), the Thousand Talents program is likely to be scaled back quickly under the onslaught of U.S. Congressional hearings and negative publicity.

Yet, as the Belt and Road continues to expand, China has stepped up its charm offensive in the educational realm, moving far beyond talent acquisition into a concerted effort to foster the next generation of China-loving workers, scholars, and even political leaders in the developing world. The official mantra of the BRI is connectivity. In fact, China boasts "five connectivities" along the Belt and Road: policy, infrastructure, trade, capital, and heart. The last of these—heart, or *minxin* 民心—is perhaps the most curious of policy goals. Under the rubric of "heart-to-heart connectivity" on the Belt and Road, China has ramped up its efforts to, on the one hand, bring students from BRI partner countries to live and study in China, and, on the other hand, export China's educational model to the world through institutional platforms such as the Confucius Institutes and Luban Workshops. The crucial extractive moment here is the summoning of talents across the Belt and Road to serve the constitutive elements of the BRI. Together, these state-sponsored educational efforts aim at "cultivating a global force that is knowledgeable about, friendly to, and fond of China" (Zheng and Ma, 2016).

The number of international students in Chinese universities has seen stunning growth in the last decade or so. In fact, the rapidity of this growth even surpassed China's own expectations. The Ministry of Education in 2010 set the goal of enrolling a total of 150,000 foreign degree students by 2020, but the goal was met six years ahead of schedule in 2014 when 164,394 international students reportedly enrolled in degree programs on Chinese college campuses. In 2017, the most recent year for which data is available, 241,543 foreign students sought a Chinese degree—a 21-fold increase from the number of 11,475 in 1999, when record-keeping started. Much of this growth has been driven by students from BRI partner countries, who totaled 144,956 in 2017—a 45-fold increase from the number of 3,244 in 1999 (Zong and Li, 2020). While the vast majority of these students are from South or Southeast Asia, rapidly growing contingents of African and European students have quickly diversified international student populations.

These figures might give the impression that higher education in China is opening up to the outside world and that the world is exhibiting growing interest in studying in China. Yet, this impression is at best a partial story, when one considers the Chinese state's aggressive efforts in luring students from its BRI allies with handsome scholarships and stipends, segregation of BRI students on satellite sites hundreds of miles from main campuses, and the wholesale export of everything from Chinese values to China's technical standards through educational programs.

Even though college admission is highly competitive for domestic students, Chinese universities are extraordinarily accessible to students from outside the country. In 2017 international applicants enjoyed an acceptance rate of 91.59 percent. Roughly two-thirds of for-degree graduate students from BRI countries were on Chinese government scholarships that covered tuition, room, and board, in addition to medical insurance and a living stipend (Hu et al., 2020). Host universities are lavishly compensated by the government for accepting Belt and Road students. In fact, state subsidies are so exceptionally generous that some Chinese universities have provided ethically questionable perks to lure international applicants. High-profile scandals have enveloped Shandong University, which assigned three Chinese female students to each international male student as "partners," and Shenyang City College, which forced Chinese students to clean foreign-student dormitories.

Perhaps in response to the negative press about these cases, a recent move on the Educational Belt and Road is to establish segregated satellite sites that are removed from the main campus of host universities. Examples include Renmin University's Silk Road College in Suzhou in Jiangsu province (roughly 700 miles from the main campus in Beijing) and Beijing Normal University's Belt and Road School in Zhuhai in Guangdong province (approximately 1,200 miles away). Targeting master's degree-seeking students from BRI countries, these schools aim to train the next generation of leaders along the Belt and Road by inculcating them with official histories and theories about China's politics, economy, society, and culture. These schools openly acknowledge their mission in serving the Chinese state's BRI expansion strategy—an acknowledgment that earns them sustained government funding. In a nutshell, such programs seek to produce graduates who are not only sympathetic to the "China model" of development—characterized by strong state leadership, press censorship, widespread surveillance and policing, and limited civil society—but also active in bringing "the Chinese miracle" to their home countries.

Outside China, the Educational Belt and Road boasts a network of "Luban Workshops" in countries as near as India and Thailand and as far as Zambia and England. (The namesake—Lu Ban—was a legendary inventor and craftsman from ancient times.) Luban Workshops are said to be "Confucius Institutes for technical education," reflecting the ambition to be as pervasive as the Confucius Institutes, and as loyal to China's overseas political interests (Lü et al., 2017). Scattered along the Belt and Road, the Workshops are technical educational centers that prepare locals for employment in the overseas arms of Chinese state-owned companies. Luban programs feature fully translated textbooks that follow Chinese national standards in areas from energy to automation, thus serving China's strategic "Made in China 2025" goal of globalizing Chinese manufacturing and technical standards. Students enrolled in Luban programs are expected to seek China's national professional certification in their technical area before they can be employed by Chinese companies, thus forgoing the professional certification system of the host country. Offering training in an expanding range of areas such as digital manufacturing and computer engineering, Luban Workshops have effectively become enclaves of technical education in BRI countries, promoting a suite of Chinese standards, businesses, interests, and values.

Educational globalization as a means for global dominance is a familiar trope. There is, for example, clear evidence of how Mexico's economic and political systems have been profoundly transformed by an elite class of U.S.-trained Mexican economists. Well-versed in the neoliberal economics of the Chicago School, these students molded Mexico into a textbook example of neoliberalism, rendering the local educational institutions and knowledge systems outdated, if not irrelevant (Babb, 2001). The BRI educational programs have followed the same playbook, except that China is far more methodical in its pursuit of global influence. Two salient qualities of Chinese educational globalization stand out. First, the Educational Belt and Road, including BRI-facing programs within China and Chinese-serving curricula overseas, enjoys substantial state endorsement and financial support from Beijing. In fact, it is common for BRI-related educational offerings to boast state sponsorship as a seal of authenticity. Ribbon-cutting ceremonies for such programs prominently feature the personal attendance of high-level Chinese officials such as ambassadors and Communist Party secretaries-general. This level of state involvement and dominance gives these educational programs a particularly Chinese flavor. Second, the Educational Belt and Road is concerned with spreading the entire complex of Chinese-approved knowledge, from details of technical specifications and coding conventions to overarching economic theories and political ideologies. In other words, the Middle Kingdom is not content with exporting its general model of "socialism with Chinese characteristics" as the United States has done with the neoliberal development model, but also is intent on controlling the underlying technical specifications, standards, and operating requirements in multiple economic sectors and social realms.

Taken together, the state-led, all-encompassing rollout of the Educational Belt and Road is reflective of the hyper-extractive logic in the age of global China. The previous strategy of overt extraction through the Thousand Talents Program and the purchase of corporations that hold intellectual property rights has become less attractive in light of the global backlash in recent years. Some programs have gone underground, while others have subsided (Mallapaty, 2018). Instead of these earlier efforts, China has successfully assembled a series of less visible extraction tactics, drawing top candidates from BRI countries into China's sphere of influence. The examples reviewed above are part and parcel of the Chinese state's quiet indoctrination campaign to normalize its "official" knowledge system. They serve to defend China's social, economic, and political agendas on the global stage.

In many regions of this Sinicized world, China has become the most obvious route to developmental success. To borrow a concept from Noam Chomsky, in the name of education, China is "manufacturing consent" on the Belt and Road (Herman and Chomsky, 2002). Through its concerted efforts, China has prepared the necessary conditions to elicit compliance, thus making alternative options increasingly unattractive. Indeed, BRI host countries such as Djibouti must wrestle with the reality that their own citizens have become skilled workers certified by Chinese authorities to operate Chinese-made bullet trains that run on tracks that comply with Chinese standards, with the help of operations manuals in Mandarin. The most deeply traumatic—but least physically tangible—consequence of the BRI is that when countries go "China-lite" (Reilly, 2013), their totalized acceptance of Chinese products, technologies, economic approaches, and governance tools seems utterly compelling to their governments, technocrats, and citizens. This is the payoff for China of its secret weapon of "heart-to-heart connectivity."

Emerging Resistance in a COVID-19 World

The coronavirus (COVID-19) pandemic has, unexpectedly, changed the terms of the debate and the intensity of the BRI's ambitions, and it has strengthened the voices of BRI skeptics. Along with the poorest members of wealthy Western democracies, much of the developing world suffers the greatest burden of the pandemic, with health care and testing inadequate, social distancing often impossible in crowded living and laboring conditions, and the poorest unable to self-quarantine when they need to work in order to eat. In this context, many blame the Chinese state for its lack of transparency and outrageous mishandling of the initial outbreak in Wuhan. The negative perception of China has simmered for a long time because of labor and environmental disputes and loan practices that lock recipients into dependent relationships with the donor. Now it has been solidified by knowledge of the grievances of Chinese citizens and the government's efforts to suppress the courageous warnings of local doctor-heroes. Criticism of the Chinese government's initial handling of the pandemic has led mainstream politicians around the world more openly to question Belt and Road propagandists' promises of co-benefits and shared prosperity—the same promises that they previously endorsed without blinking an eye.

In some places along the Belt and Road, the aggregated grievances against China's extractive activities are being woven into a wholesale reevaluation of the BRI's claims of mutual benefits and win-win outcomes. The Hamrawein coal-fired power plant in Egypt has been postponed. Pakistan has requested a more lenient repayment agreement. African leaders have banded together to call for emergency debt forgiveness. "Only a drunkard would accept these terms," declared Tanzania President John Magufuli on April 25, 2020 in regards to a proposed BRI loan of US $10 billion. Such high-level and angry rejections underscore the profound discontent with the extractive logic of the BRI, even as concerns about the spread of the virus have put many Chinese-funded and -led infrastructure projects on hold. Globally, the pandemic has given rise to rampant expressions of racism, discrimination, and xenophobia, especially against individuals of Chinese descent (Devakumar et al., 2020). Even before the wave of coronavirus infections arrived in Africa, the spread of Sinophobia had resulted in escalated tension between the locals and nearly one million Chinese nationals who live and work on the continent (Solomon, 2020). In this context, a top-level Chinese "internal report" warned of a global backlash similar to that which followed the 1989 Tiananmen Square protests (Hirschberg, 2020). Perhaps in response, China has ratcheted up its coronavirus-related aid in the form of donated ventilators, protective gear, and masks. China has even loaned medical assistance teams and built new coronavirus testing labs along a so-called Health Belt and Road in places like Angola and Gabon. (China made the same offer to California, which rejected it on the grounds of national security.)

The timing of such public outcries in the midst of a pandemic raises critical questions about accountability, subaltern voices, and political hegemony in our increasingly unequal world. The risks of working with China on the BRI are many, including the unfavorable collaterals of BRI loans, ecological consequences of massive ports and roads, and secretive bilateral negotiations that do not include those most affected by such projects. In fact, citizens of BRI recipient countries have been aware of these risks all along. Many constituencies have voiced concerns, from indigenous groups to legal activists and scholars in universities in the developing world. However, in the pre-coronavirus world, cautions and grievances were swept under the rug by political elites. Governments across Eurasia and beyond signed up for BRI partnerships with China and issued high-level endorsements for the initiative. United Nations agencies jumped on the same bandwagon.

There is a glaring contrast between the elite politics of silence and support in the pre-pandemic age and the populist politics of China-shaming in the pandemic-stricken world. The nearly universal acquiescence during the pre-pandemic era—bolstered by China's secretive bilateral agreements with 137 sovereign states and 30 international organizations—reflects a collective self-deception on the part of international political elites in their dance with the Chinese state. In embracing China's discourse of win-win development, they failed to protect the wellbeing of their own citizens, resources, and landscapes. The sudden shift to the populist politics of China-shaming in the midst of the coronavirus pandemic, however, suggests a denial of responsibility on the part of political leaders in China's BRI partner countries. China is indeed the obvious culprit in the sufferings and injustices along the Belt and Road, but the Chinese state alone could not have brought such havoc.

The double extraction and double violence on the Belt and Road could—and should—have been prevented. As we show elsewhere, China depends on a constellation of coercive controls and authoritarian techniques to achieve its model of governance domestically. Meanwhile, however, on the global stage, China's high-minded notions of co-benefits and shared prosperity depend on the willful acquiescence of foreign governments and international organizations (Li and Shapiro, 2020). The BRI would not have grown so rapidly, were it not for the implicit, complicit, and explicit support of powerful actors who might have been blinded by the payoffs in supporting China's extractive frontier. The desperate needs of developing countries for additional capital investment, and by the growing perception that China offers the only game in town, also owe significantly to the resurgence of Western isolationism. A tragic aspect of the Belt and Road's hyper-extractive violence is that it has taken one of the deadliest pandemics in human history for powerful political actors to arrive at a long-overdue reevaluation of their partnerships with China.

The global rise of populist resistance to the Belt and Road, fueled by racist and xenophobic sentiments, misses the opportunity for genuine accountability in international development. Vilifying China as the inscrutable or nefarious dragon behind the world's troubles is as misguided as eulogizing the country as a selfless Good Samaritan or Johnny Appleseed spreading the seeds of development along the Belt and Road. The reality of our highly interconnected and interdependent global commons calls not for blind trust or reactive hostility, but for transparency and accountability. At this moment of global resource scarcity, the stakes are higher, and the consequences of hyper-extractivism more destructive than ever. Acceptance of development assistance should never come with hidden strings attached. The need for equitable development is more pressing than ever. On the Belt and Road, a necessary first step would be to narrow the gap between the official win-win discourse and the lived social experience of dispossession and exploitation. If BRI investments are held up to the standards of China's own professed commitments to mutual gains, the social and economic benefits could indeed make a huge contribution to poverty alleviation and environmental protection in the developing world. However, narrowing the gap between promises and realities on the Belt and Road requires rigorous policy evaluations and evidence-based studies that can rise above the cacophony of propaganda and sloganeering. We hope this chapter has contributed to that conversation.

Chapter Acknowledgments

We are grateful to Isabella Baranyk and Jisho Warner, who provided valuable comments, and to Michelle Huang and Zhaolei Huang, who provided research assistance.

References

Andersen, R. (2013) *Governing Agrobiodiversity: Plant Genetics and Developing Countries.* Farnham: Ashgate.

Babb, S.L. (2001) *Managing Mexico: Economists from nationalism to neoliberalism.* Princeton, NJ: Princeton University Press.

Boyer, D (2017) 'Energo Power: An Introduction' in Boyer, D. and Szeman, I. (eds.) *Energy Humanities: An Anthology.* Baltimore, MD: John Hopkins University Press.

Brautigam, D. (2015) *Will Africa Feed China?* Oxford: Oxford University Press.

Brown, L.R. (1995) *Who Will Feed China? Wake-Up Call for a Small Planet.* New York, NY: W.W. Norton & Co.

Browne, A. (2015) 'Milking New Zealand's Way of Life', *The Wall Street Journal*, February 18, 2015.

China Mengniu Dairy Company Limited. (2020) *Annual Report 2019.*

Devakumar, D., Shannon, G., Bhopal, S.S., and Abubakar, I (2020) 'Racism and discrimination in COVID-19 responses', *The Lancet*, 395 (10231), p. 1194.

Escobar, A. (1984) 'Discourse and Power in Development: Michel Foucault and the Relevance of his Work to the Third World', *Alternatives*, 10 (3), pp. 377–400.

French, H.W. (2017) *Everything under the Heavens: How the Past Helps Shape China's Push for Global Power.* New York, NY: Knopf Doubleday.

George, A.L. and Bennett, A. (2005) *Case Studies and Theory Development in the Social Sciences.* Cambridge, MA: MIT Press.

Gould, K.A., Pellow, D.N., and Schnaiberg, A. (2004) 'Interrogating the Treadmill of Production: Everything You Wanted to Know about the Treadmill but Were Afraid to Ask', *Organization & Environment*, 17 (3), pp. 296–316.

Herman, E.S. and Chomsky, N. (2002) *Manufacturing Consent: The Political Economy of the Mass Media.* New York, NY: Pantheon Books.

Hirschberg, P. (2020) 'Exclusive: Internal Chinese report warns Beijing faces Tiananmen-like global backlash over virus', Reuters, May 4, 2020.

Howard, R. (2016) 'China's milk stockpile leaves New Zealand dairy farmers struggling', Reuters, March 28, 2016.

Hu, R. (胡瑞), Yin, H. (尹河), and Zhu, W. (朱伟静). (2020) 'Oversea Postgraduates Education from Belt and Road Countries in China: Status, Dilemmas and Strategies ("一带一路"沿线国家来华留学研究生教育现状、困境与策略)', *Modern Education Management (*现代教育管理*)*, (5), pp. 51–57.

Leightner, J. (2017) *Ethics, Efficiency and Macroeconomics in China: From Mao to Xi.* Abingdon: Taylor & Francis.

Li, H. (李菡静). (2015) 'Localization and globalization of the performance industry in Yunnan: Case of Smile of Angkor (云南演艺产业走出去的"在地化"与"国际化"——以《吴哥的微笑》为例)', *Studies in National Art (*民族艺术研究*)*, 28 (2), pp. 143–150.

Li, K. (李克强) (2017) Speech at the Twentieth China-ASEAN Leadership Summit (第20次中国东盟领导人会议上的讲话).

Li, Y. and Shapiro, J. (2020) *China Goes Green: Coercive Environmentalism for a Troubled Planet.* Cambridge: Polity.

Lim, T.W. (2016) 'Narratives Related to Zheng He: Explaining the Emergence of Ethnic Chinese Communities Overseas and the Rise of a Regional Trading Network' in Lim, T.W. et al. (eds.) *China's One Belt One Road Initiative.* Singapore: World Scientific, pp. 63–111.

Lü, J. (吕靖泉) et al. (2017) 'Luban Workshops: New pivot of vocational training globalization ("鲁班工坊"——职业教育国际化发展的新支点)', *Chinese Vocational and Technical Education (*中国职业技术教育*)*, (1), pp. 47–50.

Mallapaty, S. (2018) 'China hides identities of top scientific recruits amidst growing US scrutiny', *Nature News*, October 24, 2018.

Mayer, M. (2018) 'China's historical statecraft and the return of history', *International Affairs, Oxford Academic*, 94 (6), pp. 1217–1235.

Naidu-Ghelani, R. (2015) 'China's farm-buying runs into opposition', BBC News, October 29, 2015.

Nixon, R. (2011) *Slow Violence and the Environmentalism of the Poor.* Cambridge, MA: Harvard University Press.

O'Sullivan, F. (2019) 'China's growing NZ footprint raises big questions', *NZ Herald*, 2 April, 2019.

Osnos, E. (2015) *Age of Ambition: Chasing Fortune, Truth, and Faith in the New China.* New York, NY: Farrar, Straus, & Giroux.

Reilly, B. (2013) 'Southeast Asia: In the Shadow of China', *Journal of Democracy*, 24 (1), pp. 156–164.

Salois, M. (2016) 'Global dairy trade situation and outlook', *International Food and Agribusiness Management Review*, 19 (B), pp. 11–26.

Shiva, V. (2016) *Biopiracy: The Plunder of Nature and Knowledge.* Berkeley, CA: North Atlantic Books.

Solomon, S. (2020) 'Coronavirus Brings "Sinophobia" to Africa', VOA News, March 4, 2020.

State Council (国务院). (1996) *The Grain issue in China (*中国的粮食问题*).* Beijing: Information Office of the State Council (国务院新闻办公室).

Storozum, M.J. and Li, Y. (2020) 'Chinese Archaeology Goes Abroad', *Archaeologies: Journal of the World Archaeological Congress*, pp. 1–28 (Advance preview).

Subbaraman, N. (2020) 'Harvard chemistry chief's arrest over China links shocks researchers', *Nature News*, February 3, 2020.

Wang, K. (2018) 'Ancient silk road port found in Saudi Arabia', *China Daily*, March 24, 2018.

Winter, T. (2020) 'Silk road diplomacy: Geopolitics and histories of connectivity', *International Journal of Cultural Policy*, Routledge, pp. 1–15 (Advance preview).

Wu, J. (2013) 'Xi "travels in time" along the ancient trade route', *China Daily*, September 10, 2013.

Xu, X. (2019) 'Feature: Yashili factory helps reshaping Pokeno, a small town of New Zealand', Xinhua English News, October 14, 2019.

Yeniseyev, M. (2017) 'Uzbekistan launches "Samarkand City" project', *Caravanserai*, August 3, 2017.

Zhang, R. (张锐鑫) (2016) 'Xinhua Agency "Belt and Road World Tour" Reporting Group Arrives in Samarkand in Uzbekistan (新华社"一带一路全球行"报道团抵达乌兹别克斯坦撒马尔罕等城市)', *Luoyang Evening (*洛阳晚报*)*, September 5, 2016.

Zhang, Z. (张哲浩) and Yang, Y. (杨永林) (2016) *Northwest University archaeologists revisit the Silk Road in search for the Yuezhi people* (西北大学考古队重走丝绸之路⊠寻找大月氏遗迹), Guangming Daily (光明日报), 25 August.

Zhao, S. (赵珊) (2019) 'BRI Tourism Boom ("一带一路"带火沿线游)', *People's Daily* (Overseas edition) (人民日报海外版), April 26, 2019.

Zheng, G. (郑刚) and Ma, L. (马乐). (2016) 'The Belt and Road Strategy and Education of Overseas Student in China: Based on the Data from 2004 to 2014 ("一带一路"战略与来华留学生教育: 基于2004–2014的数据分析)', *Education and Economy (*教育与经济*)*, (4), pp. 77–82.

Zong, X. (宗晓华) and Li, T. (李亭松). (2020) 'Trend and future of international students from Belt and Road countries ("一带一路"沿线国家来华留学生分布演变与趋势预测)', *Higher Education Exploration (*高教探索*)*, (4), pp. 91–99.

8

GRANTING RIGHTS TO RIVERS IN COLOMBIA

Significance for ExtrACTIVISM and Governance

Whitney Richardson and John-Andrew McNeish

Introduction

Colombian courts have issued ruling recognizing 14 distinct eco-regions as rights-holders since 2016. Ten of these rights-bearing ecosystems are river basins. According to the rulings, the rights of the rivers are to be actively *protected, maintained, conserved,* and *restored.* These cases are significant concrete manifestations of a broader trend toward assigning legal rights to nature as scientists, environmentalists, indigenous communities and policymakers have come to recognize the power of strategic litigation. To date, Colombia is the country with the largest number of nature's rights court rulings worldwide (United Nations, n.d.; Radicado, 2019; Bustos and Richardson, 2020). Similar legal cases can, however, be found across the world in countries as diverse as India and New Zealand.

In this chapter, we detail the Atrato River case, the first Colombian river to be recognized as a rights-holder by a 2016 ruling. We discuss whether it, and (by association) the others that followed, represents a convincing attempt to establish a new mode of environmental governance. The rulings were primarily issued as corrective measures to redress harms due to extractive projects, and as a means to restore and protect ecological conditions that guarantee inter-dependent human rights (United Nations, n.d.; Rama Judicial del Poder Púb-lico, 2019). In this chapter we explore available empirical evidence that shows how the Atrato River ruling could help to protect the river, as well as the practical challenges it has created for riverine guardianship. We also consider the value of the Atrato case as a source of inspiration for wider environmental governance and extrACTIVISM, i.e. activism opposing the impacts of resource extraction (Willow, 2018).

The Atrato River Case and Mining in the Chocó

In 2015, the social justice research center *Tierra Digna* filed a *tutela* (a legal writ based on a claim of a breach of 'fundamental' constitutional rights) on behalf of an alliance of organizations based in Colombia's department of the Chocó. The *tutela* was directed against 26 responsible government agencies for failing to stop well-documented illegal mining throughout the Atrato River Basin (Defensoría del Pueblo, 2014a; Defensoría del Pueblo, 2014b; Corte Constitucional, 2016). Plaintiffs claimed that this failure had led to the systematic violation of their rights, i.e. rights to life, dignity, health, a healthy environment, freedom of movement, water, food security—and those of specially protected "ethnic" groups to autonomy, culture, and territory (Corte Constitucional, 2016).

The Chocó department sits in the northwest part of Colombia and is highly regarded for its high levels of cultural and biological diversity. Ninety-seven percent of the Chocó's residents are protected "ethnic" groups (87 percent Afro-descendant and 10 percent indigenous) (Corte Constitucional, 2016). Ninety-seven percent of the Chocó's surface area is made up of collective territories under common ownership, including 600 Afro-descendant communities governed by 70 community councils and 120 indigenous reserves including those of the Embera-Chamí, Embera-Dobida, Embera-Katío, Tule, and Wounan (Corte Constitucional, 2016; Macpherson, 2019). Furthermore, 90 percent of Chocó is protected forest—including a vast range of eco-systems, endemic species, and watersheds including the Atrato River. The Atrato River Basin spans 60 percent of the Chocó and more than fifteen rivers and 300 streams run through it. The Atrato is the longest river in Colombia and third most navigable (Macpherson, 2019).

Prior to Spanish colonization, indigenous communities in the Chocó region had a long history of artisanal gold mining. When Spain colonized Chocó in the 1500s, they trafficked enslaved Africans into the area and forced indigenous peoples to mine gold for the Spanish crown. As a result, the Chocó became the largest gold producer worldwide, yet almost all the generated wealth was exported, with little reinvestment in the area. After independence, slavery was abolished and many Afro-descendants settled along the coastal regions alongside indigenous groups. Mining remained the primary economic activity. Since then, administrative authorities have continued to receive royalties for mining concessions without reinvesting socially or environmentally, as demonstrated by the Chocó's high rate of unmet basic needs and deteriorating ecological conditions (Corte Constitucional, 2016).

Today, four types of mining occur in the Chocó department: 1) *artisanal mining* that is carried out manually using ancestral techniques at a small scale; 2) *semi-mechanized mining* that incorporates small equipment like motor pumps, hydraulic elevators, and small dredges; 3) *mechanized mining* that uses backhoes, dredges, bulldozers, hoses, dump trucks, high capacity motor pumps, and toxic chemicals (particularly mercury and cyanide); and 4) *mega-mining* that requires a lot of land, water, and energy and includes open pit mining. Though mega-mining can pose grave consequences to the environment, mechanized mining is considered as the most dangerous for both humans and natural entities (Corte Constitucional, 2016).

Semi-mechanized mining in the Chocó began in the 1980s when an influx of foreign actors and armed groups—paramilitaries, BACRIM (criminal) organizations, cartels, and guerilla organizations—began illegal mining operations. These actors sought to extract gold buried in the river using high-impact equipment and toxic chemicals which are cheap and portable, and ease the process of gold extraction (Güiza and Aristizabal, 2013; Corte Constitucional, 2016). Since then, there has been a proliferation of illegal mining in the region. Drug traffickers are known to launder cocaine profits by smuggling gold in and out of Colombia, by actively taxing and coercing local governments that benefit from mining, and by running shell companies that attribute gold discoveries to fictitious mines (Tubb, 2020). According to available data, in 2011 a total of 99.2 percent of the Chocó's 527 registered Mining Production Units had no mining titles or licenses, making it the region with the highest concentration of illegal mining operations (Corte Constitucional, 2016).

The proliferation of illegal mechanized mining has caused severe socio-ecological consequences. High-impact mining has resulted in the loss and contamination of water and food supplies. It has devastated subsistence and livelihoods and gravely impacted residents' health. A resident cited in the court ruling stated, "…before mechanized mining, the river was crystalline, healthy, with clear waters, and that local populations were dedicated to fishing, agriculture and artisanal mining. These were core subsistence activities for local residents and at the center of cultural life" (Corte Constitucional, 2016, p. 70). Bereft of alternatives, many residents have had to turn to illegal mining themselves, rent land to miners, or engage in sex work (T-622/16, 2016). Loss and contamination of food and water due to high impact mining have also led to the deaths of more than 30 children, impaired child development, and caused miscarriages, skin diseases, malaria outbreaks, malnourishment, and dehydration (Corte Constitucional, 2016; Comité de Seguimiento, 2018a).

High-impact mining has also had devastating consequences for the Atrato River Basin's ecological health. The use of heavy machinery and toxic chemicals has destroyed water supplies, impairing hydrological cycles and leading to increased sedimentation. In some places, little identifiable water flow remains. Mechanized mining has destroyed habitats, leading to biodiversity loss, deforestation, and the loss of genetic diversity within species. Even if high-impact mining were to cease, toxic contamination can persist for long periods of time (Corte Constitucional, 2016). By the time the Constitutional Court ruling was issued in late 2016, ecological damage to the Atrato River Basin was estimated to cover hundreds of thousands of hectares, the full extent unknown (Corte Constitucional, 2016; Delgado-Duque, 2017; OECD, 2017).

The Court affirmed that violation of plaintiffs' rights had occurred as a result of government failure to confront the proliferation of illegal mining. As a basis to remedy these complex, interdependent issues and restore conditions that guarantee plaintiff rights, the Constitutional Court issued a set of mandates tied to recognizing the Atrato River Basin as a rights-holder. Among these were provisions to increase the participation of local residents in decision-making processes with implications for the health and well-being of local residents and the Atrato River Basin (Corte Constitucional, 2016; Macpherson, 2019). In this way, the Court came to strengthen legal protection for an important element of nature as a means to guarantee the human rights which rely on the Atrato's ecological health and functioning (Corte Constitucional, 2016).

The Court named the state and local community representatives to be co-guardians of the Atrato River Basin. It ordered the Presidency to assign a state official as co-guardian, while plaintiffs were to elect a local Guardian as the official representative of the river. The Court also ordered plaintiffs to elect a body of River Guardians—composed of representatives from various resident communities—and called for a Panel of Experts to assist the River Guardians and help to ensure their participation was guaranteed in all processes (Corte Constitucional, 2016).

Furthermore, the Court issued several more mandates to help cumulatively to restore conditions to guarantee plaintiffs' rights, assigning responsible authorities to each one. Orders required that assigned authorities collaboratively develop and implement 1) short-, medium-, and long-term plans to decontaminate and restore the Atrato River Basin; 2) a comprehensive plan to neutralize and eradicate illegal mining in the region within six months; and 3) a comprehensive plan to recuperate plaintiffs' traditional livelihood and subsistence models, also within six months. These action plans were to be informed by epidemiological and toxicological studies. The Court also ordered a Follow-Up Committee, led by the Attorney General of Colombia, to oversee implementation efforts and evaluate compliance. Lastly, the Court ordered the state to ensure the *Intersectoral Commission for Chocó* to comply with the Ombudsman's 2014 Resolution 064, which was issued to address the socio-ecological humanitarian crisis in Chocó (Corte Constitucional, 2016).

A Trend towards the Rights of Nature

In 2018, just over a year after the Constitutional Court's 2016 Atrato ruling, the Supreme Court recognized the Colombian Amazon as a rights-holder, granting the Colombian Amazon region the same rights as the Atrato River Basin. The decision came in response to a *tutela* made by 25 Colombian youths, arguing that government omission to combat rampant deforestation in the region exacerbated climate change and, thus, threatened their future rights contingent on a healthy environment. As background, it is important to note that, after the signing of the 2016 Peace Deal, deforestation in the Colombian Amazon increased owing to the departure of the FARC-EP (a left-wing guerrilla group) from a region that they previously controlled. In line with the Atrato decision, the Amazon decision argued that until nature's right to exist is legally recognized, human rights will remain threatened (Corte Suprema de Justicia, 2018; Bustos and Richardson, 2020).

Later in 2018, the first regional court issued a decision demanding that the Páramo de Pisba be recognized as a legal subject. Since then, many more Colombian court decisions have recognized other rivers and ecosystems as rights-holders. These include the La Plata River; Coello, Combeima, and Cocora Rivers; Cauca River; Pance River; Otún River; Magdalena River; and Quindío River. All of these court decisions adopted the same rights as recognized by the Atrato decision; however, each decision issued a unique set of mandates intended to guarantee nature and human rights concurrently (United Nations, n.d.; Corte Constitucional, 2019). In most cases, the court deemed nature's rights as an appropriate remedy for the conflict highlighted by the lawsuit; though, in the case of the Pance River, plaintiffs filed suit on direct behalf of the river, arguing its intrinsic rights and citing the Atrato River case as precedent (Desplazada, 2019).

The courts are not the only government body acknowledging the rights of ecosystems in Colombia. In 2019 the governors of Nariño and Boyacá departments also pledged to recognize the nature's rights in their administrative proceedings within their departments (El Gobernador del Departamento de Nariño, 2019; Gobernación de Boyacá, 2019a; Gobernación de Boyacá, 2019b). Furthermore, the Jurisdicción Especial para la Paz (Special Jurisdiction for Peace) has continued to declare nature as a silent victim of the armed conflict, demanding that nature be a subject for restitution for harms done (Jurisdicción Especial para la Paz, 2019).

While Colombian courts mobilized recognition for the rights of specific river basins and ecosystems, this personification of nature as a legal subject had prior national basis. In 2011 Colombia's Law of the Victims recognized the land as a victim of the armed conflict, legally enabling the 'land' to seek restitution for harms done (Congreso de Colombia, 2011). A year before the Atrato decision, a 2015 Constitutional Court decision called for Tayrona Park's protection because of nature's intrinsic value beyond its instrumental value, paving the way for future jurisprudence to build further on this notion (Corte Constitucional, 2015).

International jurisprudence and arguments for nature's rights have also had a direct impact on the Colombian cases. The arguments and decisions issued by Colombian courts mirror designs for nature's rights governance frameworks internationally. Of note, United States based legal scholar, Christopher Stone, issued the first developed legal argument in favor of legally recognizing nature's rights in 1972. Stone's argument called for particular governance mechanisms to help to guarantee nature's rights. Stone advocated for recognizing distinct natural entities—i.e. rivers, mountains, animals, etc.—as right-holders, to help to identify and uphold their unique interests and needs to maintain ecological health (Stone, 1972).

The Atrato decision's guardianship mechanism also drew heavily from New Zealand's co-guardianship model, which named the State and the local Maori people (The iwi) as the official representatives of the Whanganui River (Te Awa Tapua) in 2014. The Whanganui River Settlement was devised as a form of restitution for colonial harms against the local Maori. Interestingly, a clerk with the Colombian Constitutional Court had conducted research on indigenous rights in New Zealand and found similarities between the Whanganui River and Atrato River Basin cases (Magallanes, 2015; Macpherson, 2019).

Prior to Colombian courts' recognition of nature's rights, Ecuador and Bolivia had passed nature's rights legislation. In 2008 Ecuador passed a new constitution that included the protection of the rights of nature, and in 2010 Bolivia adopted a constitution and "Law of Mother Earth" with similar protections. In these cases, nature's rights were positioned as a means to reflect indigenous cosmologies in order to advance the good life (*buen vivir*) and live in harmony (*Sumak Kawsay*) with Mother Earth (*Pachamama*). Such laws and promulgations have done little to change realities on the ground; both national governments have continued to move forward with an extractive agenda as their primary economic driver (Lalander, 2014). Nonetheless, the extractive agenda has more readily been challenged in court, citing nature's rights as a basis, with some wins (Kauffman and Martin, 2017).

This increased presence of ecocentrism in law can also be observed in the development of recent social theory, some of which might have played into the legal recognition of the rights of nature detailed above. In recent years, important trends have destabilized earlier anthropocentric understandings of man's domination over and separation from the natural world. Latour has, for example, influentially argued for scientific accountability to be expanded to include the human and the nonhuman (Latour and Porter, 2017). Other social theorists suggest a "post-humanist" turn aimed at further reworking our understanding of human-nature relations. Haraway (2017) suggests that we need to relearn that humans are not separate but, rather, a "companion species with a complex" assemblage of natural relations. Ingold, together with other anthropologists of the "ontological turn" (Holbraad and Pedersen, 2017), proposes the foundation of a more-than-human anthropology (Ingold, 2011). As Tsing (2017) suggests, in this anthropology we do not merely identify non-humans as static others but, instead, learn about them and ourselves in action through common activities.

Across Colombia today, proposals to recognize different forms of nature's rights remain under active consideration. In 2018, several non-governmental organizations (NGO) have filed legal petitions to recognize the rights of all rivers in Colombia (Earth Law Center *et al.*, 2018). Perhaps most notably, in the summer of 2019 Colombian legislators proposed a constitutional amendment to include a provision which recognizes nature's rights as a whole within Article 79. Article 79 affirms the human right to a healthy environment (Lozada Vargas, n.d.).

An Inspiration to Governance and ExtrACTIVISM?

The extreme levels of natural resource extraction taking place across the globe have not only caused unprecedented environmental damage, but have also stimulated sharp political, social, and cultural conflicts.

Resource extraction has been tightly connected to the histories of human development, civilization and empire, and to the processes of modernization and expectations of modernity (Harvey, 2013). Current "extractivism" has been distinguished by its single-minded disregard for environmental consequences in favor of profit and *externalization,* i.e. all costs—economic, social, and environmental—are internalized and disproportionately borne by citizens of extraction zones (Veltamayer and Petras, 2015). The borders of far-flung extractive enclaves or sacrifice zones (Lerner, 2010) have become harder to identify, as all of nature has become a commodity and earlier geographical separations reduced by technology and the concurrent higher velocity of globalization. As extractive frontiers expand ever further across the world and encroach on urban and disenfranchised populations, soft and coercive governmentalist techniques are employed by government and industry. Hearts and minds are won over by promises of jobs, investments in local services, corporate social responsibility schemes etc. The remaining uncooperative population are forcibly controlled through technologies of social pacification, including surveillance, militarized policing, and the deployment of counter-insurgency tactics branding environmentalists and land defenders as terrorists (Dunlap, 2019).

Of equal importance to the current character of extractivism is that scholars studying the growing levels of socio-environmental conflict resulting from expanding extractive frontiers have recognized that this mindset and set of practices are not free to operate with impunity. For extraction zone residents, the battling of industrial encroachment through direct action or legal challenges, lobbying government and international organizations, multi-scale alliances, media drives and targeted campaigns have become necessary for survival. Willow (2019) captures these activities intending to question, confront and tame extractivism with an antithetical term: extrACTIVISM. Her book, *Understanding ExtrActivism: Culture and Power in Natural Resources Disputes*, surveys how the contemporary resource extraction industry works and the multiple responses or extrACTIVISM it inspires to "counter extractivist development and domination" (Willow, 2018, p. 3).

ExtrACTIVISM in the Atrato case was expressed by Chocó resident groups banding together with the NGO Tierra Digna to take legal action. It was their filing of the *tutela* that mobilized the process to confront illegal mining and government inaction (Defensoría del Pueblo, 2014a; Defensoría del Pueblo, 2014b; Corte Constitucional, 2016). The rights of nature are positioned as a transformational alternative to the proliferation of illegal mining in the Atrato River Basin (Corte Constitucional, 2016; Willow, 2018).Taking legal form, extraACTIVISM importantly also moved into the state apparatus itself.

The Colombian Constitutional Court's constitutionally assigned role is to uphold the constitutional rule of law and guarantee rights. The Constitutional Court has, however, also demonstrated a willingness to carry out this role in a manner that consistently challenges the state rather than only act in its bureaucratic defence. In the Atrato case, the Constitutional Court chose to significantly push the boundaries of existing protections. In the first move of its kind in Colombia, the Court modelled the Atrato nature's rights governance approach as a means to remedy socio-ecological problems generated by extractivism and the concomitant armed conflict. It is worth noting that—while common law systems do not have the power to establish new laws—an influential 1992 Constitutional Court decision permitted the Court to grant new rights in order to uphold existing rights, recognizing the need for rights to adapt to changing conditions (Corte Constitucional, 1992). By naming the Atrato River Basin as a rights-holder, the Court drew from its available juridical tools as a means to strengthen existing rights tied to a healthy environment (Corte Constitucional, 2016).

As a primary objective, the decision required the national police and armed forces to help to develop and execute a comprehensive plan to "neutralize and eradicate" illegal mining—thereby further enforcing the goals of extrACTIVISM. Recognizing the proliferation of illegal mining as the most problematic form of mining facing residents, eradication of this harmful extractivist practice is considered necessary to restore required socio-ecological conditions. However, the Court also acknowledged that legal mining could pose severe socio-ecological risks that should also be evaluated to ensure the rights of residents and the Atrato River Basin be upheld. In this way, the Court went beyond the conflict framed by the *tutela* lawsuit to suggest that other extractivist projects be assessed and confronted in relation to both human rights and riverine rights (Corte Constitucional, 2016).

The Court's characterization of co-guardianship between Atrato residents and the state aims to strengthen the long-neglected rights of Afro-Colombians and indigenous people in the region (Macpherson, 2019). To be considered in compliance, the ruling demands that River Guardians' participation is guaranteed in all decision-making processes and that their wishes are central to the finalized comprehensive plans established to guarantee their own rights—for example, those tied to restoring traditional livelihoods and food, conditions for health—and the rights of the Atrato simultaneously. Furthermore, the ruling and follow-up reports urge the sector to incorporate River Guardians in the planning of their defense policy and to ensure differentiation between locals engaged in illegal mining due to lack of alternatives and those higher on the criminal supply chain. The ruling and subsequent compliance reports also state that the Atrato's rights are meant to uphold the rights of Atrato residents and cannot supersede them (Corte Constitucional, 2016; Comité de Seguimiento, 2018a).

Traditionally, government agencies have focused on their specific areas of mandate without sustained coordination between them. The Court saw this as contributing to the complex socio-ecological problem and ruled to require inter-institutional collaboration on all issues under shared jurisdiction regarding the Atrato case, further arguing that all agencies must seek to uphold the rights detailed by the Constitution. Therefore, all actions required by the ruling require coordination among agencies and across regions in order to be considered in compliance (Corte Constitucional, 2016; Comité de Seguimiento, 2018a).

The Court decisions were also labelled *inter comunis* (between the commons), paving the way for other individuals and communities in a position similar to the plaintiffs to cite the ruling's guarantees to defend and restore their own rights as tied to nature's integral functioning (Corte Constitucional, 2016). This feature opened the door for a domino effect, giving both courts and civilians new means to confront extractivism using the court system. It is difficult to determine if this feature alone opened the door for other courts to adopt a similar framework in their own decisions. It has, however, certainly played a role in how courts formulate their own decisions in response to similar conflicts (United Nations, n.d.).

New lawsuits seeking to guarantee nature's rights as a means to guarantee contingent human rights continue to cite the Atrato case as precedent. Other communities in Chocó are also in the process of seeking to apply the ruling's guarantees to other rivers in the region (Comité de Seguimiento, 2018b; Comité de Seguimiento, 2019; Tierra Digna, 2019).

As a form of effective resistance to ongoing extractive violence in the Atrato region, little appears to have changed on the ground. However, extrACTIVISM in the form of strategic litigation has provided an effective alternative to taking up arms, and the conceptualization of nature's rights as a tool for strategic litigation remains under active development (Comité de Seguimiento, 2019). In a country that has experienced more than 55 years of armed conflict, this is a significant development. While nature's rights have not been explicitly conceptualized as an avenue for environmental peace, Colombian nature's rights approaches can be viewed within this broader context—seeing nature as a subject of and tool for restitution.

Challenges to Guardianship and Governance

Despite the ruling's stated intentions and spreading influence, the implementation of the Atrato nature's rights governance approach has struggled to meet its stated aims. Since the ruling there have been reports of low levels of compliance with the ruling, and no sanctions of noncompliant parties have been reported. Among these myriad challenges include operationalizing co-guardianship, ongoing violence and armed conflict, inter-institutional coordination, and ongoing conflicts of interest (Comité de Seguimiento, 2018a; Comité de Seguimiento, 2018b; Comité de Seguimiento, 2019; Tierra Digna, 2019; Richardson, 2020).

The court adopted a guardianship mechanism but designed it as a shared role between an elected body of Atrato residents and the state—by contrast, for example, to designating local River Guardians as the exclusive legal guardians of the Atrato River Basin. In this shared arrangement, the Ministry of the Environment and Sustainable Development (the state guardian) was deemed responsible for coordinating inter-institutional collaboration across regions and ensuring the River Guardians' input was central to all plans ordered by the ruling (Corte Constitucional, 2016).

While co-guardianship implies shared responsibilities and powers, some power imbalances have been identified. On the one hand, state agencies receive government funding through established modes of resource allocation—for example, the National Development Plan. On the other, River Guardians are permitted to raise funds through local, national and international means. Moreover, River Guardians have become elected officials without the resources afforded elected officials, including a sustained security presence, though their work is often dangerous due to its relation to armed actors (Corte Constitucional, 2016).

Power imbalances embedded in established governance structures remain significant challenges to the implementation of the Atrato ruling. For example, the incoming president issues a National Development Plan every four years. This plan determines the administration's priorities and economic interests, and it allocates national resources to mobilize this agenda. This process is what determines what resources are distributed, and to which ministries (Aguilar-Støen *et al.*, 2016).

In effect, this resource distribution reflects the power inherent in a Ministry to carry out its defined agenda. Recent National Development Plans have continued to prioritize extractivist interests over environmental protection, allocating more resources to the mining agencies than, for example, the Ministry of Environment and Sustainable Development. Thus, resource allocation influences ministerial capacity to meet environmental goals and uphold the rights of residents, especially those of Afro-descendants and indigenous groups (Morales, 2017; Restrepo Botero and Peña Galeano, 2017; Paz Cardona, 2018; Diaz Parra, 2019; Pardo, 2019). Meanwhile, President Duque continues to open up land to private foreign investors (King and Wherry, 2020), and some recent Court decisions have sought to speed up mining agendas in the territories, bypassing the rights of indigenous groups to *consulta previa* (prior consultation) in collective territories and overturning decisions based on previous decisions stemming from prior consultation retroactively by five years (Corte Constitucional, 2018; Paz Cardona, 2020). A court decision in May 2020 cited the recent coronavirus (COVID-19) pandemic as creating conditions requiring a move to online prior consultations. Anti-extractivist campaigners have claimed that this is yet another move to weaken local rights and further speed up the process of licencing concessions (Observatorio de Derechos Territoriales de los Pueblos Indígenas, 2020).

Human, Nature, and Indigeneity

An essentialist perspective of indigenous culture can be identified in many nature's rights arguments, including the Atrato ruling (Movement Rights et al., 2015). This essentialism has been encouraged by the strategic reduction of identities by indigenous peoples themselves in order to gain recognition of their distinct identities, by similarly reductive human rights and environmental campaigns, and by particular perspectives in more-than-human theory such as *Amerindian perspectivism* (Vivieros de Castro, 2012). At its most narrow, this perspective depicts indigenous peoples as living in a pre-modern society that is in a pact with nature, and that is both at odds with (and seriously threatened by) cultural and territorial encroachment (Tuck and Yang, 2012; Barcan, 2019).

Through the Atrato case, an alliance of so-called "ethnic" Afro-descendant and indigenous conveyed a special relationship with the Atrato River Basin. The River provides them with all of their essential needs, including water for drinking and bathing, food, recreation, and spiritual and cultural needs. They depend on the river for their physical and spiritual sustenance, and have distinct relationships with the river not just as their ancestral territory, but as a "space to reproduce life and recreate culture" (Macpherson, 2019, p. 143). In this way, the claimant communities sought to emphasize that their well-being relies on the integral functioning and ecological health of the Atrato River Basin—and that they have a shared interest in protecting it (Corte Constitucional, 2016; Richardson, 2020).

The Court aimed to respect this special relationship and viewed the Atrato River Basin as an extension of these resident communities—much like the notion suggested by New Zealand's Maori who declared, *I am the river, and the river is me,* as well as other nature's rights legislation which claimed to reflect indigenous cosmologies (such as in Ecuador and Bolivia) (Lalander, 2014; Gudynas, 2015; Magallanes, 2015; Macpherson, 2019). Having adopted this understanding of the river as an extension of the Atrato "ethnic" residents, the Court introduced a conception of *biocultural rights* into Colombian jurisprudence (Bavikatte and Bennett, 2015; Corte Constitucional, 2016).

While clearly well-intended, the Court's application of the hybrid notion of biocultural rights assumes that certain perspectives exist in all indigenous and Afro-descendent Atrato communities. This assumption has positive political potential, but also some potential risks and contradictions (Offen, 2003; McNeish, 2012; Tuck and Yang, 2012; Barcan, 2019; Macpherson, 2019).

The influence of other nature's rights cases (e.g. the Whanganui case in New Zealand in particular) along with the pre-existing Colombia legislative frameworks to protect cultural diversity and biological diversity led the Court to think that a bio-cultural approach would positively afford already categorized "ethnic" communities greater political agency, by offering both nature and culture greater protection (Barcan, 2019; Macpherson, 2019; Richardson, 2020). As a tool, nature's rights when backed in practice by a model for river guardianship, were also thought to provide an additional means to confront harmful extractivist interests in addition to already available legislative tools (Alvarado and Rivas-Ramírez, 2018; Macpherson, 2019).

Indigenous and Afro-descendent organizations in the region and throughout Colombia have without exception been in strong agreement with the Court ruling on the Atrato and the idea of nature as a subject of rights (CRIC, 2017). However, drawing from earlier scholarship on the subject, Offen (2003) suggests this might also create a situation in which respect for ethnic rights becomes contractual—for example, whereby ethnic groups are made responsible for stewarding the land in exchange for territorial rights. Some comparative legal scholars (O'Donnell, 2017; MacPherson, 2019) have, however, questioned whether the formalization of the biocultural rights of the Rio Atrato might overly formalize local communities' use of and access to the river.

Essentialist attitudes toward local indigenous and Afro-descendent communities assume that these communities are only concerned with protection and an ontological connection with the river and not its practical and commercial use. As MacPherson (2019) highlights, this kind of assumption has proven problematic in many contexts, including the case of the Whanganui case in New Zealand. Maori water rights claimants before the Waitangi Tribunal expressed a desire to "walk in two worlds: to resist assimilation and protect their knowledge and law but also to benefit commercially from development" (Macpherson, 2019, p. 223). The way in which indigenous peoples choose to "use" natural resources might not, in fact, coincide with Western notions of indigenous culture.

Although this has so far not been an explicit problem in the Atrato River case, MacPherson asks whether such legal essentialization can cause similar complications to existing territorial claims:

> We know from the legal pluralism literature that when states recognize indigenous rights and interests, there is an inevitable process of translation, accommodation and mediation. Legal personality is a mechanism used to recognise indigenous and tribal relationships and jurisdictions to manage the natural world. However, the indigenous rights are not recognized in their complete form, and are actually limited via the process of recognition. As an example of this, while the Atrato communities' biocultural rights are positioned as being territorial in nature, and although the indigenous and Afro-descendent communities successfully claimed a failure to protect their right to "territory," the Court does not recognize a right to property for the communities in the river, nor for the river to own itself.
>
> *(MacPherson, 2019, p. 156)*

Such a legal reduction also presents a possible loophole through which state authorities might feasibly contest claims to territory in the interest of pursuing ongoing extractive interests in the region, thus contributing to ongoing territorial tensions. Despite outward-facing emphasis on the need to protect the Atrato River Basin and residents and the development of nature's rights as a strategic tool (while also calling into question the impact of "legal" extractive activities), the State's legal ownership over the subsoil continues to permit the State to extract nonrenewable resources for its own interests nationwide. Therefore, it is worth noting that the rhetoric surrounding upholding the rights of ethnic communities and the Atrato remain subject to legal contestation and disregard by authorities with extractive interests.

As is characteristic of reductive understandings of indigeneity and "ethnic" identities, essentialist views might contribute to an erasure of the complexity of indigenous and Afro-descendant communities and their interest in the right to not only protect, but make use of their natural wealth (Bicker *et al.*, 2003; Hooker, 2005; Tuck and Yang, 2012; Ojulari, 2015; Blaser and de la Cadena, 2018; Barcan, 2019; Ramírez, 2019).

Complex Political Ecologies: Governance in Social Minefields

River governance of the Atrato River Basin confronts a complex political ecology. In an August 2020 conversation with John-Andrew McNeish, Ximena González, one of the founders of Tierra Digna and a lawyer behind the Atrato case, commented that it was this complex reality that posed the greatest threat and challenge to the success of the Atrato ruling and the continued work of the river guardians.

Positioned as an "activist" decision by the Court, MacPherson questions why the unorthodox decision has *not* met with significant resistance by the executive government (MacPherson, 2019). Given the initial resistance of the government agencies targeted by the *tutela* to admitting responsibility, the same agencies have formally embraced the decision. There has been no move to nullify the decision, despite their legal ability to do so (Corte Constitucional, 2016; MacPherson, 2019). Macpherson suggests this should raise suspicion among activists and analysts that the decision and guardianship model might be "without teeth," i.e. incapable of deterring the government from its plans for economic development or holding it accountable for the river's protection in the face of its ongoing illegal use.

Gaps between alleged ambitions and effective implementation are observed in other regions claiming to seek guarantees for nature's rights. In both Ecuador and Bolivia, protections for nature's rights and concepts of *buen vivir* remain poorly applied, manipulated by the government and erratically implemented by the courts (Lelander, 2014). The ongoing expansion of extractive frontiers are also observed to have continued in these areas (Bury and Bebbington, 2013; Göbel and Ulloa, 2014; Revelo-Rebolledo, 2019).

A common criticism of the Río Atrato decision is that the model of river guardianship it introduces is overly ambitious, idealistic, and impractical (MacPherson, 2019). For example, effective river guardianship requires communication and collaboration across multiple riverine communities spread over a vast region with varying degrees of mobility. To succeed, river guardianship must operate across a vast and complex topography. The Atrato watershed covers 40,000 sq km and stretches 750 km from the Andes to the Gulf of Urabá on the Caribbean Sea, in which there is a rich but also diverse set of ecosystems. Each group and its members have diverse experiences and interests and, prior to the Atrato decision, many of these groups had limited prior communication. The river guardianship mechanism required collaboration for the first time, generating a unique opportunity for "ecopolitical imagination" at a scale previously unknown (Cagüeñas *et al.*, 2020).

With that said, there are only fourteen official River Guardians (with equal male and female representation), representing select groups. Therefore, many interests and voices might not be part of the conversation (Comité de Seguimiento, 2018a; MacPherson, 2019). Some tensions within the group have emerged regarding conflicting interests. For example, some groups continue to have an interest in traditional mining in their territories, while others are concerned that this could prolong a problematic extractive economy (Cagüeñas *et al.*, 2020). Disagreements and tensions within the group might prevent the formation of a unified vision for representing the Atrato River Basin and its many inhabitants.

Furthermore, each group has a unique relationship with the vast nature of this river, as the river presents itself distinctly across the breadth of the basin. As Cagüeñas, Galindo Orrego, and Rasmussen note:

> Making the Atrato a subject of rights implies telling new stories, weaving new relationships and inventing practices that must arise from a close relationship

with the nature of this river. This represents a challenge for the eco-political imagination, as it requires the creation of translation mechanisms that allow the behavior of all beings that make up the Atrato basin, both human and non-human, to be covered by the legal logic that encourages the sentence.

(Caguenas et al., 2020, p. 171)

Different components of the Atrato River Basin have often-competing interests that must be understood and represented. This task falls squarely onto the River Guardians.

It is worth emphasizing that the *tutela* was filed to confront a demonstrably noncompliant government body, and today compliance with the ruling remains low (Corte Constitucional, 2016). While a ruling implies required actions, disciplinary measures in the event of noncompliance must hold noncompliant actors accountable. Furthermore, to be effective, these disciplinary measures must be proportional to the impact of noncompliance, and these disciplinary measures must be issued in a timely manner to facilitate corrective measures. Early 2018 compliance updates indicate active discussions around appropriate sanctions due to low levels of compliance; however, by the 2019 report, mention of disciplinary action is weak to non-existent (Comité de Seguimiento, 2018a; Comité de Seguimiento, 2018b; Comité de Seguimiento, 2019). To date, there has been no clear indication of sanctions being issued for noncompliance.

The Atrato ruling was introduced in a region that lacks the sustained presence of the national police and security forces, and illegal armed groups have taken advantage of the security vacuum. Therefore, the implementation context is rife with conflict and room for error, risking failure to confront violence and harboring potential to increase it. To a high degree the Atrato ruling operates within what Rodríguez-Gavarito (2010) terms "social minefields."

Writing with a focus on the Colombian government's implementation of prior consultation, Rodríguez-Gavarito suggests that social minefields:

are true social *fields*, characterized by the features of enclave, extractive economies, which include grossly unequal power relations between companies and communities, and a limited state presence. They are *mine*fields because they are highly risky; within this terrain, social relations are fraught with violence, suspicion dominates, and any false step can bring lethal consequences.

(Rodríguez-Garavito, 2010, p. 5)

Reflecting on the particular context of Colombia, Rodríguez-Garavito also observes that these fields of negotiation are also minefields in a very literal sense given that they correspond to territories that are in dispute that are plagued by anti-personnel mines planted by illegal, armed groups as a strategy of war and for obtaining territorial control (Rodríguez-Garavito, 2010). This analytic description can shed light on the context of the Atrato ruling.

Violence and the threat of assassination against individuals involved in the process of confronting illegal mining remains a significant concern in the Atrato watershed, and high levels of confrontation and violence between actors in the region remain high (Comité de Seguimiento, 2018a; Comité de Seguimiento, 2018b; Friedman, 2018; *Redacción Colombia, 2020*, 2019; Tierra Digna, 2020). The defense sector has failed to produce concrete comprehensive plans to eradicate illegal mining. While reports indicate that security forces have "eradicated" some illegal mining machinery along the River Basin (by blowing it up), many machines have been repaired and remain in use. Those that have not been repaired have fallen into the river, causing further ecological damage. Still, indicators for total progress remain unknown and some figures submitted as evidence of compliance were inconsistent (Comité de Seguimiento, 2018a; Comité de Seguimiento, 2018b; Comité de Seguimiento, 2019).

In recent years, Colombia has had the second highest rate of assassinations against human rights and land defenders worldwide, making it an issue of particular concern to governance approaches which seek to achieve human rights and environmental goals (Global Witness, 2019). Human rights and environmental activists, indigenous and Afro-descendent leaders, receive daily threats of assassination by letter or SMS on a daily basis in Colombia. By increasing the visibility of human rights and land (or river) defenders as legal guardians of rights-bearing natural entities, the risk and threat of violence against these guardians will potentially increase (*Redacción Colombia 2020*, 2019). Without significant political will and backing to support these defenders, increased visibility might also become matched with a security and economic deficit, essentially immobilizing and threatening effective action.

Conclusions: Possibility and Pessimism

The Atrato River decision together with the ruling on the Whanganui River in New Zealand, represent significant developments in environmental jurisprudence, inspiring a raft of similar efforts of governance, protection, and extrACTIVISM across the world. Although of clear importance, as we have demonstrated, the existing evidence from the watershed area reveals there has, so far, been little meaningful change in the governance and socio-ecological conditions within the Atrato River Basin. We conclude this chapter observing possibilities but also with a sense of pessimism, given the complexities of the political ecology in which the ruling must function.

The Colombian state is determined to persist with a plan for economic development based largely on the extraction of natural resources despite the adverse socio-ecological impacts and increasing jurisprudence for recognizing nature's rights. Although the Atrato decision on the rights of rivers has garnered significant national and international attention as a novel approach to environmental protection, significant conflicts of interest remain cemented in local and national governance structures. The national extractive agenda continues to accelerate, and illegal armed actors continue to hold significant power throughout the entire watershed of the Atrato River. The River remains a minefield, both social and ecological.

While the Atrato decision has further inspired an international movement to reimagine human-nature relations and become a mechanism of extrACTIVISM, the depth and breadth of local work required to operationalize the eco-political visions and confront the magnitude of the socio-ecological devastation remain daunting barriers to achieving the stated aims of the *tutela* action and court decision. The value of the Atrato approach as an effective life jacket for vulnerable human and non-human natural communities remains in question, owing to a lack of political will, legal loopholes, armed illegal actors, a defense sector that defies legal norms, power imbalances, and a paradigm of governance reliant on expanding extractive frontiers.

An initial examination of the Atrato approach suggests that, although a new eco-political imagination has been activated, nature's rights have yet to crystalize fully in practice as a significantly different approach to environmental governance in the region. While the Atrato River now has formal rights, its health and the reliant interests of Afro-descendant and indigenous groups throughout the watershed remain in grave doubt.

References

Abate, R. (2019) *Climate Change and the Voiceless: Protecting Future Generations, Wildlife, and Natural Resources.* Cambridge: Cambridge University Press.

Aguilar-Støen, M., Toni, F., and Hirsch, C. (2016) 'Forest governance and REDD' in De Castro, F., Hogenboom, B., and Baud, M. (eds.) *Environmental Governance in Latin America.* New York, NY: Palgrave Macmillan.

Alvarado, P.A.A. and Rivas-Ramírez, D. (2018) 'A Milestone in Environmental & Future Generations' Rights Protection: Recent Legal Developments Before the Colombian Supreme Court', *Journal of Environmental Law*, 30 (3), pp. 519–526.

Atapattu, S. and Schapper, A. (2019) *Human Rights and the Environment: Key Issues.* New York, NY: Routledge.

Barcan, R. (2019) 'The campaign for legal personhood for the Great Barrier Reef: Finding political and pedagogical value in the spectacular failure of care', *Nature and Space*, pp. 1–23.

Bavikatte, S.K. and Bennett, T. (2015) 'Community stewardship: The foundation of bio-cultural rights', *Journal of Human Rights & the Environment*, 6 (1), pp. 7–29.

Bicker, A., Ellen, R., and Parkes, P. (eds) (2003) *Indigenous Environmental Knowledge and its Transformations: Critical Anthropological Perspectives.* Amsterdam: Overseas Publishers Association.

Blaser, M. and de la Cadena, M. (2018) 'Pluriverse: Proposals for a World of Many Worlds' in Blaser, M. and de la Cadena, M. (eds) *A World of Many Worlds.* Durham, NC: Duke University Press.

Bugge, H.C. (2013) 'Twelve fundamental challenges in environmental law: An introduction to the concept of rule of law for nature' in Voigt, C. (ed.) *Rule of Law for Nature: New Dimensions and Ideas in Environmental Law.* Cambridge: Cambridge University Press.

Burdon, P.D. (2012) 'A Theory of Earth Jurisprudence', *Australian Journal of Legal Philosophy*, 37, pp. 28–60.

Bury, J. and Bebbington, A. (eds) (2013) *Subterranean Struggles: New Dynamics of Mining, Oil, and Gas in Latin America*. Austin, TX: University of Texas Press.

Bustos, C. and Richardson, W. (2020) 'Nature's Rights in Colombia: An Emerging Jurisprudence' in Zelle, A.R., Wilson, G., Adam, R., and Greene, H.F. (eds.) *Earth Law: Emerging Ecocentric Law-A Guide for Practitioners*. New York, NY: Wolters Kluwer.

Cagüeñas, D., Galindo Orrego, M.I., and Rasmussen, S. (2020) 'El Atrato y sus guardianes: Imaginación ecopolítica para hilar nuevos derechos [The Atrato River and Its Guardians: Ecopolitical Imagination for Weaving New Rights]', *Revista Colombiana de Antropología*, 56 (2).

Chapin, M. (2003) 'A Challenge to Conservationists', *World Watch*. Available at: www.questia.com/magazine/1G1-124444744/a-challenge-to-conservationists.

Comité de Seguimiento. (2018a) Tercer Informe de Seguimiento Sentencia T-622 de 2016 [Third Follow-Up Report Ruling T-622 from 2016].

Comité de Seguimiento. (2018b) Cuarto Informe de Seguimiento Sentencia T-622 de 2016 [Fourth Follow-Up Report Ruling T-622 from 2016].

Comité de Seguimiento. (2019) Quinto Informe de Seguimiento Sentencia T-622 de 2016 [Fifth Follow-Up Report Ruling T-622 from 2016].

Congreso de Colombia. (2011) Ley 1448 de 2011 [Law 1448 of 2011]. Available at: www.unidadvictimas.gov.co/es/ley-1448-de-2011/13653.

Corte Constitucional. (1992) Case T-406/92, Estado Social de Derecho/Juez de tutela [Social Rule of Law/Judge of Tutela]. Available at: www.corteconstitucional.gov.co/relatoria/1992/t-406-92.htm.

Corte Constitucional. (2015) Case T-606/15, La Sala Sexta de Revisión de la Corte Constitucional. Available at: www.corteconstitucional.gov.co/relatoria/2015/t-606-15.htm.

Corte Constitucional. (2016) Case T-622/16, La Sala Sexta de Revisión de la Corte Constitucional. Available at: www.corteconstitucional.gov.co/relatoria/2016/t-622-16.htm.

Corte Constitucional. (2018) Case SU095/18, La Sala Plena de la Corte Constitucional. Available at: www.corteconstitucional.gov.co/relatoria/2018/SU095-18.htm.

Corte Suprema de Justicia. (2018) Case STC4360–2018, Sala de Casación Civil. Available at: http://files.harmonywithnatureun.org/uploads/upload605.pdf.

CRIC. (2017) Corte Constitucional declara río Atrato como sujeto de derechos. Available at: www.cric-colombia.org/portal/corte-constitucional-declara-al-rio-atrato-como-sujeto-de-derechos.

Defensoría del Pueblo. (2014a) Crisis humanitaria en Chocó: Diagnóstico, valoración y acciones de la Defensoría del Pueblo [Humanitarian crisis in Chocó: Diagnostics, assessment and actions of the Ombudsman's Office, Bogotá.

Defensoría del Pueblo. (2014b) Resolución Defensorial No. 064: Crisis humanitaria en el Departamento del Chocó 2014 [Ombudsman Resolution No. 064: Humanitarian crisis in the Department of Chocó 2014]. Available at: www.defensoria.gov.co/es/public/resoluciones/2552/Resolución- Defensorial-064-de-2014-Defensorial.htm.

Delgado-Duque, L. (2017) 'El papel de los grupos ambientalistas contra la minería ilegal en Chocó; más allá del lobby' ['The role of environmental groups against illegal mining in Chocó: beyond the lobby'], *Revista Estrategia Organizacional*, 6 (1).

Desplazada. (2019) N.U.R. 2019-00043-00, República de Colombia Juzgado Tercero de Ejecución de Penas y Medidas de Seguridad. Available at: www.desplazada.co/wp-content/uploads/2019/07/19-07-12-JUZ.-EJECUSION-DE-PENAS-Tut.-2019-00043-00-Rio-Pance-1.pdf.

Diaz Parra, K. (2019) Extractivismo en la brava en el Plan Nacional de Desarrollo del gobierno del Duque [Like it or not, extractivism in the National Development Plan of the Duque government], Semana Sostenible. Available at: https://sostenibilidad.semana.

com/impacto/articulo/extractivismo-a-la-brava-en-el-plan-nacional-de-desarrollo-del-go
bierno-de-duque/44087.

Dunlap, A. (2019) *Renewing Destruction: Wind Energy Development in Oaxaca, Mexico*. New York, NY: Rowman & Littlefield.

Earth Law Center, International Rivers, and RIDH. (2018) Amicus Brief Urges Fundamental Rights for the Anchicaya and All Colombian Rivers. Available at: www.earthla wcenter.org/elc-in-the-news/2018/8/amicus-brief-urges-fundamental-rights-for-the-anc hicay-and-all-colombian-rivers.

Earth Law Center. (2016) 2016 Update: Fighting for Our Shared Future: Protecting Both Human Rights and Nature's Rights. Available at: www.earthlawcenter.org/co-violation s-of-rights/?utm_content=LDF%20tweet%20co-violations%20report.

El Gobernador del Departamento de Nariño. (2019) Decreto No. 348 [Decree No. 348]. Available at: https://servicio.xn--nario-rta.gov.co/DespachoGobernador/Normatividad/a rchivos/Decretos/2019/Decreto-348-2019-07-15.pdf.

Friedman, J. (2018) 'The Only Protection Is God: Negotiating Faith and Violence in Chocó', The Pulitzer Center. Available at: https://pulitzercenter.org/reporting/only-p rotection-god-negotiating-faith-and-violence-choco.

Global Witness. (2019) Enemies of the State: How governments and businesses silence land and environmental defenders. Available at: www.globalwitness.org/en/campaigns/envir onmental-activists/enemies-state/.

Göbel, B. and Ulloa, A. (eds) (2014) *El extractivismo minero en Colombia y América Latina*. Bogota: Biblioteca Abierta.

Gobernación de Boyacá. (2019a) Boyacá Sigue Avanzando [Boyacá Moves Forward]. Available at: http://sedboyaca.gov.co/wp-content/uploads/2020/01/RamiroBarragan_Programa DeGobierno_2020-2023.pdf.

Gobernación de Boyacá. (2019b) Gobernadores de Boyacá y Nariño, reafirmando la fuerza de las regiones, firman pacto para implementación de decretos por la vida [Governors of Boyacá and Nariño, reaffirm the strength of the regions, sign pact to impact decrees for life]. Available at: www.boyaca.gov.co/gobernadores-de-boyaca-y-narino-reafirmando-la -fuerza-de-las-regiones-firman-pacto-para-implementacion-de-decretos-por-la-vida-2.

Gordon, G. (2018) 'Environmental Personhood', *Columbia Journal of Environmental Law*, 43 (1), pp. 49–91.

Gudynas, E. (2015) *Derechos de la Naturaleza: Ética Biocéntrica y Políticas Ambientales*. Buenos Aires: Tinta Limón Ediciones.

Güiza, L. and Aristizabal, J.D. (2013) 'Mercury and gold mining in Colombia: A failed state', *Universitas Scientiarum*, 18 (1), pp. 33–49.

Haraway, D.J. (2017) *Staying with the Trouble: Making Kin in the Chthulucene*. Durham, NC: Duke University Press.

Harvey, D. (2003) *The New Imperialism*. Oxford: Oxford University Press.

Harvey, D. (2005) *A Brief History of Neoliberalism*. Oxford: Oxford University Press.

Holbraad, M. and Pedersen. M. (2017) *The Ontological Turn: An Anthropological Exposition*. Cambridge: Cambridge University Press.

Hooker, J. (2005) 'Indigenous Inclusion/Black Exclusion: Race, Ethnicity and Multicultural Citizenship in Latin America', *Journal of Latin American Studies*, 37 (2), pp. 285–310.

Ingold, T. (2011) *Being Alive: Essays on Movements, Knowledge and Description*. Abingdon: Routledge.

Jurisdicción Especial para la Paz. (2019) Unidad de Investigación y Acusación de la JEP, "Reconoce Como Víctima Silenciosa el Medio Ambiente" [Investigation and Indictment Unit of the JEP, "Recognizes the Environment as a Silent Victim"]. Available at: www.jep. gov.co/SiteAssets/Paginas/UIA/sala-de-prensa/Comunicado%20UIA%20-%20009.pdf.

Kauffman, C.M. and Martin, P.L. (2017) 'Can Rights of Nature Make Development More Sustainable? Why Some Ecuadorian lawsuits Succeed and Others Fail', *World Development*, 92, pp. 130–142.

King, E. and Wherry, S. (2020) 'Colombia's Environmental Crisis Accelerates Under Duque', NACLA. Available at: https://nacla.org/news/2020/04/20/colombia-environmental-crisis-duque.

Lalander, R. (2014) 'Rights of Nature and the Indigenous Peoples in Bolivia and Ecuador: A Straitjacket for Progressive Development Politics?', *Iberoamerican Journal of Development Studies*, 3 (2), pp. 148–172.

Latour, B. and Porter, C. (2017) *Facing Gaia: Eight Essays on the Climatic Regime*. Oxford: Polity Press.

Lerner, S. (2010) *Sacrifice Zones: The Front Lines of Toxic Exposure in the United States*. Cambridge, MA: MIT Press.

Lozada Vargas, J.C. and Congreso de la República de Colombia: Cámara de Representantes. (n.d.) Proyecto de Acto Legislativo "por el cual se modifica el artículo 79 de la Constitución Política de Colombia" [Legislative Act Project "to modify article 79 of the Colombian Constitution"].

Macpherson, E. (2019) 'Rivers as subjects and indigenous water rights in Colombia' in Massoud, M.F., Meierhenrich, J., and Stern, R.E. (eds.) *Indigenous Water Rights in Law and Regulation: Lessons from Comparative Experience*. Cambridge: Cambridge University Press.

Magallanes, C.J.I. (2015) 'Nature as an Ancestor: Two Examples of Legal Personality for Nature in New Zealand', *VertigO*, 22.

McNeish, J.A. (2012) 'More than Beads and Feathers: Resource Extraction and the Indigenous Challenge in Latin America', in H. Haarstad (ed), *New Political Spaces in Latin American Natural Resource Governance: Studies of the Americas*. New York, NY: Palgrave Macmillan.

Morales, L. (2017) 'Peace and Environmental Protection in Colombia: Proposals for Sustainable Rural Development Report', *Inter-American Dialogue*. Available at: www.thedialogue.org/wp-content/uploads/2017/01/Envt-Colombia-Eng_Web-Res_Final-for-web.pdf.

Movement Rights, Indigenous Environmental Network, and Global Exchange. (2015) *Rights of Nature & Mother Earth: Sowing Seeds of Resistance, Love & Change*. Oakland Movement Rights.

Neopolitanos. (n.d.) Aire de Bogotá: sujeto de derechos [Air of Bogotá: subject of rights]. Available at: https://neopolitanos.org/proyectos/aire-de-bogota-sujeto-de-derechos.

Ng'weno, B. (2008) 'Can Ethnicity Replace Race? Afro-Colombians, Indigeneity and the Colombian Multicultural State', *The Journal of Latin American and Caribbean Anthropology*, 12 (2), pp. 414–440.

Observatorio de Derechos Territoriales de los Pueblos Indígenas [Observatory of Territorial Rights of the Indigenous Peoples]. (2020) 'Impactos del Covid-19 en los Derechos Territoriales de los Pueblos Indígenas en Colombia' [Impacts of Covid-19 on the Territorial Rights of Indigenous Peoples in Colombia]. Available at: www.ohchr.org/Documents/Issues/IPeoples/SR/COVID-19/IndigenousCSOs/COLOMBIA_Observator_de_Derechos_Humanoa_y_Secretar%C3%ADa_Técnica_Ind%C3%ADgena.pdf.

Ødemark, J. (2010) 'Timing Indigenous Culture and Religion: Tales of Conversion and Ecological Salvation in the Amazon' in Johnson, G. and Kraft, S.E. (eds.) *Handbook of Indigenous Religion(s)*. Leiden: Brill.

OECD. (2017) 'Due Diligence in Colombia's Gold Supply Chain: Gold Mining in Chocó', Available at: https://mneguidelines.oecd.org/Choco-Colombia-Gold-Baseline-EN.pdf.

Offen, K.H. (2003) 'The Territorial Turn: Making Black Territories in Pacific Colombia', *Journal of Latin American Geography*, 2 (1), pp. 43–73.

Ojulari, E. (2015) 'The social construction of Afro-descendant rights in Colombia' in Contemporary Challenges in Securing Human Rights. London: Institute of Commonwealth Studies.

Pardo, A. (2019) El Plan Nacional de Desarrollo profundiza el modelo extractivista [The National Development Plan deepens the extractivist model], *Razón Pública*. Available at: https://razonp ublica.com/el-plan-nacional-de-desarrollo-profundiza-el-modelo-extractivista.

Paz Cardona, A.J. (2018) 'Colombia's new president faces daunting environmental challenges', Mongabay. Available at: https://news.mongabay.com/2018/08/colombias-new-p resident-faces-daunting-environmental-challenges/.

Paz Cardona, A.J. (2020) 'For Colombia, 2019 was a year of environmental discontent', Mongabay. Available at: https://news.mongabay.com/2020/01/for-colombia-2019-was-a -year-of-environmental-discontent.

Rama Judicial del Poder Público. (2019) Radicadión 63001–2333–000–2019–00024–00, Tribunal Administrativo del Quindío Sala Cuarta de Decisión.

Ramírez, M.C. (2019) 'Militarism on the Colombian Periphery in the Context of Illegality, Counterinsurgency and the Post-Conflict', *Current Anthropology*, 60 (19), pp. S134–S147.

Redacción Colombia 2020. (2019). '"¿Nos van a matar a todos por defender el río Atrato?": Líder social del Chocó ["They're going to kill us for defending the Atrato River?": social leader from Chocó]', *El Espectador.* Available at: www.elespectador.com/colombia2020/pais/nos-va n-matar-todos-por-defender-el-rio-atrato-lider-social-del-choco-articulo-857687/.

Restrepo Botero, D.I. and Peña Galeano, C.A. (2017) 'Territories in Dispute: Tensions between 'Extractivism', Ethnic Rights, Local Governments and the Environment in Bolivia, Colombia, Ecuador and Peru', *Alternative Pathways to Sustainability*, 9, pp.] 269–290.

Revelo-Rebolledo, J. (2019) 'The Political Economy of Amazon Deforestation: Subnational Development And The Uneven Reach Of The Colombian State', PhD Dissertation, University of Pennsylvania.

Richardson, W. (2020) 'Nature's Rights in Colombia: An Exploration of Legal Efforts to Secure Justice for Humans and Nature', Master's thesis, Norwegian University of Life Sciences.

Rodríguez-Gavarito, C. (2010) 'Ethnicity.gov: Global Governance, Indigenous Peoples and the Right to Prior Consultation in Social Minefields', *Indiana Journal of Global Legal Studies*, 18 (1), pp. 263–305.

Rounds, K. (2019) *Why Colombia's deforestation spiked after the FARC's demobilization*, Colombia Reports. Available at: https://colombiareports.com/why-colombias-deforesta tion-spiked-after-the-farcs-demobilization.

Stone, C. (1972) 'Should Trees Have Standing? Towards Legal Rights for Natural Objects', *Southern California Law Review*, 45, pp. 450–501.

Stone, C. (2010) *Should Trees Have Standing? Law, Morality, and the Environment.* Oxford: Oxford University Press.

Tierra Digna. (2019) Risas, Sueños y Lamentos del Río [Laughter, Dreams and Regrets of the River]. Available at: https://tierradigna.net/pdfs/web2019.pdf.

Tsing, A. (2017) *Mushroom at the End of the World: On the Possibility of Life in Capitalist Ruins.* Durham, NC: Duke University Press.

Tubb, D. (2020) *Shifting Livelihoods: Gold Mining and Subsistence in the Chocó, Colombia.* Seattle, WA: University of Washington Press.

Tuck, E. and Yang, K.W. (2012) 'Decolonization is not a metaphor', *Decolonization: Indigeneity, Education & Society*, 1 (1), pp. 1–40.

United Nations. (n.d.) Rights of Nature Law, Policy and Education, Harmony with Nature Law List. Available at: www.harmonywithnatureun.org/rightsofnature.

United Nations Environmental Program. (2019) *Environmental Rule of Law: First Global Report.* Available at: www.unenvironment.org/resources/assessment/environmental-rule-law-first-

global-report#:~:text=NAIROBI%E2%80%94%2024%20January%202019%20%E2%80%9
3%20The,over%20the%20last%20four%20decades.

Veltamayer, H. and Petras, J. (2015) *The New Extractivism: A Post-neoliberal Development Model or Imperialism of the 21st Century?* London: Zed Books.

Vivieros de Castro, E. (2012) *Cosmological Perspectivism in Amazonia and Elsewhere*. Manchester: Journal of Ethnographic Theory.

Voigt, C. (2013) *A Rule of Law for Nature: New Dimensions and Ideas in Environmental Law*. Cambridge: Cambridge University Press.

Whyte, K. (2017) 'Is it colonial déjà vu? Indigenous peoples and climate injustice', in J. Adamson and M. Davis (eds), *Humanities for the Environment: Integrating Knowledge, Forging New Constellations of Practice*. Abingdon: Routledge.

Willow, A.J. (2019) *Understanding ExtrActivism: Culture and Power in Natural Resource Disputes*. Abingdon: Taylor & Francis.

9

EXTRACTIVISM AT YOUR FINGERTIPS

Christopher W. Chagnon, Sophia E. Hagolani-Albov, and Saana Hokkanen

Introduction

The twenty-first century has seen a meteoric rise in the use and availability of technology aimed at individuals, by which we mean technology that is developed and deployed to be used by individual consumers. This technology includes personal computers, smartphones, tablets, and other handheld digital devices. Veiled by entertainment, interpersonal communication, and quick or convenient access to products and knowledge, an underlying and ever-present agenda involves collecting data about the individual using the device. The consumer becomes both the resource for collecting data and the target of the potential uses and abuses of the data collected. In this chapter we explore the infiltration of extractivist logic into the relationship between those providing the digital infrastructure and consumers in the digital realm. Extractivist logics are inextricably bound up with capitalism and other configurations of modernity—and with extractivism comes violence.

There are distinct modes of violence that unfold throughout the digital realm that are directly related to violence perpetrated in natural resource extraction, for example effects of mining lithium for the batteries used in digital devices. By drawing extractivist logic into the digital realm, new forms of violence are unleashed, that are often insidiously indirect and even manifestly unrecognizable, but are no less damaging on the socio-spiritual and physical levels. There are many unknowns in regard to effects or even potential violence that could be perpetrated against individuals when their personal data is accumulated in mass and deployed against them or monetized (Segura and Waisbord, 2019).

In this chapter we contribute an analysis of an ever more complex web of extractivisms. Here different forms of digital and data extractivism are observed to intersect with natural resource and financial extractivisms in their underlying logic and processes. We highlight how this complex web needs to be analyzed in the modern era, to uncover the linkages and extensions of extractivist violence. The extractivist logic continues to expand into arenas where the extent of the infiltration of extractivist modes of operation has only recently been recognized.

Expanding Extractivisms

Not all scholars and activists are in accord with the push to expand understandings of extractivism. For example, Gudynas (2018) maintains that expanding the concept of extractivism beyond the realm of natural resources—to finance, or additional forms of development—is detrimental to the analytical and descriptive power of the concept, and thus undermines the search for alternatives. However, from an historical-ontological perspective the concept of extractivism rests upon a universalizing "natural law" in which the exploitation of "nature" features as an ontological prerequisite to the forms that European modernity developed over the last 500 years (see Chapter 1). As Mezzadra and Neilson (2017) note, new forms of financial and digital processes facilitate the expansion of resource extraction in the global economic system. The digitization of finance and data render these sectors of the global economy dependent on one another in increasingly complex ways. Monetarily, the most significant extractions currently take place on the digital platforms of global financial speculation, largely run by algorithms, through a computerized system with vast violent consequences for the everyday lives and livelihoods of beings around the world. The links to this digital realm and the rise of non-productive capital as the key sectors of capitalist expansion since 1990 are often hard to discern (Dowbor, 2018). What matters here are the logics, mindsets, and ideologies that stem from extractivist ontological dispositions (see Chapter 1), rather than the particular resource or technology. Moore (2018) argues this in his critique of Eco-Marxist theories (e.g. Malm, 2016) that place the most emphasis on coal in the surge of industrial capitalism. Indeed, the existence and prominence of less directly visible or tangible extractivist thrusts behind all sorts of tangible and mindset transformations fit in neatly with Dunlap and Jakobsen's conceptualization of "total extractivism," which is "centered on the deployment of violent technologies aiming at integrating and reconfiguring the earth and absorbing its inhabitants, meanwhile normalizing its logics, apparatuses and subjectivities, as it violently colonizes and pacifies various natures" (2020, p. 6).

This expanded and deepened understanding of extractivism guides attention towards the centrality of extractivist practices and mentalities within the broader modern world-system, and even during prior millennia of empire and civilization-buildings. This conceptualization also uncovers the expansionary and totalizing nature of extractivist thrusts. A central aspect of this global extractivism emphasized by Dunlap and Jakobsen (2020) is the centrality of coercion and social pacification, which enables rolling out and continuation of extractivist practices and the resulting environmental degradation. Violence and militarization are identified as the main mechanisms of coercion and social pacification (Dunlap and Jakobsen, 2020). However, there are types of violence(s) that play out against the human psyche, which are also central to the overarching violences associated with extractivism. In data extractivism, these assaults to the psyche occur through increased exposure to algorithms and programs designed to make users dependent and catch their attention repetitively in digital realms. This results in the parallel process of data extractivism via extraction of knowledge of personal and human tendencies of behavior, and other processes that could be likened to digital colonialism (Thatcher *et al.*, 2016).

As forms of social control, data extractivism and data violence are becoming ever more necessary for extractivism, as they are used to discipline, to convert the subjectivities of people, and to supersede alternative relations between people and their environments. In addition, pro-corporate digital campaigns and resistance campaigning are becoming ever more central in politics, including electoral politics and contentious politics around natural resources (Kröger, 2013; 2020). These sorts of "positive mechanisms" of control (following Foucault, 1978/2007) are integral in social pacification and the creation of docile masses, as they legitimize the continuation of extractivist practices. This subtle aspect of violence, which is especially present in the realm of data extractivism, is crucial as "extractive violence does not always involve armored vehicles, riot police and helicopters" (Dunlap and Jakobsen, 2020, p. 9).

For these reasons, it is important to look at expanded concepts of extractivism to better understand new encroachments that destroy or radically alter lived environments. In this chapter we contemplate the forms of violence that result from the progressively intricate knots that digital technologies weave into different formations of extraction and accumulation. We are sympathetic to the proliferation in the use of the concept of extractivism, as scholars and activists seek to better understand new encroachments by a variety of actors, including: corporations; old and new elites; the multi-billionaires of the digital and financial spheres; progressive governments; actors behind complex investment tools such as churches and pension funds; and even environmental non-governmental organizations engaged in green-grabbing conservation initiatives.

Extractivisms: Digitized and Datafied

The collection, manipulation, and deployment of data are excellent examples of how extractivist processes are useful to describe practices beyond direct natural resource extraction. Data extractivism is a part of a wider self-reinforcing total extractivism that operates at multiple levels within the modern world system, connecting extractivism of natural resources to the extractivism of our thoughts and identity through data (see Figure 9.1).

Before looking at the direct link to natural resources, and the ways extractivism and violence express themselves at different levels of data collection and usage, it is worthwhile briefly to review the terminology. As this is a burgeoning area of study, it is easy to conflate the terms "data" and "digital." As a result, it is important to take a moment to differentiate data collection from other types of digital extractivisms.

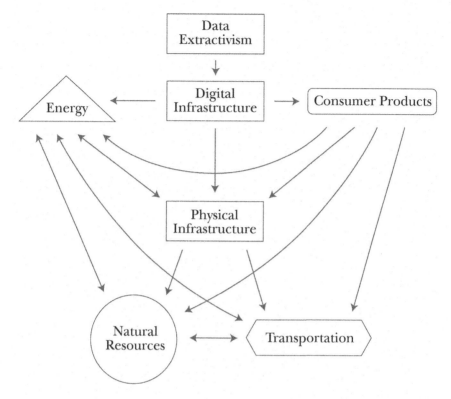

FIGURE 1 This figure illustrates our conceptualization of the web that connects data and natural resource extractivisms. The arrows indicate the lines or directions of dependence; for example Data Extractivism is dependent on Digital Infrastructure.

The definition of 'digital' in the Cambridge English Dictionary is: *using or relating to digital signals and computer technology*, with the business definition adding in: *especially the internet*. This definition can relate to a plethora of tools, spaces, and resources that are quite varied. According to Couldry and Mejias, data is "information flows that pass from human life in all its forms to infrastructure for collection" (2019, p. xiii). So, while data extractivism certainly falls under the umbrella of digital extractivism, they are not synonymous terms. For example, cryptocurrency mining or gold farming are other varieties of digital extractivisms not directly linked to the manufacture and harvesting of personal data (see Table 9.1). Further investigation into types of digital extractivism is beyond the scope of this chapter, as they have very different processes, mechanisms, and outcomes from personal data extraction.

Data extractivism is one of the newest cogs in the self-reinforcing machine of total extractivism (Dunlap and Jakobsen, 2020). It pushes the rationales and destruction of extractivism into our daily lives, as people, their movements, thoughts, and even social connections become the product (Couldry and Mejias, 2019).

TABLE 9.1 Delineating types of digital extractivisms

Type of Extractivism	Object of Extraction	Mode of Extraction	Who Profits
Data Extractivism (Sadowski, 2019)	Personal data	Any sort of internet usage, social media, geotracking, voice pickup, among others. Data points are collected and combined to be packaged and used or sold.	Big tech companies, data brokers, social media, and the companies that produce the infrastructure for data collection.
Gold Farming (Heeks, 2008; Gago and Mezzadra, 2017)	Currency, Items, and Characters in Massively Multiplayer Online Games	Individuals working in a game for extremely long hours to collect resources and level up characters. These resources and characters are then sold directly to people for real money.	A boss, company, or government keeps most of the profits.
Cryptocurrency Mining (Smith, 2019; Rosales, 2019)	Cryptocurrency	Large numbers of energy-intensive computer banks working constantly on extremely complicated algorithms in order to produce cryptocurrency "coins."	Owner(s) of the computer banks. This could be individuals, companies, governments, or other organizations.

Note: There are at least three extractivisms that are digital in nature but utilize extremely different modes of extraction for their respective resources. This is not meant to be exclusive, but rather is a starting point for further exploration.

Data extractivism has a fundamental connection to natural resource extractivism. The growing use of the digital infrastructure for harvesting data, like Google, WeChat, and other social media, drives demand for the physical infrastructure and energy required to utilize those platforms; this, in turn, drives other extractivisms (Dunlap and Jakobsen, 2020). The manufacture of the consumer products relies on the extraction of rare earth elements and other raw materials. In addition, the movement of the parts and finished products across the globe relies largely on fossil-based energy. Beyond the active life of the products needed to interact with the digital infrastructure, there are the issues of the waste, pollution, and human exploitation that attend the disposal of old and broken devices. This is a fundamental, though broad, connection to the violences against the environment, humans, and non-human-beings arising from other forms of resource and energy extraction and extractivism. There is also the material aspect of the ever-expanding physical infrastructure that is needed to keep the digital infrastructure operational (e.g. fiber optic cables, power transmission lines, towers, data farms, and satellites, among the myriad other physical items) and their knock-on impacts on life and the environment (Parks and Starosielski, 2015). As a result, digital infrastructures depend on natural resource extraction, while at the same time natural resource extraction is increasingly driven by the digital—especially data.

Data extractivism involves a type of violence associated with even the most basic collection of data, namely lack of consent. One of the major hallmarks of data extractivism is that there is no meaningful agreement to the harvesting of information. The most common way companies exploit this is the end-user licensing agreement or the Terms of Service of most programs, websites, and devices. These agreements are often designed to be long and difficult to read, and can hide clauses that revoke the rights of users to use or be compensated for their data. The complicated nature of these agreements effectively leaves the user with no power and few alternatives. One must either agree to the terms, and thus give up rights to the data generated by use of the product, or not use the product at all (Sadowski, 2019). This can be seen as a form of accumulation by dispossession, which is based on appropriating resources at zero or very low costs (Harvey, 2003). Couldry and Meijas (2018) even go so far as to suggest parallels between required consent in a website's Terms of Service and the Spanish empire's *Requerimiento*, in which the conquistadors recited an incomprehensible dictum— in the presence of a notary—demanding the acceptance of colonial rule or face violence (see de Vitoria, 2010). Both situations, they argue, require a legally recognized monopoly of force. In the *Requerimiento* it was physical force, whereas with data it is a concentration of economic power, in that, "Whatever the form of force used, its effect now, as then, is through the discursive act that accompanies it to embed subjects inescapably into relations of colonization" (Couldry and Meijas, 2018, p. 341). In this process of legally coerced consent, the conditions for various manifestations of violence are established.

The potential for new forms of extractivist violence is spreading exceptionally fast precisely because these forms are not direct, explicit, or widely recognized. Rather, they are based on a logic of alluring hegemonic expansion wherein the subjects give consent to being targets of extraction, in exchange for using the digital infrastructure, whether it be for work, entertainment, or communication, among the myriad other uses of the digital infrastructure (Van Dijck, 2014). To date, much of the literature on extractivism has overlooked extraction that occurs in the digital realm. This could be due to the notion that extraction is an act that occurs only with and in the material realm, and the digital realm operates apart from the material realm. However, it is convincingly argued that the digital realm and the material realm (or non-digital realm), are in practice, one and the same (see, for example, Horst and Miller, 2012; Pink *et al.*, 2016). In understanding the digital as an extension of the life-world rather than as a separate sphere "out there," the types and possibilities of violence are found to increase in complexity, often becoming obscured or latent, and showing up in ways seemingly far removed from a colloquial conceptualization of the digital.

Moving beyond the ways that infrastructures drive other extractivisms and the ways in which violences are inflicted on the creators of data by simply interacting with the system, data extractivism leads to other socio-environmental damage. There are pollution-like effects on the broader social fabric connected to the way people communicate and how communication is shared in the digital era. Online environments are constructed to a certain extent solely to extract data; for example, social media has been found to be addictive, and former employees of social media companies have claimed they are designed to be addictive (Andreassen *et al.*, 2012; Andersson, 2018; Schwär and Moynihan, 2020). The fundamental design of these digital interactive spaces makes it easier to create an echo chamber and cut out people who disagree with or are different from the user. This turns dissenting voices into faceless "others." Violence is laced in multifarious ways through these processes and while not immediately apparent, it is always immanent. In order to explore these myriad effects and their accompanying violences, it is worthwhile to look at the resources and processes through which data extraction occurs.

Subtle but Violent

The confounding aspect of data extractivism is that a single piece of data is virtually worthless, but the more that pieces of data are combined, the more valuable the data. The products that follow from the data grow exponentially, allowing a new configuration of information (Sadowski, 2019). One of the most common uses of data—and one of the biggest drivers of its potential violences—is the creation of what are called "data doubles." These abstracted versions of people are created using pieces of data collected from one or a variety of sources through a process that Haggerty and Ericson (2000) describe as a surveillant assemblage. An individual will generally have multiple data doubles, each created by different companies and networks, using data both proprietarily extracted and purchased. Although attached to individuals, the use of the data double is not strictly tailored to the individual— instead it is cross-referenced using Artificial Intelligence (AI) with other data doubles to come up with recommendations and ideas based on probability (Couldry and Mejias, 2019). For example, if you search for a video on YouTube, the suggestions for following videos will be based on what data doubles similar to your own search for or click on next.

The pervasive use of this system—and companies' increasing reliance on the system—can lead to a variety of violences. Some are deeply personal, but hard to predict, because they can impact the growth and development of individuals, and impacts could theoretically be greater on younger generations who might grow up more dependent on this technology. This relates to potential loss of autonomy through a greater dependence not only on technology, but also on AI to handle basic tasks even within technology. For example, finding new music by listening to the radio compared with Spotify with custom playlists, or learning about politics or science by talking with different people and going to lectures compared with an infinite list of suggested videos on YouTube. While it is not always obvious in the face of being fed a seemingly endless stream of content, this dependence could hinder the ability to find new things and escape algorithmically created echo chambers. Data doubles can also relate directly to discrimination and violence, such as with the phenomenon of cybervetting, which occurs when companies examine data doubles from individuals as part of a hiring process, including going into personal data unrelated to the position. This has led to some expectations of individuals to discuss, unprompted, past issues which could be discernible from their data double. While some companies hail this technology as a boon for streamlining, the ability to allow for stronger gatekeeping and discrimination based on unrelated activities is clear (Hedenus and Backman, 2017). In this way, the data revolution of past decades has ushered in a new era that permeates different spheres of life, extracting knowledge through an extractivist logic imbued with multiple forms of violence.

The interplay between AI and data doubles gives rise to most of the uses of data in data extractivism. Data doubles, once compiled, are used and referenced by AI as the informational basis for completing tasks. However, different AIs work with different types of data, depending on the task. It should be noted that AI is not inherently nefarious; it depends on the intentions of the people and corporations creating the AI. As a tool of extraction in the accumulation, processing, circulation, and usage of data, AI has resulted in variegated forms of violence, giving rise to concepts like 'data violence' (Hoffmann, 2018) and "algorithmic violence" (Onuoha, 2018). These concepts are related to Galtung's concept of structural violence, wherein social structures and institutions perpetuate a form of violence by preventing people from meeting their fundamental needs (2018). Data and algorithmic violence center around how the algorithms that drive automated AI decision-making can perpetuate and deepen violences such as inequalities, segregation, racism, and sexism. This is not *necessarily* intentional—although it can be—but at the very least it occurs because the people designing the AI have underlying structural biases they are unaware of—or do not have a good grasp of the issues they are programming into the AI—and do not understand the best methods and sources for gathering relevant data.

There are already numerous examples of data and algorithmic violences, whether intentional or unintentional. Eubanks (2018) discusses how the automation of decision-making can impact access to life-saving health and social support, which disproportionately hurts impoverished communities. Safransky (2019) argues that data-driven city planning in "smart cities," brought in to make decision-making seem politically unbiased, has in effect recreated the racially discriminatory practice of redlining and unwittingly enforced informal segregation. There is the example of crime prediction software, which tries to foresee the likelihood of crimes occurring in different places in order to inform police patrols. However, they often use datasets that are of poor quality and racially biased. As such, these measures have not been linked to more efficient policing. Rather they have been linked to racial profiling and police harassment of minorities (Mooney and Baek, 2020).

These violences are not limited to the governmental sphere, but also go into the tools of everyday digital life. Facebook AI has a history of discriminating against Native American users by flagging their names as fake, banning them, and requiring the banned individuals to provide multiple forms of identification to customer service before they are reinstated (Holpuch, 2015). In a gross example, Google AI has projected racism by incorrectly automatically tagging pictures of black people as gorillas (Guynn, 2015). Amazon was found to be using AI to identify impulse buyers and charge them more than non-impulse buyers for the same products, because it was assumed that they were less likely to do research on prices or notice a price hike (Zittrain, 2008). When Amazon's foray into facial recognition AI was turned to photos of members of the U.S. Congress, it misidentified 28 of the congress people as being people from publicly available police mugshots. In this incident, the AI disproportionately misidentified the Black and Latino congress people (Singer, 2018).

Overall, data extractivism has a strong connection with a variety of violences. In the way that it drives other types of extractivism by increasing demand for energy and resources, it drives and exacerbates the violences of those extractivisms. There is violence in the way that companies force data creators to "consent" to their data being extracted, or else be unable to use these vital systems. There is damage and violence in the way that data doubles are used to limit our interactions, opportunities, and choice. There is data/algorithmic violence built into AI that informs our governments and drives our engagement in digital spaces. These violences and depletions are insidious; they grow in impact as technology embeds itself deeper into our lives, and generations begin to grow up with no conception of what life could be like without these intrusions.

Digital Violence IRL

For proponents of limiting the lens of extractivism strictly to natural resources, one of the major criticisms of including the resource of data is that the associated/caused violences are only online and do not spill over IRL (to use the internet parlance, "In Real Life" or the everyday physical world). Although the previous section touched on ways that data/algorithmic violence can easily leap over into physical violence, it is worthwhile to touch on some more concrete examples of the intrusion, manipulation, and literal violence that have grown from the products and methods of data extractivism, as well as the toxic social environment that it creates.

The Chinese context provides some interesting examples, as Chinese companies have been at the forefront of developing and rolling out facial recognition infrastructure and AI (Simonite, 2019). This context provides some of the most famous and extensive examples of how facial recognition technology can spread into many facets of public life. Issues of consent, collection, and usage of data have mixed the digital with the physical world via the usage of facial recognition technology. The people who are having their faces recognized and processed while they walk down the street have little idea of where the data is going, and give no direct consent. There are even government mandated regulations that require facial recognition scans to be able to engage with certain technologies and products, for example signing up for a sim card or internet service (Kuo, 2019). In many workplaces, employees are required to clock in using facial recognition with little or no knowledge of where that data goes (Borak, 2019). Facial recognition can even be used to order and pay for fast food (Hawkins, 2017).

Stepping out from consent, the consequences of facial recognition come into the real world. In some Chinese cities, facial recognition technology has been installed to prevent jaywalking—by effectively doxing, or collecting transgressors' personal information with malicious intent. This is done by using facial recognition technology to project the faces of jaywalkers on billboards as well as showing their pictures, names, and partial identification (ID) numbers on a traffic police website. There is also discussion of expanding the system to automatically text fines to the mobile phones of jaywalkers via social media platforms (Li, 2018). While the thought of official doxing might be unnerving, the case gets far more intrusive and dystopian when looking at the usage of surveillance cameras in Xinjiang (where the Uighur minority group makes up a majority of the population), where facial and ID recognition, as well as mandatory checkpoints, follow people wherever they go. An unsecured database of a surveillance company in the city Urumqi, Xinjiang was found to have facial recognition records and ID scans for 2.5 million of the 3.5 million inhabitants of the city (Buckley and Mozur, 2019). Given the rollout of this level of surveillance via facial recognition and the start of reeducation camps, detaining up to 1 million Uighurs, it is hard to ignore how data can create violence outside the confines of the purely digital realm (Mozur, 2019).

This is not to say that this spillover is a uniquely Chinese issue; it is a global one. Beyond the examples of the previous section, the pervasiveness of the QAnon conspiracy theory and actions inspired by it show how the addictive infrastructure for data extraction and the socially toxic environment it creates can have ramifications outside of the digital realm. This includes in 2016 when a man was inspired by the conspiracy and online echo chambers to drive hundreds of kilometers with an assault rifle, handgun, and knife to a Washington, DC pizza restaurant. His aim was to free victims of left-wing elite child trafficking that the conspiracy said were being held and ordered there; he held people hostage at gunpoint for hours and discovered that there were no secret passages before being arrested (Robb, 2017). We also see U.S. politicians making references to the conspiracy and the spread of the conspiracy to other parts of the world (Stanley-Becker, 2020; Bradley et al., 2020).

Although these are some quick snapshots, there are innumerable examples of how data extractivism and the toxic environment that it creates can cause violence to spill over into the physical realm in a visceral way. These examples are only likely to increase as tools of data extractivism push further into our lives, and the digital and non-digital realms become increasingly—perhaps inextricably—enmeshed.

Conclusion

This chapter has outlined how the lines between realms of extraction have become blurred. As a result, and as the literature cited in this article shows, there is a clear effort by a rising number of scholars to understand the entanglements of datafied and digitized formations of extractivism as they bind with more established notions, processes, and practices of extractivism. There is a need for a deeper and critical analysis of the rich complexities of the interface of natural resource, digital, and intellectual extractivisms to unveil the complex web of extractivisms in this era. This chapter has provided some initial thoughts on the violences manifest in and through these newer configurations of extractivism(s). There is still much ground to cover in utilizing extractivism(s) as a tool to provide systemic understandings of our extractive age, and much additional research needs to be done, but as the other chapters in the volume demonstrate, the conceptual work is already well underway.

References

Andersson, H. (2018) 'Social Media Apps Are "Deliberately" Addictive To User', BBC News. Available at: www.bbc.com/news/technology-44640959.

Andreassen, C., Torsheim, T., Brunborg, G., and Pallesen, S. (2012) 'Development of a Facebook Addiction Scale', *Psychological Reports*, 110 (2), pp. 501–517.

Borak, M. (2019) 'Man Mistaken For His Co-Workers Illustrates The Flaws Of Facial Recognition', *South China Morning Post*. Available at: www.scmp.com/abacus/tech/a rticle/3029424/man-mistaken-his-co-workers-illustrates-flaws-facial-recognition.

Bradley, M., Angerer, C., and Suliman, A. (2020) 'Qanon Supporters Join Thousands At Protest Against German Coronavirus Rules', NBC News. Available at: www.nbcnews. com/news/world/qanon-supporters-join-thousands-protest-against-germany-s-corona virus-rules-n1238783.

Buckley, C. and Mozur, P. (2019) 'How China Uses High-Tech Surveillance To Subdue Minorities', *The New York Times*. Available at: www.nytimes.com/2019/05/22/world/a sia/china-surveillance-xinjiang.htm.

Couldry, N. and Mejias, U. (2018) 'Data Colonialism: Rethinking Big Data's Relation to the Contemporary Subject', *Television & New Media*, 20 (4), pp. 336–349.

Couldry, N. and Mejias, U. (2019) *The Costs Of Connection: How Data Is Colonizing Human Life And Appropriating It For Capitalism*. Stanford, CA: Stanford University Press.

de Vitoria, F. (2010) *Political writings* (A. Pagden and J. Lawrance, eds). Cambridge: Cambridge University Press.

Dowbor, L. (2018) *The Age of Unproductive Capital: New Architectures of Power*. Cambridge: Cambridge Scholars Publishing.

Dunlap, A., and Jakobsen, J. (2020) *The Violent Technologies of Extraction: Political Ecology, Critical Agrarian Studies and the Capitalist Worldeater*. London: Palgrave MacMillan.

Eubanks, V. (2018) *Automating Inequality*. New York, NY: St. Martin's Press.

Foucault, M. (1978/2007) *Security, Territory, Population: Lectures at the Collège de France 1977–78*. London: Palgrave Macmillan.

Gago, V. and Mezzadra, S. (2017) 'A critique of the extractive operations of capital: Toward an expanded concept of extractivism', *Rethinking Marxism*, 29 (4), pp. 574–591.

Gudynas, E. (2018) *'Extractivisms: Tendencies and Consequences'* in Munck, R. and Wise, R.D. (eds.) *Reframing Latin American Development*. New York, NY: Routledge.

Guynn, J. (2015) 'Google Photos Labeled Black People "Gorillas"', Eu.usatoday.com. Available at: https://eu.usatoday.com/story/tech/2015/07/01/google-apologizes-after-p hotos-identify-black-people-as-gorillas/29567465.

Haggerty, K. and Ericson, R. (2000) 'The surveillant assemblage', *British Journal of Sociology*, 51 (4), pp. 605–622.

Harvey, D. (2003) *The New Imperialism*. Oxford: Oxford University Press.

Hawkins, A. (2017) 'KFC China Is Using Facial Recognition Tech To Serve Customers – But Are They Buying It?', *The Guardian*. Available at: www.theguardian.com/technol ogy/2017/jan/11/china-beijing-first-smart-restaurant-kfc-facial-recognition.

Hedenus, A. and Backman, C., (2017) 'Explaining the Data Double: Confessions and Self-Examinations in Job Recruitments', *Surveillance & Society*, 15 (5), pp. 640–654.

Heeks, R. (2008) 'Current Analysis and Future Research Agenda on "Gold Farming": Real-World Production in Developing Countries for the Virtual Economies of Online Games', *Development Informatics Working Paper*, (32), Available at: http://dx.doi.org/10.2139/ssrn. 3477387.

Hoffmann, A. (2018) 'Data Violence And How Bad Engineering Choices Can Damage Society', *Medium*. Available at: https://medium.com/s/story/data-violence-and-how-ba d-engineering-choices-can-damage-society-39e44150e1d4.

Holpuch, A. (2015) 'Facebook Still Suspending Native Americans Over "Real Name" Policy', *The Guardian*. Available at: www.theguardian.com/technology/2015/feb/16/fa cebook-real-name-policy-suspends-native-americans.

Horst, H. A. and Miller, D. (2012) *Digital Anthropology*. London: Berg.

Kröger, M. (2013) *Contentious Agency and Natural Resource Politics*. London: Routledge.

Kröger, M. (2020) *Iron Will: Global Extractivism and Mining Resistance in Brazil and India*. Ann Arbor, MI: University of Michigan Press.

Kuo, L. (2019) 'China Brings In Mandatory Facial Recognition For Mobile Phone Users', *The Guardian*. Available at: www.theguardian.com/world/2019/dec/02/china-bring s-in-mandatory-facial-recognition-for-mobile-phone-users.

Li, T. (2018) 'Just Jaywalked? Check Your Mobile Phone For A Message From Police', *South China Morning Post*. Available at: www.scmp.com/tech/china-tech/article/ 2138960/jaywalkers-under-surveillance-shenzhen-soon-be-punished-text.

Malm, A. (2016) *Fossil Capital: The Rise of Steam Power and the Roots of Global Warming*. London: Verso Books.

Mezzadra, S. and Neilson B. (2017) 'On the multiple frontiers of extraction: Excavating contemporary capitalism', *Cultural Studies*, 31 (2–3), pp. 185–204.

Mooney, T. and Baek, G. (2020) '*Is Artificial Intelligence Making Racial Profiling Worse?*', Cbsnews.com. Available at: www.cbsnews.com/news/artificial-intelligence-racial-profi ling-2-0-cbsn-originals-documentary.

Moore, J.W. (2018) 'The Capitalocene Part II: Accumulation by appropriation and the centrality of unpaid work/energy', *The Journal of Peasant Studies*, 45 (2), pp. 237–279.

Mozur, P. (2019) 'One Month, 500,000 Face Scans: How China Is Using A.I. To Profile A Minority', *The New York Times*. Available at: www.nytimes.com/2019/04/14/technol ogy/china-surveillance-artificial-intelligence-racial-profiling.html.

Onuoha, M. (2018) 'Notes On Algorithmic Violence', GitHub. Available at: https://github. com/MimiOnuoha/On-Algorithmic-Violence.

Parks, L. and Starosielski, N. (2015) *Signal Traffic: Critical Studies of Media Infrastructures*. Champaign, IL: University of Illinois Press.

Pink, S., Ardevol, E., and Lanzeni, D. (eds.) (2016) *Digital Materialities: Design and Anthro-pology*. London: Bloomsbury Academic.

Robb, A., (2017) '*Pizzagate: Anatomy Of A Fake News Scandal*', *Rolling Stone*. Available at: www.rollingstone.com/feature/anatomy-of-a-fake-news-scandal-125877.

Rosales, A. (2019) 'Radical rentierism: Gold mining, cryptocurrency and commodity col-lateralization in Venezuela', *Review of International Political Economy*, 26 (6), pp. 1311–1332.

Sadowski, J. (2019) 'When data is capital: Datafication, accumulation, and extraction', *Big Data & Society*, 6 (1), pp. 1–12.

Safransky, S. (2019) 'Geographies of Algorithmic Violence: Redlining the Smart City', *International Journal of Urban and Regional Research*, 44 (2), pp. 200–218.

Schwär, H. and Moynihan, R. (2020) 'Instagram And Facebook Are Intentionally Conditioning You To Treat Your Phone Like A Drug', *Business Insider*. Available at: www.businessinsider.com/facebook-has-been-deliberately-designed-to-mimic-addictive-painkillers-2018-12?r=US&IR=T.

Segura, M.S. and Waisbord, S. (2019) 'Between data capitalism and data citizenship', *Television & New Media*, 20 (4), pp. 412–419.

Simonite, T. (2019) 'Behind The Rise Of China's Facial-Recognition Giants', Wired. Available at: www.wired.com/story/behind-rise-chinas-facial-recognition-giants.

Singer, N. (2018) 'Amazon's Facial Recognition Wrongly Identifies 28 Lawmakers, A.C.L.U. Says', *The New York Times*. Available at: www.nytimes.com/2018/07/26/technology/amazon-aclu-facial-recognition-congress.html.

Smith, H. (2019) '*The Shady Cryptocurrency Boom On The Post-Soviet Frontier*', Wired Available at: www.wired.com/story/cryptocurrency-boom-post-soviet-frontier.

Stanley-Becker, I. (2020) 'How The Trump Campaign Came To Court Qanon, The Online Conspiracy Movement Identified By The FBI As A Violent Threat', *The Washington Post*. Available at: www.washingtonpost.com/politics/how-the-trump-campaign-came-to-court-qanon-the-online-conspiracy-movement-identified-by-the-fbi-as-a-violent-threat/2020/08/01/dd0ea9b4-d1d4-11ea-9038-af089b63ac21_story.html.

Thatcher, J., O'Sullivan, D., and Mahmoudi, D. (2016) 'Data colonialism through accumulation by dispossession: New metaphors for daily data', *Environment and Planning D: Society and Space*, 34 (6), pp. 990–1006.

Van Dijck, J. (2014) 'Datafication, dataism and dataveillance: Big Data between scientific paradigm and ideology', *Surveillance & Society*, 12 (2), pp. 197–208.

Zittrain, J. (2008) *The Future of The Internet – And How to Stop It*. London: Penguin UK.

10

CARBON REMOVAL AND THE DANGERS OF EXTRACTIVISM

Simon Nicholson

Introduction

Climate change is many things. A predominant but incomplete way to think about climate change is that it is the problem of an overconcentration of greenhouse gases in the atmosphere. Responding to climate change is, from this starting point, about cleaning up energy, transportation, and food systems to limit the emission of additional greenhouse gas molecules.

In recent years, climate models meant to help policymakers to understand the greenhouse gas concentration implications of policy choices have been indicating that emissions reductions alone are likely not enough to keep global warming from crashing through dangerous temperature thresholds. This insight has driven a growing cadre of scientists and policymakers to exhort the need for, in addition to greenhouse gas emissions abatement, the *removal* of carbon dioxide from the atmosphere. Carbon removal (also sometimes termed carbon dioxide removal, greenhouse gas removal, or negative emissions technologies of approaches) refers to a set of current or imagined technologies and land and ocean management practices that could pull vast amounts of carbon dioxide out of the atmosphere. The idea is to take carbon dioxide into long-term storage or direct it to beneficial use, lessening the atmospheric concentration of this most important greenhouse gas (Morrow *et al.*, 2018).

In this chapter I consider how carbon removal has found its way into the international climate response conversation and the dangers inherent in carbon removal done badly. This is a story of the interplay between computer-based climate models and policy development. Some of the discussion below gets a bit technical, looking at the ins and outs of how carbon removal has come to be seen as an essential part of international climate change response based on the workings of climate models. The basic message, though, is straightforward. Computer-based models are a sophisticated tool to help us peer into plausible futures. The models themselves and the uses to which they are put, though, have quirks, some of which have to do with the assumptions upon which they are based. I show below how an *extractivist* orientation is present in the models in ways that shape the results the models display and the policy conversations that the models facilitate. The chapter unpacks how extractivism is at play and how it might be guarded against in future consideration of carbon removal options.

The conceptual leaping-off point for the chapter is, then, "extraction." The term extraction connotes a set of processes whereby materials are taken from the earth and put to human use. Extraction in itself is a neutral notion. All individual members of all species make use of the world around them. Biological systems operate via flows of nutrients. Solar energy is converted into biologically available chemical energy through photosynthesis and the material wastes generated by one individual or species are utilized as inputs by other individuals or species, resulting in great chains of biological connectedness (McDonough and Braungart, 2002). Non-biological systems on earth similarly operate in cycles—think carbon or nitrogen or water cycles—such that the extraction of materials and the passing of materials through different bodies and states is very much the natural order. By this reckoning, the fact that human beings engage in extractive processes is not necessarily a harmful or problematic thing in environmental terms.

"Extractivism," by contrast, connotes a logic. Though extraction can be considered a neutral set of processes and practices, modern-day human beings tend not to extract in the manner of other species. Instead, our economies, societies, and individual lifestyles depend on the extraction and utilization of materials in ways and at scales that distort, disturb, and often destroy natural processes. An extractivist mindset or pervading set of understandings opens the whole world to human exploitation, justifying taking with too little regard for the environmental and social consequences.

Such a logic informs not just traditional extractive processes and industries—mining of minerals and fossil fuels, taking of timber and other products from the world's forests, removal of fish from the world's oceans, and the like. In addition, in these early days of the Anthropocene, the logic of extractivism can be discerned in a much wider variety of human practices. As John-Andrew McNeish and Judith Shapiro point out in this volume's Introduction, an extractivist logic serves to justify the violence of removal and exploitation that are hallmarks of our present hyper-extractive age, noting further that green technologies and other contemporary efforts to tackle environmental ills can themselves depend on the continuance of social exploitation and the contaminating practices of non-renewable extraction.

What, then, of carbon removal? Carbon removal can be viewed, in a positive light, as potentially righting a prior extractive wrong. By this reckoning, carbon removal could operate as a kind of *de*-extraction, allowing the carbon pollution associated with the earlier mining of oil, gas, and coal deposits or the plundering of the world's soils and forests to be pulled from the atmosphere, where in large quantities it poses dangers to all life, and returned to underground reservoirs or placed in other forms of safe storage. This understanding of carbon removal is apparent in what some call carbon removal approaches "negative emissions technologies"—a reversal of prior emissions ills.

I hold the position that carbon removal must now be considered a necessary component of climate change response. I have been convinced, by the scientific assessments that I examine with something of a critical eye in this chapter, that carbon removal must now be a major component of humanity's responding effectively to climate change, and my research and public policy group, the Institute for Carbon Removal Law and Policy, aims to assess and advocate for sustainable approaches to carbon removal. However—if carbon removal is undertaken according to an extractivist logic or if the promise or practice of carbon removal are used to perpetuate reliance on extractivist ways of being, this could have dire consequences. Carbon removal, even if it is countering past extractive undertakings, can be an extractivist activity like any other. Much depends on how carbon removal as an enterprise is understood and how particular forms of carbon removal are ushered into being.

The development and deployment of large-scale carbon removal options is now being touted in mainstream climate change response quarters as essential. Yet as my colleague David Morrow would say, even if carbon removal is important, not all forms of carbon removal are created equal. Some forms of carbon removal utilized in particular ways will be needed and will be beneficial for people and the planet. Others, though, could entrench or perpetuate the very social processes that are driving climate change. Carbon removal, if poorly considered or poorly implemented, could worsen rather than combat the very problem that it is designed to address.

Climate change is about concentrations of greenhouse gases in the atmosphere. It is also, though, about much else besides. It is about the social, political, and economic arrangements that have driven carbon from terrestrial storage into the atmosphere. It is about the systems of belief and understanding that mediate humanity's collective relationship with the other-than-human world. And it must fundamentally be about the environmental and social implications of efforts to respond. Climate change, said differently, is not just a technical problem amenable to straightforward technical answers. A focus on extractivist logics reminds and teaches us that climate change is a complex *social* challenge. Responding to climate change requires the best of us, not just in terms of ingenuity applied to development of technical responses but in terms of how people, other species, and places are attended to in efforts to comprehend climate change and in the actions taken to address it.

The chapter is organized as follows. The next section offers a brief overview of what is meant by carbon removal and how the consideration of carbon removal has gained traction in the space of a handful of recent years. Then, we dig more deeply into the workings of climate models—and particularly integrated assessment models (IAMs)—to look at how an extractivist logic is already at play in the scientific and policy assessment of carbon removal options. We follow by considering what it will take to guard against an extractivist carbon removal. We then provide a brief conclusion.

What is Carbon Removal?

The idea of removing carbon from the atmosphere and putting it into storage has been a part of the international policy consideration of climate action since its earliest days (Carton *et al.*, 2020). Augmentation and protection of forests for the role they play as carbon sinks is captured via the Reducing Emissions from Deforestation and Forest Degradation (REDD+) platform. Soil carbon sinks have long been a focus of land use analyses and were pushed as an important component of climate change response by the French government during its hosting of the 2015 United Nations Framework Convention on Climate Change conference of the parties.

In more recent years, the international carbon removal conversation has moved far beyond consideration of such "natural" carbon repositories. The impetus for an expansion of what is needed from carbon removal and the options available for carbon removal has come from two main quarters. First, a range of recent scientific assessments based principally on computer-based climate models, discussed in more detail below, have suggested the need for staggering amounts of carbon drawdown to prevent the crossing of critical temperature thresholds. Second, new potential opportunities for carbon removal have emerged because of scientific and engineering investigations that posit chemical or technological pathways for the removal and storage of atmospheric carbon.

BOX 10.1 FORMS OF CARBON REMOVAL

There are a wide variety of potential ways to draw carbon dioxide down from the atmosphere and direct it to storage or put it to use. The acronym for this expanding field is CCUS, for "carbon capture, utilization, and storage," indicating a two-step process: 1) Capture carbon dioxide from the atmosphere; and 2) Do something with the captured carbon dioxide to prevent it from reentering the atmosphere.

When it comes to step 1 (the carbon capture step), it is helpful to think of a spectrum of options. At one end of the spectrum is a set of biological pathways to carbon drawdown—carbon dioxide pulled down into growing trees, plant and microbial material in soils, or life growing in oceanic ecosystems. At the other end are chemical or engineered pathways—direct air capture (using a chemical membrane and mechanical systems to separate carbon dioxide directly from the open air) and enhanced mineralization or enhanced weathering (grinding and spreading certain minerals to speed the process by which atmospheric carbon finds its way into rock formations). In the middle, between these two poles, is bioenergy with carbon capture and storage (BECCS). BECCS involves growing biomass as a way to pull carbon dioxide out of the atmosphere, turning the biomass into a liquid fuel or burning it directly, and then capturing the carbon dioxide released during combustion.

For step 2, the carbon storage or utilization step, there is again an array of options. The fully biological approaches store carbon in biological systems, turning carbon into the stuff of trees or seagrasses or soil deposits. There is vast potential for additional carbon storage by these "natural climate solutions," as some call them. Carbon held in biological systems, though, must be monitored and governed closely to avoid return to the atmosphere. Forest fires, farmers deciding to once again plough their fields, or the destruction of coastal ecosystems can all release stored carbon.

Engineered carbon removal approaches and BECCS can store carbon by having the carbon dioxide rendered into a liquid form and then pumping the liquid underground into so-called "legacy" oil and gas wells (places from which some or a majority of oil and gas has already been removed) or natural rock formations. The idea is then to trap the carbon dioxide underground for decades or centuries. Enhanced weathering offers still another route to carbon storage, aiming to hold carbon in solid mineral deposits.

There are also ways to utilize captured carbon dioxide. One company, Carbon Engineering, for instance, is using direct air capture to pull carbon dioxide from the atmosphere and is then converting the captured carbon into a liquid fuel. Across the full lifecycle of the fuel much less carbon dioxide finds its way into the atmosphere per unit of energy. This approach reduces flows of carbon dioxide into the atmosphere but, importantly, does not potentially lessen atmospheric concentrations in the same way as underground or biological storage, meaning that those kinds of uses are not really "negative emissions technologies." Taking carbon dioxide into long-lived products like cement, meanwhile, is a use that is more analogous to long term underground storage methods.

Let us look in turn at the two main drivers of an expanded role for carbon removal in climate response. On the first point—that authoritative reports have been making the case that carbon removal will be essential to avoid dangerous climate futures—a pair of reports from the Intergovernmental Panel on Climate Change (IPCC) are most telling and have had the most impact. The IPCC's Fifth Assessment Report (AR5), released in 2014, indicated, by way of climate modeling, the likely need to draw down billions of tons of carbon dioxide by 2100, ratcheting up rapidly from 2050, to have a good chance of staying beneath a 2°C threshold of warming above pre-industrial averages (Intergovernmental Panel on Climate Change, 2014). Mid-range estimates from the report suggested a need to remove 670 billion metric tons (Gt) of carbon dioxide during this century. The implication is that something like 10 Gt of carbon dioxide or more would need to be removed each year by century's end. To put this into perspective, all current global anthropogenic emissions of carbon dioxide amount to about 40 Gt per year. The scale at which carbon removal is now being contemplated is simply gargantuan.

The more recent IPCC special report on 1.5°C (noting that 1.5°C is the aspirational target from the Paris Agreement) featured a recalculation of the global carbon budget, suggesting a slightly expanded window for climate action and the potential for traditional emissions abatement activities alone to be enough to keep warming below 2°C (Intergovernmental Panel on Climate Change, 2018). Still, the report suggests that the more ambitious 1.5°C target appears, according to the models, to be almost impossible to reach without the utilization of massive amounts of carbon removal.

These kinds of scientific assessments, based on climate models and scenario-based projections of human actions under conditions of climate change, have shifted the conversation about carbon removal. Carbon removal used to hover on the fringes of the climate change response conversation. Biological carbon removal pathways were seen as useful augmentations to efforts to keep greenhouse gases from entering the atmosphere and technological carbon removal approaches were seen as overly costly or unable to be scaled. Now, biological carbon removal and, in a widening set of circles, technological and engineered carbon removal options are being talked about in the same breath as emissions abatement, with carbon removal now being looked at either as a form of climate change mitigation or at the very least as something that will be critically important alongside traditional climate change response activities (Lomax *et al.*, 2015; Waller *et al.*, 2020). This shifting dynamic is captured and was foreshadowed in the Paris Agreement's call, in Article 4, for a balancing of greenhouse emissions from sources and capture by sinks by mid-century.[1]

The climate models have made clear that carbon removal will be needed. Much hinges, then, on *how* carbon removal appears in the climate models and how those models themselves serve as a guide to social and political consideration of climate change and climate change response options. As it stands, representation of carbon removal in climate models and the broader climate modeling universe itself have problematic dimensions that need to be unpacked. The links between the structure of model-informed scientific assessments and an extractivist logic are explored in the section that immediately follows.

On the second point noted above—the fact that scientists and engineers have now posited a wide array of possible modes of carbon removal that go far beyond forest and soil sinks—there are a couple of things to call attention to.

The first is that there is a proliferating array of potential options for large-scale carbon removal. Some, like forests and soil sinks, are reasonably well understood in their basic technical elements, though there are still big estimate ranges even for these more established pathways and much still to learn about whether and how such options can be scaled to provide a meaningful contribution to the levels of carbon removal called for in the models and reports referenced above. Some of the more speculative chemical or engineered forms of carbon response – options like direct air capture with carbon storage or enhanced weathering (also known as enhanced mineralization) (see Box 10.1 above)—seem to offer large potential for carbon removal and storage, but have an array of potential risks or current cost impediments that again makes unclear the potential scalability of the approaches (Fuss *et al.*, 2018).

The second is that all forms of carbon removal currently being discussed would take a good deal of time, effort, and money to develop, and would only operate over very long timescales to bring down atmospheric levels of carbon or to offset the emissions from hard-to-mitigate societal activities. Carbon removal is almost certainly still needed, but I point out these challenges to make clear that carbon removal cannot be seen as any kind of *replacement* for emissions abatement. Crucially, it is also not clear how receptive people would be to the land use changes that, say, massive BECCS operations would entail, or the pipelines needed to move captured carbon dioxide to storage locations that would accompany development of direct air capture facilities, or the potential negative implications for human and environmental health entailed by grinding up and spreading minerals for enhanced weathering. Technical response options are also social and political responses, in that there will be contestation over various pathways and projects, and there are myriad different ways to do something like BECCS of direct air capture, producing differing environmental and social impacts and differing constellations of winners and losers. So while climate models tend to give some time for the bringing online of large-scale carbon removal—the currently accepted wisdom is that carbon removal will have to be scaled up over the second half of the century—there is in fact some urgency associated with the investigation of carbon removal, to see whether or not any of the contemplated options can pan out and to make sure that carbon removal plans proceed along socially and environmentally desirable paths.

The phrase "pan out" obscures a lot. Most of the analyses to this point of carbon removal options have focused on absolute or relative cost and technical potential and have paid too-little attention to social dimensions. In the concluding section of the chapter I will turn back to consider the kinds of analyses of carbon removal needed now to escape the trap of an extractivist logic.

Carbon Removal and the IPCC: Extractivism in the Climate Models

The above section indicated that the mainstreaming of carbon removal has come largely on the back of assessments of climate futures derived from computer models. In this section I consider how carbon removal has emerged from recent authoritative reports and how the computer models themselves operate to shape the consideration of carbon removal alongside other forms of climate change response. A big piece of the story has to do with the structure of the scenarios that undergird modern climate change computer modeling, and especially the assumptions that are captured in those scenarios. What we'll see by way of a shallow dive into climate modeling is the extent to which an extractivist logic is transported into climate models, and particularly the IAMs that are the backbone of climate change projections, via a set of underlying assumptions about societal futures.

Climate modeling is something of an arcane and opaque endeavor. Some of this has to do with the nature of the enterprise itself. The work of climate modeling is technical and highly specialized. This makes it hard for non-experts to discern how climate modelers do their work and, moreover, how to fully and robustly interpret the results of that work. Climate modelers also tend to operate in tight teams around tight projects (although there is much collaboration between and among teams), while facing a set of professional pressures to make their results politically and socially relevant. Yet given the importance of climate models in policy and broader consideration of what climate change is, what it means, and how to respond to it, peeking behind the climate modeling veil is an important exercise.

The focus here is IAMs. They are the backbone of model-based analysis of the social dimensions of climate change. They are "integrated" in the sense that they bring together different modules that represent energy systems, the land sector, typically a simplified representation of the climate system (simplified as compared to massive Earth System Models that aim for much more complete representation of the climate), and other physical and social systems of interest to researchers. The characterization here is a little loose because there are a number of different major IAMs that operate in slightly different ways. Of most interest to what follows is that the major integrated assessment modeling teams have come together around a common scenario architecture, described in more detail below, and operate according to a range of common assumptions.

Scenarios have been basic to consideration of climate response for decades. The kinds of scenarios that are captured by modern IAMs can be called "exploratory scenarios", in that they are designed to indicate in a systematic fashion an array of potential and plausible futures, as a tool for scientific assessment and policy development (Nikoleris *et al.*, 2017). Scenarios are imperfect tools, however. Any scenarios, climate scenarios included, are open to an obvious line of attack: it is really hard to predict the future. Those who work with scenarios seek to guard against such criticisms by asserting that scenarios are images of the future that do not seek to be either *projections* or *forecasts*. The scenarios played out in computer-based simulations are meant to guide interrogation of possible futures, not to act as perfect representations or foretellings of those futures.

However, despite those much-repeated caveats, it is easy to *read* the outputs of IAMs as concrete representations of the future. Model outputs lead to numbers and graphs in authoritative reports stamped with the imprimatur of science, such that when a model output reports, "based on a whole bunch of assumptions about how the world works and how the future might pan out it sure looks like we are going to need a lot of carbon removal," it can look like what the model is conveying is, "carbon removal is now essential for climate change response." The difference is subtle but it is important, particularly given the ways that climate modeling and policy development operate in a kind of paired dance, as described in more detail below.

The state of the mainstream integrated assessment modeling art is informed and is captured in the IPCC's AR5 and Special Report on 1.5°C, along with the scoping work already underway for the sixth assessment report (AR6). I will focus here particularly on AR5. In advance of AR5, a collection of modelers set about determining two different linked scenarios to replace the earlier "SRES" (Special Report on Emissions Scenarios) scenario pathways. The two different elements of this new scenarios architecture are as follows:

1. The "representative concentration pathways" (RCPs) originally spanning four different possible greenhouse gas concentration futures and their concomitant radiative forcings (that is, factors that alter the amount of the sun's energy absorbed by the Earth or the atmosphere); and

2. The "shared socioeconomic pathways" (SSPs) offer five different packets of socioeconomic factors—population, economic growth, education, urbanization, and generalized rates of technological development—indicating how the social world might evolve in the absence of climate policy.

Together, the RCPs and SSPs now constitute an agreed-upon set of starting points for the work of the major IAM modeling teams that provide input into the IPCC processes (van Vuuren *et al.*, 2014). The RCPs offer projections of how greenhouse gas concentrations might grow across this century and what that could mean for average levels of atmospheric warming by century's end. The SSPs then describe a range of different socially defined present and future conditions, with the understanding that each of those futures drives innately towards higher or lower greenhouse gas concentrations even in the absence of policies designed to respond to climate change. The SSPs and RCPs are meant to operate together in the work of the modeling teams, so that a given scenario will show how a particular RCP is generated from a particular SSP. That is, the modeling teams are able to use this architecture to ask, can I get my model to produce a certain amount of warming by the end of the century (the RCP component) based on a particular set of social conditions as a starting point (the SSP component)?

To provide more clarity, let's now bring carbon removal into the picture. The RCP2.6 scenario is the "low radiative forcings (that is, relatively low climate change) by the end of the century" pathway that, in model runs, offers the best chance of keeping atmospheric temperatures beneath 2°C. In order to achieve the RCP2.6 pathway, the vast majority of model runs that were reported in IPCC AR5 utilized carbon removal (or "negative emissions") at vast quantities by mid-century.

If all of that seems a bit technical and arcane, that is because it is. At the same time, it is an immensely important result. It was this finding that drove carbon removal firmly onto the international climate change response agenda. For some critics, though, this was a case of, as Matthew Nisbet has put it, "legitimating the unbelievable" (Nisbet, 2019). Carbon removal options at the kinds of scales called for in AR5 do not exist and might never exist. The model results say that large scale carbon removal is needed, but no such large-scale carbon removal might ever materialize. This has led the IPCC to be accused of inadvertently or willfully justifying rounds of political target-setting followed by inaction. The international community agreed to act to limit warming to 2°C (and more recently, via the Paris Agreement, to aim to limit warming to no more than 1.5°C); the IAMs then showed that the target is possible, so long as there is reliance on an imagined set of possible future technologies and land and oceanic management options. This gives rise to what Oliver Geden has called "magical thinking" out of the IAMs—where the models appear to promise a magical fix for climate change and the policy community willingly reaches for it (Geden, 2015).

An additional dynamic was at play in AR5. In the IAMs only two negative emissions options were present—afforestation/reforestation and BECCS. The upshot is that it was not carbon removal as a big suite of potential options that were shown to be needed by the IAMs, but rather the models could be read as suggesting that massive amounts of afforestation and BECCS would be essential to meeting temperature targets.

BECCS, at the scales called for in the models, if it were to do alone all of the carbon removal work that the models are calling for, would require up to twice the land area of India to be planted in new agricultural feedstocks; would require a new pipeline and transportation infrastructure at least as big as the current global oil, natural gas, and coal infrastructure to move carbon dioxide from point of capture to point of sequestration; and would in a variety of other ways amount to a massive agricultural and industrial undertaking with associated social and environmental impacts (Burns and Nicholson, 2017). The IAMs suggested that large-scale BECCS would be needed. But physics, biology, and the operation of the social world suggest that large-scale BECCS (or, at least, BECCS at the scale suggested in the climate models) is almost certainly not possible.

It was widely reported out of AR5 that BECCS would be needed to save the world (or some variant). The modelers intended no such message through their summarizing of model results. Instead, BECCS was a stand-in in the models for a whole potential suite of carbon removal options. Moreover, BECCS arose from the models out of a set of assumptions that *bake in an extractivist logic*. The assumptions captured in the RCPs and now (for AR6) in the RCP and SSP architecture *assume that the social world is basically going to carry on as it is now*. There are big differences, to be sure, between SSP1 (an imagined "sustainability" world where there has been some embrace of lower consumption pathways and greener tech) and, say, SSP3 (an imagined "regional rivalry" world where countries retreat into nationalistic or bloc positions and there is low international appetite for environmental action) (O'Neill *et al.*, 2017). All of the scenarios, though, turn on quite simple relationships between a very short list of variables. The variables are also conservative in the sense that they do not vary a great deal from the world of today. Even SSP1, the "sustainability" world, has increasing GDP through the end of the century and a set of additional assumptions that make it incompatible with, say, a de-growth reading of sustainability or a sustainability premised on something other than what Jennifer Clapp and Peter Dauvergne have characterized as "market liberal" environmentalism (Clapp and Dauvergne, 2011).

The IAMs, in turn, do not factor in the potential for big sudden changes, for big surprises, for massive positive or destabilizing social, economic, political, or technological shocks. They look, instead, to produce useful representations of plausible futures extrapolated in a linear fashion across a narrow range of variables. This means that when the models report that BECCS is needed at large scales to keep warming beneath 2°C, what the models are really saying is that in their very narrow and flawed representations of the world, the limited toolkit available to the models requires utilization of BECCS to reach the temperature target determined by policymakers.

The previous sentence sounds harsh. It is not meant to. All models are narrow and flawed representations of the world. The IAMs are incredibly useful as one way to explore climate futures. That said, we should always be cognizant that because the IAMs basically work with straight-line extrapolations from the world as it is today, they bake in core features of the present world. They bake in features of the global economy and technological development, for instance, making it appear that these features will be stable and consistent moving forward. Because there is no room for discontinuities or abrupt changes, an extractivist logic is at work in the IAMs, just as it is in the present hyper-extractive real world. What I mean by this is that the IAMs, working now from the RCP + SSP architecture, assume a set of relatively stable relationships that hold the current world intact. In such a world, when the global political economy cannot undergo radical change, when societies cannot experience sudden and far-reaching values shifts, and political power relations are hard and fast, the models struggle to represent worlds with limited temperature increase absent carbon removal.

Escaping an Extractivist Logic

The scenarios utilized by the IPCC and other scientific assessments of climate change are immensely important. They help, as Edward Parson has put it, to "make ... required speculation more disciplined" by making more explicit the underlying assumptions about economic and other social processes (Parson, 2008). However, the mainstreaming of carbon removal into the climate change response conversation points to some ways in which the influence of climate models on policy processes can be problematic.

One thing that has happened with the mainstreaming of carbon removal is that policy incentives have started to be developed in some places to encourage research and development. In the United States, for instance, a number of different federal policy instruments have been advanced, among them the 45Q tax credit and the USE IT Act.

The 45Q tax credit provides an incentive to store captured carbon geologically. As it currently operates, the provision in the U.S. tax code offers $20 per ton (increasing over 10 years to $50 per ton) for carbon pulled from the atmosphere and sequestered in saline aquifers. In addition, and importantly, the provision offers $10 per ton (increasing over 10 years to $35 per ton) for carbon dioxide that is injected underground in support of what is known as "enhanced oil recovery" (EOR). EOR is a practice that allows fossil fuel companies to inject CO_2 into unproductive wells, to retrieve oil and gas that would otherwise be unavailable via traditional methods. The USE IT Act is also meant to drive, though in a somewhat different fashion, the development of carbon capture, utilization, and sequestration (CCUS) activities. The Act, if it becomes law, would provide money for U.S. government research into CCUS and would also support public-private investigation into and development of new CCUS technologies and pipelines to transport captured carbon.

In both instances, these efforts have received bipartisan support, which is a rarity in relation to any feature of contemporary U.S. politics and particularly rare in relation to energy and climate policy. In addition, both efforts have received broad support from a coalition of environmental NGOs and fossil fuel industry groups that work as the Carbon Capture Coalition.

The carbon storage step is essential to any implementation of direct air capture or BECCS technologies, in addition to the potential use of carbon capture methods at sources of fossil fuel combustion (e.g. coal-fired power plants). Those environmental groups that have publicly supported 45Q and the USE IT Act have mostly done so on the grounds that the world will need carbon removal technological systems and that anything that can help to develop those systems should be supported. Yet the fact that such systems are even being considered is because of a failure of humanity to this point to effectively wrestle with the existential challenge posed by climate change and because of the stubbornness and intransigence of existing power structures. Tax credits directed to enhanced oil recovery on the back of modeling efforts that nudge towards the status quo seem the very essence of, well, a status quo response. The very companies that have profited via fossil fuel extraction are now being handed additional money to pull additional hydrocarbons out of the ground, in pursuit of (or perhaps for some under the guise of) working on a key element of the response to climate change.

Guarding against carbon removal extractivism will take concerted work. We need, for one thing, work within the IAMs to better characterize a full range of carbon removal options, and, importantly, there need to be supplementary ways of rigorously exploring alternative futures that go beyond the IAMs. We also need to foster a view of carbon removal that goes beyond the highest level finding that carbon removal is an essential or useful component of climate response. As soon as the idea that carbon removal is essential becomes the norm, then established actors and power structures set about nudging carbon removal in preservation of the existing order. Carbon removal can be a part of many *different* climate responses. Some useful thinking on how to move from what I have been calling here an extractivist carbon removal to a carbon removal attentive to all the myriad needs of just climate policy has been captured in recent work by David Morrow and colleagues (see Box 10.2 below).

BOX 10.2 ONE EARTH PRINCIPLES FOR CONSIDERATION OF CARBON REMOVAL

A recent statement published in the journal *One Earth* does an effective job characterizing a set of high-level principles for an anti-extractivist carbon removal (Morrow et al., 2020). The statement was developed by a diverse group of academics, environmental policy practitioners, and civil society representatives. The four principles they arrived at via a consensus-building process are as follows:[2]

Principle 1: Don't forget the long game. Carbon removal should be thought of as one small piece of an overall response to climate change. The biggest piece of the climate response puzzle will and must continue to be emissions abatement. Carbon removal could be part of a climate response strategy in a variety of ways, from generalized carbon drawdown to compensating for harder-to-abate sectors. Even though there will be important disagreements over what kinds of carbon removal are used in pursuit of what ends, those disagreements do not take away from the need to get started now on carbon removal research, development, and deployment.

Principle 2: It's not all about the carbon. With the above said, it is too easy to get fixated on the carbon drawdown potential of big abstract response options. We cannot lose sight of the social, economic, and environmental implications of carbon removal, and those implications are best seen at the level of individual projects. In addition, notions of equity, desired futures, and transparency must be basic to consideration of carbon removal at all levels.

Principle 3: Split, don't lump. Another way in which a focus on broad categories of carbon removal fails us is by obscuring important differences between different approaches to carbon removal and *within* particular approaches. There are lots of different ways to do direct air capture or afforestation, for instance. Fine-grained analysis of particular options utilized in particular ways in particular locations is the best way to separate the carbon removal good from the carbon removal bad.

Principle 4: Don't bet it all on being right. Carbon removal options might not pan out. Nor, though, should the world count solely on rapid emissions abatement. It is important to begin work on carbon removal options now, not as a way to delay or offset needed work to keep greenhouse gases from entering the atmosphere, but because delaying work now on carbon removal could rob future generations of a needed tool.

Conclusion

Interest in carbon removal is growing. Increasingly, carbon removal options are being considered by scientists and engineers, policymakers, and populations, in response to ever-more dire projections of Earth's future under conditions of anthropogenic climate change. Still, it is impossible yet to say whether carbon removal options will ever be developed at scale or the role that they might play in climate-related actions of the future. Even with such deep uncertainty, though, carbon removal is shaping the climate conversation of today. It is important that the consideration of carbon removal be pushed beyond technical consideration of "how much carbon at what cost" and beyond the boundaries erected by too-narrow readings of climate modeling results towards much deeper examination of the social, environmental, political, and other implications of making carbon removal part of the climate response portfolio, and to avoid having carbon removal be captured irrecoverably by an extractivist logic. Carbon removal that is done well promises to be an important and useful part of humanity's climate change response. Carbon removal done poorly could simply entrench the very social dynamics and power structures that got the world into this mess in the first place.

Notes

1 Article 4(1) of the Paris Agreement reads: "In order to achieve the long-term temperature goal set out in Article 2, Parties aim to reach global peaking of greenhouse gas emissions as soon as possible, recognizing that peaking will take longer for developing country Parties, and to undertake rapid reductions thereafter in accordance with best available science, so as to achieve a balance between anthropogenic emissions by sources and removals by sinks of greenhouse gases in the second half of this century, on the basis of equity, and in the context of sustainable development and efforts to eradicate poverty [emphasis added]."
2 The headings come directly from the original article. I have then trimmed down the explanatory language. Nuance is necessarily lost in such an exercise. I urge the reading of the original statement of principles for a full account.

References

Burns, W. and Nicholson, S. (2017) 'Bioenergy and carbon capture with storage: The prospects and challenges of an emerging climate policy response', *Journal of Environmental Studies and Sciences*, 7 (4), pp. 527–534.

Carton, W., Asiyanbi, A., Beck, S., Buck, H.J., and Lund, J.F. (2020) 'Negative emissions and the long history of carbon removal', *WIREs Climate Change*, p. 671.

Clapp, J. and P. Dauvergne (2011). *Paths to a Green World: The Political Economy of the Global Environment.* Cambridge, MA: MIT Press.

Fuss, S., Lamb, W.F., Callaghan, M.W., Hilaire, J., Creutzig, F. Amann, T., Beringer, T., de Oliveira Garcia, W., Hartmann, J., Khanna, T., Luderer, G., Nemet, G.F., Rogelj, J., Smith, P., Vicente, J.L.V., Wilcox, J., del Mar Zamora Dominguez, M., and Minx, J.C. (2018) 'Negative emissions—Part 2: Costs, potentials and side effects', *Environmental Research Letters*, 13 (6): 063002.

Geden, O. (2015) 'The Dubious Carbon Budget'. *The New York Times.*

Institute for Carbon Removal Law and Policy. (2020). Homepage. Available at: www.am erican.edu/sis/centers/carbon-removal.

Intergovernmental Panel on Climate Change. (2014) *Climate Change 2014: Synthesis Report.* Contribution of Working Groups I, II and III to the Fifth Assessment Report of the Intergovernmental Panel on Climate Change. Geneva, IPCC: 151.

Intergovernmental Panel on Climate Change. (2018) Summary for Policymakers. In: Global Warming of 1.5°C. An IPCC Special Report on the impacts of global warming of 1.5°C above pre-industrial levels and related global greenhouse gas emission pathways, in the context of strengthening the global response to the threat of climate change, sustainable development, and efforts to eradicate poverty. Geneva, World Meteorological Organization: 32.

Lomax, G., Workman, M., Lenton, T., and Shah, N. (2015), 'Reframing the policy approach to greenhouse gas removal technologies', *Energy Policy*, 78, pp. 125–136.

McDonough, W. and M. Braungart (2002). *Cradle to Cradle: Remaking the Way we Make Things.* New York, NY: North Point Press.

Morrow, D.R., Buck, H.J., Burns, W., Nicholson, S., and Turkaly, C. (2018) 'Why Talk About Carbon Removal?', Washington DC, Institute for Carbon Removal Law and Policy.

Morrow, D.R., Thompson, M.S., Anderson, A., Batres, M., Buck, H.J., Dooley, K., Geden, O., Ghosh, A., Low, S., Njamnshi, A., Noël, J., Táíwò, O., Talati, S., and Wilcox, J. (2020) 'Principles for Thinking about Carbon Dioxide Removal in Just Climate Policy', *One Earth*, 3 (2), pp. 150–153.

Nikoleris, A., Stripple, J., and Tenngart, P. (2017) 'Narrating climate futures: Shared socio-economic pathways and literary fiction', *Climatic Change*, 143 (3), pp. 307–319.

Nisbet, M.C. (2019) 'The Carbon Removal Debate: Asking Critical Questions about Climate Change Futures', Washington DC, Institute for Carbon Removal Law and Policy.

O'Neill, B.C., Kriegler, E., Ebi, K.L., Kemp-Benedict, E., Riahi, K., Rothman, D.S., van Ruijven, B.J., van Vuuren, D.P., Birkmann, J., Kok, K., Levy, M., and Solecki, W. (2017) 'The roads ahead: Narratives for shared socioeconomic pathways describing world futures in the 21st century', *Global Environmental Change*, 42, pp. 169–180.

Parson, E.A. (2008), 'Useful Global-Change Scenarios: Current Issues and Challenges', *Environmental Research Letters*, 3 (4), pp. 1–5.

Van Vuuren, D.P., Kriegler, E., O'Neill, B.C., Ebi, K.L., Riahi, K., Carter, T.R., Edmonds, J., Hallegatte, S., Kram, T., Mathur, R., and Winkler, H. (2014), 'A new scenario framework for Climate Change Research: Scenario matrix architecture', *Climatic Change*, 122 (3), pp. 373–386.

Waller, L., Rayner, T., Chilvers, J., Gough, C.A., Lorenzoni, I., Jordan A., and Vaughan, N. (2020) 'Contested framings of greenhouse gas removal and its feasibility: Social and political dimensions', *WIREs Climate Change*, 11 (4), p. 649.

PART 4

Frontier Spaces

11

HYPER-EXTRACTIVISM AND THE GLOBAL OIL ASSEMBLAGE

Visible and Invisible Networks in Frontier Spaces

Michael John Watts[1]

Introduction

James Ferguson famously described African oil zones as "enclaved mineral-rich patches" where "security is provided....by specialized corporations while the... nominal holders of sovereignty...certify the industry's legality...in exchange for a piece of the action" (Ferguson, 2006, p. 204). His model of spatial mercantilism associated with "seeing like an oil company" has always struck me as out of sync with the political economic realities of both the world of oil in particular and extractive industries in general (see Ferguson, 2005). Seeing like an oil company privileges the notion that oil capital satiates its corporate appetite from its oil patches by barely touching down, alighting onto "patches" and operating through a logic of spatial confinement and enclosure. Ferguson radically confines the spaces of oil, as if all that mattered was the wellhead, the concession, or the international oil company's gated residential communities and corporate compounds. Rather, the oil well and the oilfield are planetary phenomena grounded in what Mezzadra and Neilson call the "operations of capital," an immense global assemblage of oil extraction, logistics, finance, and corporate power (Mezzadra and Neilson, 2019). The mine, the wellhead, the oilpatch—all must be deterritorialized or, to use different language, rendered planetary (Labban, 2014; Arboleda, 2020).

To take an African case, the enclaved oil-patch hardly captures the enormity of the hydrocarbon footprint across the oilfields of the Niger Delta, Nigeria, or indeed the larger oil cosmos of which it is part (Omeje, 2006; Adunbi, 2015). Virtually every inch of the region is touched by the industry, directly or indirectly. Over 6,000 wells have been sunk, roughly one well for every ten sq. km quadrant in the core oil-producing states. There are 606 oilfields (360 on shore) and 1,500 "host communities" with some sort of oil or gas facility or oil infrastructure. There are 4,315 km of multi-product pipelines and 7,000 km of crude oil pipelines, mostly owned and operated by a subsidiary of the national oil company, Nigerian National Petroleum Company (NNPC), 22 storage depots, 275 flow stations, ten gas plants, 14 export terminals, four terminal oil jetties, four refineries, and a massive LNG and gas supply complex. NNPC and its joint-venture partners (Shell, Exxon-Mobil, Total, and Eni), independents, and indigenous companies (such as Aiteo and Addax Petroleum) and a raft of related oil service companies directly employ an estimated 100,000 people—a figure that is certainly a con- siderable underestimate. It amounts to, minimally, a 65,000 sq km oil "patch." It bears repeating that what I have glossed over here is simply the logistical and infrastructural footprint of the industry.

The Niger Delta's oil frontier resembles an astonishing spatial patchwork, a quilt of multiple overlapping and intersecting spaces of territorial concessions, blocs, pipelines, risers, rigs, flowstations, and export terminals. Spatial technologies and representations are foundational to the oil industry: seismic devices map the con- tours of reservoirs, and geographic information systems monitor and meter the flows of products within pipelines. Hard rock geology is a science of the vertical, but when harnessed to the marketplace and profitability, it is the map that becomes the instrument of surveillance, control, and rule. The oil and gas industry are a cartographer's dream-space: a landscape of lines, axes, hubs, spokes, nodes, points, blocks, and flows. As a space of flows and connectivity, these spatial oil networks are unevenly visible (often subsurface and virtual) in their operations. A pipeline might run through a village alongside or even through residences and fields, only to disappear when it reaches a river or creek as engineers lay the pipelines into the river channel or into the sea; sometimes complex wellheads—Christmas trees is the professional term—might appear dramatically rising out of the water as if they were some terrifying sea-serpent. The delta is littered with plugged wellheads sitting in the middle of a cleared "pad" (often overgrown and heavily polluted), abandoned and typically oozing oil or hissing quietly (Amunwa, 2011; Amnesty International, 2015; Stakeholder Development Network, 2015).

Abandoned wells point to the larger trauma of serial oil spills dating back to the very origins of the industry in the late 1950s (Watts, 2008). The Nigerian Depart- ment of Petroleum Resources estimates that 1.89 million barrels of petroleum were spilled into the Niger Delta between 1976 and 1996 out of a total of 2.4 million barrels spilled in 4,835 incidents (see United Nations Development Programme, 2006). Data on pipeline malfunction (so-called "vandalizations" and "ruptures") provided by the NNPC for 2005–18 reveal a total of 35,670 incidents, and the volume of "petroleum products" lost over that period was 4,737,046 metric tons (33.7 million barrels) (Nigerian National Petroleum Company, 2008–2018). The recently established federal oil spill monitor agency (the National Oil Spill Detection and Response Agency, NOSDRA) identified a total of 12,628 spill events between 2006 and 2019. It is often said that the Niger Delta experiences the equivalent of an Exxon Valdez spill every year.[2]

As Gavin Bridge says, "the hole is the essential feature of the extractive land- scape, but the hole is just the start" (Bridge, 2015; no page number). The actual footprint of the oil and gas system's enclave and its logistical and other infrastructures in the Niger Delta is just the beginning of a planetary story. Big oil (i.e. national and international oil companies and so-called indigenous operators) is part of a global value chain (Gereffi et al., 2005; Tsing, 2009). At its most capacious and expansive, this extractive assemblage includes a suite of commodity trading-houses, state actors, investment banks, engineering and service companies, shipping, refining, and logistics, including state and private security forces and other forms of surveillance. Critically, the assemblage also includes a heterogenous suite of other actors: oilfield insurgents, militias, local artisanal refiners, criminal organizations, trade unions, non-governmental organizations and advocacy organizations, both local and global (such as Global Witness and Amnesty International), multilateral development institutions, development assistance agencies, and transnational regulatory institutions such as the Extractive Industries Transparency Initiative (EITI) (Appel et al., 2015).

Is this assemblage best understood as an oil patch or perhaps as an enclave? I think neither. The idea of a vast and heterogeneous oil assemblage, replete with diverse actors and agents, exhibiting spatial complexity and the varied forms of territorialization, deterritorialization, and layered sovereignties that it entails, points to a rethinking of contemporary extraction in relation to global capitalism in its various neoliberalized forms. Such a rethinking is what the concept of hyper-extraction is designed to address. In spatial terms—that is to say with a full accounting of the layered and overlapping sovereignties associated with the production and management of a multiplicity of oil and gas spaces—what is on offer is something akin to what Henri Lefebvre calls spatial hyper-complexity: it is a territory (Lefebvre's term) of nested, overlapping, and fissioned spaces (Lefebvre, 2005).[3] The enclave space—perhaps less central than often thought—is but one element of an oil and gas world constantly in the throes of de- and re-territorialization. My chapter endeavors to shed some light on the oil assemblage and its spaces by focusing on the intersections of finance, logistics, and rent (and rentier relations) as forms of value extraction. Empirically, I draw from the Arctic, Nigeria, and Mexico and focus on two entry points into the operations of the assemblage, shedding light on the porous boundaries between the licit and illicit, formal and informal, the visible and the invisible (see also Appel, 2019): the first is oil theft, piracy, and artisan refining (as instances of what I shall call the invisible supply chain), and the second is the world of commodity-trading firms and so-called "first trades" (as an exemplar of the shadow world of global oil markets).

What's Hyper About Hyper-Extractivism?

Hyper-extraction can be construed in a number of related but distinctive ways. One is simply the expanded scale and output—the basic quanta—of resources extracted and consumed. From 1970 to 2017 the annual global extraction of materials[4] grew from 27 billion tons to 92 billion tons, while the annual average material demand grew from seven tons to over twelve tons per capita, an annual average growth of 2.6 percent[5] (roughly twice the rate of population growth) (United Nations Environment Programme, 2019, p. 42). The new millennium ushered in a major increase in global material requirements, which grew at 2.3 percent per year from 1970 to 2000, but accelerated to 3.2 percent per year from 2000 to 2017, driven largely by major investments in infrastructure and increased material living standards in East Asia and the Pacific. While there was a brief slowdown in the growth rate of demand for materials between 2008 and 2010 as a result of the global financial crisis, this has clearly had a limited impact on the overall trajectory.

Over the last century, resource extraction from non-renewable stocks has grown while extraction from renewable stocks has declined as the agricultural economy has contracted in relation to manufacturing (Organization for Economic Co-operation and Development, 2015). Once accounting for some 75 percent of global material extraction, biomass today accounts for less than a third of total extraction. By 2010 non-renewable resource extraction represented over two-thirds of global material extraction, with construction minerals making up over 30 percent, fossil energy 20 percent, and metal and metal ores 13 percent. Fossil fuels—the most traded primary material accounting for half of the global total of 11.6 billion tons of direct physical exports currently—have grown in absolute terms from 6.2 billion tons to 15 billion tons, but their share in global extraction decreased from 23 percent in 1970 to sixteen percent in 2017. Natural gas, conversely, had a growth rate of 2.8 percent average yearly growth, and coal displayed 2.1 percent yearly growth, both in excess of petroleum with a 1.3 percent yearly growth. Global primary materials use is projected to almost double from 89 gigatons (Gt) in 2017 to 167 Gt in 2060 (Organization for Economic Co-operation and Development, 2018).

There are, naturally, other senses of hyper-extraction. One conjures up the speed, intensity, and energy (in the peculiar form of technological innovation) of contemporary extractive systems (Szeman, 2017). The rate and scale of extraction is one attribute of contemporary extraction's hyper qualities—the massive scars and land movement such as those entailed in the Canadian tar sands or Kennecott's Bingham Canyon copper mine—but there is relatedly the degree to which new technologies offer the possibility of enhanced recovery rates, the opening of new frontiers previously foreclosed (fracking is an obvious case), and the deployment of high-tech instruments for discovery, estimation, and surveillance of resources (three-D seismic imaging, for example, in deepwater mining). The very notion of the "digital mine," and the digital transformation of the oil industry, are cases in point (*Mining Review Africa*, 2019). An in-house industry journal puts it this way: "augmented reality, virtual reality, AI, intelligent automation, and the interconnectedness of all devices, hardware, and plant machinery will completely change the face of day-to-day oil and gas operations" (Oil and Gas IQ, 2019). The digital and the virtual point to the sector-specific interfaces between extraction and infrastructure, one expression of which is the ability to move, transform, and refine/process massive quantities of materials at unprecedented speeds for an array of novel end uses. Rare earths and their role in the informatics sector are simply one instance (Klinger, 2018).

Hyper-extraction can also be put to service as a tagline for the capaciousness—the planetary scope and scale—of the extractive supply-chain networks (Bridge, 2008). At stake is not simply the quanta of the commodity extracted but also the density, connectivity, and tensions among different but functionally related supply chains—extractive, manufacturing, finance, logistical—that intersect in extra-ordinarily complex global configurations resembling artist Mark Lombardi's global networks (Lombardi, 2003). In terms of logistical orders and global supply chains, oil and gas are arguably one of the vastest, complex, and securitized of infra-structural and logistical spaces (Cowen, 2010). In a way that other global supply chains are not, oil and gas have (since the early twentieth century) been a textbook illustration of a state-military-industrial-corporate complex. Like other infra-structures, oil and gas logistical systems are *unevenly* visible. They are both private and public (and sometimes hybrid mixes of both) and stand complexly in relation to spatial fixity: Pipelines might be fixed, but semi-submersible rigs are mobile between off-shore fields in between periods of sedentary drilling. It is often said that large-scale technical systems are a system of substrates, invisible until they malfunction; they are taken for granted and to that degree offer up an illusion of freedom. Filip de Boeck says: "[Infrastructures] are mainly present in their absence" (de Boeck, 2012). Deepwater rigs are cases in point: offshore and out of sight. Out of sight, that is, until they are not—as the massive Deepwater Horizon blowout in the Gulf of Mexico revealed.[6] But the question of visibility is largely situational seen through the lens, say, of Americans filling up their gas tanks and in any case is only a partial truth. From another vantage point (on the oil patches or fracking fields), the logistical system is *hyper-visible*—pipelines running through villages, gas flares continuously emitting startlingly harsh illumination, wellheads on farmsteads, villages cheek by jowl with massive liquefied natural gas plants—the system is unavoidable and omnipresent (Larkin, 2013).

As a hyper-extractive assemblage, the oil and gas supply chain and its logistical orders operate less across a frictionless, smooth, monochromatic abstract space (in the sense deployed by Henri Lefebvre in *The Production of Space*) than through a networked mosaic of more or less regulated, more or less ordered, more or less calculable nodes, sites, and spaces. Take for example the oil "frontier." To the petro-geologist, the frontier is a geological province—a large area often of several thousand square kilometers with a common geological history—which becomes a petroleum province when a "working petroleum system" has been discovered. The play, or collection of oil prospects, has its own unique reservoir properties, tem-peratures, flow characteristics, viscosity, and so on. All mapped, calculated and ordered as part of a technological zone (Barry, 2006). But as part of the global supply chain, these plays are often at the margins and fringes, the so-called "liminal spaces" of the unregulated fracking fields of North Dakota or the Nigerian oil fields plagued by violence. All of this is characteristic of frontiers everywhere: namely, the parts of the extractive system marked by the circumvention of infra-structural and administrative grids of the formalized economy. Without these irre-gularities and asymmetries across the supply chain, there would, of course, be no arbitrage. After all, without this linking of, as it were, the ordered and disordered, the licit and illicit, the cores with the interstitial periphery, logistics with "counter-logistics" which constitute the circulatory politics and frictions of contemporary supply chain capitalism, it is not clear what all those financiers, speculators, hedge fund and equity managers would actually do.

Finally, and for this chapter most crucially, there is the meaning of hyper-extraction as expanded, extended, or enhanced extraction. Extraction in this account has become "a generalized feature of capitalism as we know it today" (Ye et al., 2020, p. 171). It draws upon three related but slightly different strands of political economy. One is the move to deterritorialize and render "planetary" the mine, as explicated by Mazen Labban and Martin Arboleda, i.e. the idea that "capitalist urbanization secrets the planetary mine—everyday, above ground, scattered, diffuse, perpetual and swelling" (Labban, 2014, p. 564; see also Arboleda, 2020). Central to the planetary approach is not simply emphasizing scale, and interconnectivity (the city as the "inverted mine") and breaking with methodological nationalism. Rather, it is understanding extraction as a set of shifting dynamic frontiers produced and enmeshed in forms of contemporary racialized capitalism and empire. A second thread is the related work of Sandra Mezzadra and Brett Neilson in their book *The Politics of Operations* (Mezzadra et al., 2017; Mezzadra and Neilson, 2019) Their focus is on the production of multiple edges and frontiers of expanding capitalism, the layered sovereignties and variegated legal spaces of global capital, and the new spatial and temporal complexities of capitalism associated with capital's circulation and colonization of social life, or what they call the politics of operations. In particular, it is the operations of a trifecta of "sectors" and their connections that provide the core entry point: extraction, logistics, and finance.[7]

The final approach to the notion of an enlarged extraction requires a little more elaboration. I shall refer to it as extractive rents, a body of work that has collectively addressed the question of contemporary capitalism and "rule by rentiers" (Piketty, 2014; Standing, 2016; Mazzucato, 2018). Not surprisingly, financial rentiers, which is to say firms engaged primarily in financial activities and earning revenue primarily through the ownership and exploitation of financial assets, have been in the spotlight, the principal agents of what has come to be seen as the dominance of Wall Street and finance capital. As a form of critique, rents are seen as "unearned" (rather than productive as a source of accumulation). Owners of land, mineral resources, intellectual property, and a panoply of other income-generating financial and non-financial assets are seen to exercise a sort of hegemony within a neoliberalized and financialized capitalism. When economists refer to a rent-seeking political economy, they typically invoke a lack of market competition and see the source of rent as state intervention or restrictions on economic activity. Others see rent as any income derived from ownership, possession or control of assets (including financial assets) that are scarce or artificially rendered scarce. Implicit in differing explications of rent—all too complex to enter into here (see Christophers, 2019)—is the notion of both monopoly power not only of ownership or control but also in the marketplace. In this sense rent is income derived from the ownership, possession, or control of scarce assets under conditions of limited or no competition (2019).

Central to the rentier world is the determination and distribution of property rights that are not deployed to produce new commodities but rather to extract value via rent (what has been called "value" grabbing through "pseudo-commodities," see Andreucci *et al.*, 2017). There is, to take the idea of a planetary extractive system, an expanding class of rentiers operating in the interstices of, for example, the multiple agents in the oil and gas assemblage (financiers, commodity traders, oil insurgents, politicians, military, corporations and so on) who profit without producing (Harvey, 2007; Lapavitsas, 2009). Rent-bearing assets—how they are created, their opportunities to extract value, and conflicts and struggles over the property rights that underlie them—are pivotal to contemporary capitalism, and to extraction in particular. The state figures centrally in rents for a trio of reasons: it customarily creates and institutes property rights, it typically regulates, enforces and legitimates the distribution of rights and titles and their use, and not least—and this is especially so in oil state where mineral rights are nationalized—it is itself a landlord or acts like a landlord (Hausmann, 1981; Schmitt, 2003). But these rights might also inhere in international law or through the operations of multilateral development institutions. Either way, "the proliferation of private property relations over everything imaginable significantly expands the terrain for rent extraction and related struggles" (Andreucci *et al.*, 2017, p. 38).[8]

Rents (and the rentier state) have been a staple in the diet of extractive analysis for many decades (Mommer, 1990; Hertog, 2010) But planetary extraction, and the dominant forms of neoliberalized finance capital associated with it, point to the importance of the massive proliferation of rents and rent opportunities—"value grabbing"—within the operations of the oil and gas assemblage. This is no longer solely a product of corrupt rent-seeking petro-states but operates across multiple spaces and sectors, across the licit and illicit, and among cores and frontiers, a development which highlights the blurring of conventional borders in thinking about the global political economy of extraction (Ye *et al.*, 2020). One of the purposes of this chapter is to elucidate the vast proliferation of rents in extractive economies understood at the planetary levels and show how these rents blur distinctions between legal and legal, formal and informal, state and civil society, boundaries, and frontiers.

The Digital Arctic: Deepwater Oil as a Hyper-Extractive System

A vignette. On August 2, 2007 a Russian submarine carrying two parliamentarians planted a titanium flag two miles beneath the North Pole. At stake were lucrative new oil and gas fields—by some estimations 10 billion tons of oil equivalent—on the Arctic sea floor. A decade later in December 2017, the U.S. National Oceanic and Atmospheric Administration (NOAA)—significantly, an arm of the U.S. Department of Commerce—released a report proclaiming a "New Arctic," signaling massive, irreversible changes in the material composition of the Arctic

Ocean and its peripheries.[9] A world of forbidding sea ice is now re-construed through the lens of runaway melt, thaw, liquefaction, and off-gassing and a *new ocean* emerges demanding to be observed, represented, documented, exploited, and policed at multiple scales. Confronting new systems of global oceanic and atmospheric circulation, a vast constellation of satellites, drones, buoys, cables, supercomputers, servers, and sensors will give form to the New Arctic, a "digital ocean" whose geo-economic and geostrategic value rests on forms of legibility and computational calculation. A liquid Arctic is both a knowledge and infrastructural frontier, calling into being new forms of "environmental intelligence" (EI) and logistical orders of extraction, circulation, and securitization. All of this is in the service of a new frontier of accumulation, a so-called "trillion-dollar ocean." What is at stake is building a logistics space for the Anthropocene. As Kalvin Henely (2012) put it: "if you think of Wall Street as capitalism's symbolic headquarters, ... the sea is capitalism's trading floor writ large."

One part of this digital Arctic story concerns resources, especially but not exclusively oil and gas. Deepwater oil and gas production in the Arctic (and elsewhere) is nothing new of course; the logistical and infrastructural investments in the oil and gas global supply chain have already left their profound footprints not simply on the ocean floor but in and through the oceanic world in the form of pipelines, flow-stations, risers, rigs, tankers, tank-farms, gas flaring vents, semi-submersible rigs, blowout preventers, and so on.[10] It is now commonplace for test wells to delve through 7,000 feet and more of maritime waters and 30,000 feet of sea floor to tap oil in tertiary rock laid down 60 million years ago. A single test well might cost over $250 million. A great deep-water land grab is under way: primitive accumulation at significant depths. Warming wrought by global climate change has opened Arctic prospects containing an estimated one-eighth of the world's remaining oil and a quarter of its gas (according to the U.S. Geological Survey). But the arrival of peak oil has triggered increasingly high-risk techniques and geographies of extraction, especially in deepwater and the extreme environments of the Arctic now amplified under conditions of climate change.[11] NOAA has adopted Environmental Intelligence—rebranding itself as America's environmental intelligence agency—to mold the New Arctic policy narrative as a security concern through the problem of data production, management, and deployment. Adapted from long-standing military-scientific techniques of geographic, meteorological, and otherwise geophysical knowledge production, EI frames the New Arctic through an established military-industrial-academic complex operating at many levels—structural, logistical, and infrastructural.

What distinguishes the contemporary variant of EI, however, is the addition of speculative finance capital and its logics of risk (Arroyo, in progress). By changing the risk landscape, EI becomes a strategic domain of value that maps out possible scenarios and multiplies speculative opportunities by trafficking in New Arctic futures. Environmental Intelligence asserts the ascendancy of geospatial data in the valuation and evaluation of risky uncertain futures as a space of economic and political securitization—it is a sort of "emerging market." It makes use of the vast resources of Silicon Valley rather than the secret state technologies and military satellites, ships, and other sensing platforms typical of Cold War-era big science. Bay Area firms focus on small, automated, cheap systems—from Saildrone's unmanned solar and sail-equipped sensor packages to Planet Labs' CubeSat swarms—to produce data that is market ready for just-in-time maritime logistics, everywhere-war security operations, and for the extractive sector. The idea of a new Arctic Ocean endeavors to map a space of the yet-to-be observed, represented, exploited, and policed, at multiple spatial and temporal scales (see Mason, in press), as an epistemic object and a logistical order in the m making, expanding the means by which the region's strategic worth is evaluated. NOAA's coinage of the New Arctic might appear to be a predominantly American techno-political project. But it is a supranational enterprise as important to Norway or Russia as it is to China or Canada.

But data collection is the leading edge of finance capital and state-led investment. As NOAA was rebranding itself, Guggenheim Investment Partners LLC, a New York firm, offered the first Arctic-specific investment portfolio, while China published a comprehensive Arctic strategy for a Polar Silk Road. The U.S. Defense Advanced Research Project Agency seeks to deploy sensor networks of floatation devices for real-time maritime monitoring in an Ocean of Things, while U.S. defense contractor and ocean technology startup Liquid Robotics, a Boeing subsidiary, has outlined its vision for a digital ocean. The Arctic mineral and energy frontier are thus what Alexander Arroyo calls a "geography of speculation" (Arroyo, 2020).

Oceanic oil and the digital Arctic reveal how the concept of hyper-extraction offers a sort of full-screen technicolor picture of the twenty-first century extractive political economy. It points to a planetary oil and gas assemblage in which the politics of operations on the ground encompass extraction, logistics, technology, and finance. In rendering the wellhead "planetary," it offers an important reckoning: extraction is less an old-world nineteenth century industry rooted in classical imperialism than a leading edge of contemporary capitalism ceaselessly searching for new frontiers of real and formal subsumption of nature (Murray, 2004; Boyd *et al.*, 2001).

The Planetary Well: Oil Theft and Illicit Capitalism

On April 18⁻19,, 2018, a global conference, Oil and Fuel Theft 2018, was held in Geneva. Building upon the work of the Atlantic Council, the conference aimed to forge a global network of stakeholders in order to share information, expertise, and other mutual support in taking on "a worldwide threat to security and prosperity." Oil and Fuel Theft 2018 drew 140 attendees from around the world, including the leadership of national oil companies (NOCs) from Iraq, Libya, Mexico, Ghana, and Uganda, as well as government delegations from the USA and the Philippines and multilateral organizations such as the World Customs Organization and the International Maritime Organization, international oil and service companies, and other corporate actors such as Dow Chemical. Among the offerings was striking testimony by General Mahmound al-Bayati, Director-General Counter-Terrorism and National Security Advisor for the Republic of Iraq, who outlined the history and genesis of how large-scale oil and fuel smuggling took root and in his country—coming to light in the infamous corrupt Oil-for-Food Program between 1995 and 2003—including the dynamics of oil smuggling for profit by Islamic State in Iraq and Syria (Vienneast, 2016; Tichý, 2019). But the Islamic State and its oil investments are simply the tip of an iceberg, and these patterns are repeated the world over.

The Atlantic Council's three recent reports—*Downstream Oil Theft: Global Modalities, Trends, and Remedies; Downstream Oil Theft: Implications and Next Steps*; and *Oil on the Water: Illicit Hydrocarbons Activity in the Maritime Domain*—offered the first comprehensive picture of global hydrocarbon crime. The scale of the illicit oil economy is mind-boggling. Globally, it is estimated that $133 billion worth of oil and fuel annually is stolen, adulterated, or fraudulently transferred at some point in its supply chain, an estimate that includes only refined (and not crude oil) products. But this figure is a massive underestimate, as it does not include the sorts of losses associated with fraudulent oil trading contracts or oil revenues unaccounted for or "lost" through public financial institutions in oil-states like Venezuela or Nigeria. Liquefied gas is also stolen and illicitly traded. Crucially, oil theft is not simply the preserve of petro-states in the Global South marked by "poor governance." In the European Union, revenue loss caused by theft of oil and fuel is estimated to be worth €4 billion; the illicit cross-border trade in oil between Mexico and the United States involving not just Mexican cartels but American trading houses and oil companies is a multi-billion dollar business network (Reinhart, 2014; Jones and Sullivan, 2019). This illicit money machine not only turns on organized criminal gangs, terrorist groups, and insurgents but on corrupt public officials and security forces, offshore financial centers, and the global oil leviathan. What is on offer is a sort of global oil mafia operating in the interstices of the oil and gas global value chain.

Oil theft points to a larger systemic and structural pathology within the vast oil and gas complex—according to market research by IBISWorld the total revenues for the oil and gas drilling sector came to approximately $3.3 trillion in 2019, roughly four percent of global GDP[12]—namely, endemic corruption and illicit financial flows (Organisation for Economic Co-operation and Development, 2016). In resource-rich post-colonial states, somewhere between 25 percent and 55 percent of global capital flows could be illicit. It is widely acknowledged that illicit finance capital is deeply enmeshed with international crime networks (narcotics, arms, smuggling) and illicit commercial practices like tax and pricing fraud. Twenty percent of the 242 enforcement actions under the U.S. Foreign Corruption Practices Act came from the extractives sector—by far the highest for any industry, while of the 427 foreign bribery actions examined in a 2014 Organisation for Economic Co-operation and Development (OECD) report, twenty percent were lodged in the extractive sector (Organisation for Economic Co-operation and Development, 2014; Foreign Corrupt Practices Act Clearing House, 2018).

The scale of illicit financial flows (IFF) in extractive economies across the Global South is gargantuan. According to Global Financial Integrity, the real normalized cumulative IFFs from Sub-Saharan Africa (SSA) between 1980 and 2009 amounted to $846 billion (over $40 billion per year in the 2000s) (Global Financial Integrity, 2013); UNECA, 2018). Net recorded outflows from West and Central Africa—and from the trio of oil producers, Nigeria, Congo, and Angola—swamped recorded transfers into other regions over the decade ending 2009. Oil and gas exports accounted for over 55 percent of all IFFs in SSA during the same period. Data from the Brookings Institution estimate that between 1980 and 2018 SSA received nearly $2 trillion in foreign direct investment and official development assistance, but produced over $1 trillion in illicit financial flows: four of the top seven IFF African producers of illicit flows 1980–2018 (totaling almost $200 billion) are oil producers[13] (Signé et al., 2020).

Oil and gas provide the richest of soils for IFF risk. State control of the industry in producer states is widespread and provides a massive hunting ground for rents on the part of the political, military, and business classes. The global supply chain is deeply financialized, not only in the investment required for exploration and production but also and especially in the trading system, a domain marked by opacity. OECD's typology of corruption risks across the extractive sector analyzed 131 corruption cases, including oil and gas, and noted that corruption risks might arise at any point in the extractive value chain (Organisation for Economic Co-operation and Development, 2016). The award of mineral, oil, and gas rights, and the regulation and management of operations accounted for almost 75 percent of all cases, and involved bribery of foreign officials, embezzlement, misappropriation, and diversion of public funds, abuse of office, trading in influence, favouritism, and extortion, bribery of domestic officials and facilitation payments. Large-scale, so-called "grand," corruption involving high-level public officials is widely associated with the award of mineral and oil and gas rights, procurement of goods and services, commodity trading, revenue management through natural resource funds, and public spending. Sophisticated vehicles for channeling illegal payments, disguised through a series of offshore transactions and complex layers of corporate structures often involving shell companies, are recurrent features of the oil and gas sector landscape.

Perhaps no country on earth is more closely associated with large scale oil theft ("bunkering") than Nigeria (though Mexico, Iraq, and Russia follow close behind).[14] The scale and costs of hydrocarbon crime in Nigeria are notoriously difficult to quantify because of the multiplicity of points where oil in its various expressions (crude, kerosene, refined petroleum, oil revenues) is stolen, but also because Nigerian state and regulatory authorities, as well as corporate actors, lack consistent and accurate metrics. In fact, the commonly expected global standards for measuring and metering across the national supply chain are weak or absent. Estimates of crude oil and fuel stolen and revenues lost vary, often widely, as indeed does the data on pipeline sabotage and attacks. Nigeria lost approximately 204 million barrels, valued at 4.57 trillion naira (roughly $18 billion), to oil theft in the four years between 2015 and 2019, according to estimates by the Nigeria Natural Resource Charter (NNRC); that is to say, the Federal Government lost approximately 43 percent of its revenue to oil theft over four years (Nigeria Natural Resource Charter, 2018; Nasir, 2020; *Nigeria Business News*, 2020). Nigerian EITI estimated oil theft at $42 billion between 2009 and 2018 (*Nigeria Bulletin*, 2020). According to international oil company figures, Chevron, Shell, and Nigerian Agip Oil Company lost $11 billion between 2009 and 2011 owing to theft and sabotage. The NNPC, which is the parastatal charged with management of the industry, spent $2.3 billion on pipeline repairs and security from 2010 to 2012 and almost $100,000 million in the first quarter of 2019 alone. By some estimates, 500,000 people are employed in the theft business, broadly defined.[15]

Illegal bunkering of Nigerian crude oil originated in the 1960s in part during the Biafran civil war (1967–70) but subsequently expanded under military rule when top army and navy officers began stealing oil—or allowing others to steal it—to enrich themselves and maintain political stability while also busting tight OPEC quotas. Local and foreign intermediaries did much of the legwork—Lebanese and Greek enablers loomed large—but the scale was small, perhaps a few thousand barrels per day. According to some reports (Katsouris and Sayne, 2013), lower global oil prices and Nigerian output, combined with the relatively closed group of actors involved, helped contain the business. Growing involvement by the Nigerian security forces after military rule ended in 1999 and active involvement by Nigerian political and business classes (so-called political "Godfathers," well-placed political party members and high-ranking civil servants) all pointed to the rise of a well-organized "oil mafia." As militancy arose across the oilfields in the 2000s a new set of actors—insurgents, armed criminal and youth groups, local chiefs and political operatives—muscled their way into the oil theft business. All of this pointed to an oil black market of a systemic sort: the illicit capture of various assets and rents in the oil system, shady oil contracts and licenses, opaque "oil swaps," fraudulent trading deals, and the pillaging of public revenues (derived from oil) that course through the country's fiscal federal system (Watts, 2015; 2018).

Pipelines, Taps, and Topping up: The Illicit Life of a Barrel of Oil

Let's start with a wellhead in the Niger Delta, Nigeria. Nembe Creek Well 7, behind Mile 1 Community in Bayelsa State, feeds into the 97 km pipeline, Nembe Creek Trunk Line (NCTL). The trunk line is one of Nigeria's major oil transportation arteries that evacuates crude from the onshore fields to the Atlantic coast for export. Owned by Aiteo Group, NCTL was recently purchased from Shell Petroleum Development Company (SPDC) as part of the related facilities of the prolific oil block OML 29. NCTL's construction commenced in 2006 and was finally commissioned in 2010 at the cost of $1.1 billion. Billed as a replacement to the ageing and often vandalized Nembe Creek Pipeline which had suffered significant losses due to incessant fires, sabotage, and theft, SPDC made use of the pipeline to transport crude oil from the OML 29 starting at Nembe Creek to a manifold at the Cawthorne Channel field on OML 18, and finally to the Bonny Island oil terminal for export (and for liquefied natural gas). In December 2011, barely one year after the line was commissioned, the pipeline was shut down for one month to repair leaks caused by crude thieves. In early 2012 Shell claimed that crude oil valued at $16 million (over 60,000 barrels per day) was being stolen *daily* from the NCTL (Clark, 2014). Two short sections of pipelines in Brass and Nembe local government areas had over 600 attacks (theft, sabotage, operational failures) between 2006 and 2019, and over 200 oil theft events in a two-year period (2012–14) (Whanda *et al.*, 2017; Ngada, 2018). Within a year of opening, Shell discovered 17 bunkering (theft) spots along 3.8 km of the pipeline (Ogunde, 2012). The NCTL pipeline is arguably one of the most attacked, sabotaged and compromised pipelines on the face of the earth.[16] Typically—that is to say when it is not shut in by *force majeure*—the pipeline seems to be losing more oil than it is transporting.

In national terms, the collective assault on the integrity of Nigeria's pipelines is staggering: according to government data (the accuracy of which is open to question), in the oil producing Niger Delta alone between 2006 and 2019 there were over 12,000 spill events, 75 percent of which were located in three states, Bayelsa, Rivers, and Delta, and over 35,000 pipeline "incidents" (National Oil Spill Detection and Response Agency, 2020; Nigerian National Petroleum Company, 2020). Officially, the government record says that over three-quarters of all spills and incidents were due to "sabotage," which includes acts of oil theft and attacks on infrastructures by anti-state militant groups. Pipeline ruptures and oil theft in particular have waxed and waned over time; since the return to civilian rule in 1999 the incidence, regularity, and quantity of oil theft has fluctuated shaped in part by the electoral cycle, by the price of oil, and by shifting patterns of criminal and militia activity. Theft reached the staggering height of around 350,000 barrels per day and between 2006 and 2009, when an armed insurgency threw the oil fields into disarray and oil theft was funding the rebel cause (see Watts, 2007; Obi and Rustad, 2011; Nwajiaku-Dahou, 2012; Adunbi, 2015). However, since the signing of the government amnesty with 30,000 militants in 2009, oil theft has only increased (Rexer, 2019).

The Ontological World of the Illegal Tap

Let's now follow that barrel of crude[17] as it passes from the wellhead into the trunk pipeline (the story might be slightly different if the pipeline is carrying refined fuels or gas). Within a short distance of the well, the crude flow is compromised by a "hot" or "cold tap," either on land or, if the pipeline is running along the floor of the creeks and estuaries of the delta, underwater. Hot tapping involves creating a branch connection to a pipeline in which the oil is flowing under pressure. To access lines running underwater and to conceal the tap, a small area of swamp around the pipe might be cordoned and drained and an isolating valve is welded or fitted mechanically to the pipe. After fitting—and with the valve open—the pipe is drilled to the maximum size through the valve, or the pipe is drilled part-way through and doused with sulfuric acid to complete the job once the line is in place.

Exceptional skill and knowledge is required of oil infrastructure—the sparks from drilling easily ignite the fuel—and tapping is typically undertaken by corrupt or former oil-industry technicians and engineers, or increasingly by a class of professional "bunkerers" who are part of "unions" or small corporate groups. In more elaborate (and large scale) bunkering, crude oil might be diverted from manifolds or flow stations[18] rather than individual pipelines, operations that require complicity and corrupt behavior from both local security forces and company operators.

In cold tapping, criminal or armed militias (sometimes referred to as oil gangs or oil mafia) blow up a pipeline, putting it out of use long enough for them to attach a spur pipeline. Many but not all pipeline bomb attacks appear to be linked to oil theft in order to enable a spur pipeline to be fitted, but during the period of armed insurgency in the Niger Delta (2005–2006, and earlier periods of activity by armed militias (2003–04, for example), attacks were launched in retaliation for military operations or as a way of extorting payments from transnational oil companies anxious to avoid *force majeure* (these were typically cash payments disguised as community development made to community "youth organizations"). The illegal spur pipeline transports the crude, often over several kilometers, to a convenient creek, where it is released into flat bottomed loaders (barges) or wooden 'Cotonou boats" and then transshipped to differing locations, local, regional, and international.

All stolen oil that is taken out of Nigeria for sale elsewhere—probably about 80 percent of all stolen oil until recently—appears to be initially transported in surface tanks or barges. Much of the oil to be distributed within Nigeria—depending upon the quantity either to local refiners or to major state-run refineries—appears to be transferred into drums, generally transported by trucks, or after local refining by canoe to remote creek communities. The ability to tap with impunity requires the complicity of and the payment of rents by the tappers to local military and security forces (the Joint Task Force), local police, coastguards, security, and low-level technical operators working for oil companies, local militants (they can be so-called secret societies, vigilante groups, ethnic militias or anti-state insurgents such as the Movement for the Emancipation of the Niger Delta (MEND) who "tax" the movement of stolen crude near their creek encampments, and village chiefs and other youth groups in oil "host communities."

Arranged in and around the tap and its installation, in short, is an ensemble of actors and agents held together by patterns of value extraction and rent, at once a sort of ontology of infrastructures and a political order of invisible supply chains (see Østensen and Stridsman, 2017). The local tappers (skilled welders with experience in the industry as opposed to those hacksaw or puncture siphons) typically work in teams of 3–6 people. Their proliferation, particularly post-2009, and their networked relations with security forces and actors within the industry (the oil company community liaison officers, flowstation technicians, oil service companies) has resulted in the rise of informal "unions" (Stakeholder Democracy Network, 2017; 2018). These unions can arrange for a tap placement (a recent report says the fee is roughly $6,200—in a country where per capita income is $2,300) arranging often for the reduction in pressure on the pipelines by having company officials in the control rooms in the flow stations on their payroll. The unions provide security by paying off local security forces and "settle" with local community leaders.

Operating the tapping points, irrespective of the scale of the tap, entails a "consortium" of security (by local youth and payoff to local state and federal security forces), technical capability and operational access capable of earning around $1 million a month; the monthly costs entail a union fee ($500), security payoffs ($1,200) and labor and equipment ($4,500). Payments to local community leaders in the oil host communities and to company technicians constitute additional expenditures. The union, provided with protection by local armed youth groups, delivers the oil to barges that either move offshore or arrange for local delivery to artisanal refiners. All along this local tapping and transporting supply chain are points of value extraction through rents which might take the form of extortion as much as formalized bribes.

All of the oil companies, including the national oil company, in theory provide surveillance and security to manage pipelines for reasons of safety and security. The costs of spills and explosion in and around communities, farms and fishing grounds are especially high. Yet the massive proliferation of what the companies and government agencies see as sabotage and theft suggests that either these systems are weakly enforced, or there is widespread collusion. Along some of the major trunk lines there are serial taps—in some sections there could be literally 100 or more taps per year (Ngada and Bowers, 2018). Both the companies and the federal government, moreover, have made use of local unemployed youth (as an employment strategy for a massive wageless class of alienated and frustrated youth across the region) to protect pipelines. But this in turn, building upon long standing grievances between communities and the companies, has simply provided yet more avenues for value extraction and rent-seeking.

Since 2009, as part of the Amnesty Plan, the federal government essentially has placed some 30,000 amnestied militants and their commanders on their payroll. This demobilization strategy provided security contracts ("surveillance contracts") to the most powerful commanders—and to oil host community chiefs and local contractors—to protect infrastructure. Not surprisingly, the amnesty program created tensions and conflict between former militias and their leaders squabbling over payments, while in practice superintending over the criminal oil theft enterprise. These surveillance functionaries were in effect "ghost workers" rarely carrying out any work while the amnesty program provided expanded opportunities to extract rents while converting commanders into local "businessmen" and indirectly funding private armies.[19]

Oil theft's world of rents, "taxes," and extortion is not confined to the world of tapping, shipping, and security, but also extends to environmental clean-up and restoration. In view of the overwhelming number of spills and pipeline interdictions each year, both corporations and federal and state regulatory agencies are liable for remediation. A number of federal and state agencies have jurisdiction over the oil and gas spillage response system, but for regulatory purposes it is the

Joint Investigation Process (JIV)—which entails the submission of four forms to the national regulatory agency (NOSDRA) and a similar JIV independent assessment by the company—that offers inordinate space for graft (Amnesty International, 2015; 2018). The framework under which such assessments are conducted—contained in the Oil Spill Recovery, Clean-up, Remediation and Damage Assessment Regulations, of 2011 ("Clean-up")—requires that a joint investigation team (JIT), comprising of the owner or operator of the facility from which oil has spilled, community and state government representatives, and NOSDRA, be constituted immediately after an oil spill notification is made, which ordinarily should be within 24 hours of the occurrence. The JIT is required to visit the spill site and investigate the cause and extent of the spillage and under regulation 40 of the Clean-up regulations, it entails a physical evaluation of the soil and surrounding environment in order to determine the impact of the incident, proper remediation procedures, and monitoring the remediation progress.

The horizons for value extraction when oil companies, the state and impoverished poor communities are brought together in a "stakeholder process" are legion. Amnesty International noted that "[t]he process is heavily dependent on the oil companies: they decide when the investigation will take place; they usually provide transport to the site of the spill; and they provide technical expertise, which the regulatory bodies lack" (Amnesty International, 2013, p. 14). Participatory involvement is relatively limited and tokenistic: few members of the community are able to participate; typically, the oil companies deal with chiefs—or those they designate—and male youth leaders. Not infrequently, community representatives are asked to sign incomplete forms and communities denied a copy of the form after signing it. Individuals are frequently paid to sign a JIV and company contractors in turn pay to get the clean-up contract. For example, a spill in Bayelsa State at Ikarama in 2011 at a Shell facility illustrates intersecting forces of lack of transparency, of an inadequate response system capable of effectively responding to conditions in the Delta, and the corruption of the JIV process itself (Olawuyi and Tubodenyefa, 2018). Pressures and payoffs exerted by the operators including threats by security agencies resulted in the coercion of the community stakeholders to acquiesce and agree to the finding of sabotage even though the communities believed it was operational failure.

When viewed through the prism of regulation and surveillance, the oil theft, the spill and clean-up system is one in which an array of foxes (regulators, companies, military, chiefs, militant commanders) are deployed to guard the henhouse. The assemblage is a shadow world of bribes, intimidation, extortion, fraud and illicit finance. The federal military forces provide protection for the major actors while offering a veneer of state legitimacy (taking the problem of oil theft seriously) by arresting, without necessarily prosecuting, low level barge operates and large numbers of small artisanal refineries, all the while leaving the black market operations intact. As amnesty payments dried up, or were absconded with by the commanders, and as employment opportunities through government programs declined as oil prices collapsed in 2014, many of the former militants had incentives to turn to artisanal refining and expanded tapping of pipelines.

Topping Up and the Piratical World

Tapping—hot or cold—is only one among a number of means to steal crude oil. There are others. One is "topping up" at the export terminal. Oil company employees can be bribed into allowing unauthorized vessels to load. Authorized vessels can be topped—filled with oil beyond their stated capacity—and the excess load sold. Oil revenues can also be embezzled, or money made through the sale of export licenses, credentials, bills of lading, and so on. This "white collar" branch of oil theft allegedly involves pumping illegally obtained oil onto tankers already loading at export terminals, or siphoning crude from terminal storage tanks onto trucks. Bills of lading and other shipping and corporate documents might be falsified to paper over the theft. Some topping off might also happen at sea via ship-to-ship transfer when barges holding up to 3,000 metric tons of oil unload onto smaller tankers with a capacity of 10,000 metric tons anchored offshore. Thieves generally use these small tankers to store and transport oil locally, although a few of the more seaworthy vessels might carry stolen oil to refineries or storage tanks within the Gulf of Guinea. Several small tankers can service a single oil theft network. Once the crude stored in them builds to a certain level, crews transfer it to a coastal tanker or an international class "mother ship" waiting further offshore. These ship-to-ship (STS) operations, typically occurring at night, can involve topping up a legal cargo of oil or filling an entire mother ship.

Oil theft from export terminals entails a different set of actors from within the upper echelons of the industry as well as a set of international agents—the shipping companies and a network of commodity traders and financiers—who can arrange for the international transfer and sale of oil products in China, North Korea, Israel, and South Africa. Political actors have a key role "due to their formal role in Nigeria's economy, as government regulators of the oil and maritime industries in Nigeria, or as businesspeople who process oil, provide support services to oil firms, and ship oil" (Hastings and Phillips, 2015, p. 457). They are enablers and intermediaries standing between local economic and political networks and international actors operating in the global oil assemblage.

The other means of stealing crude oil is piracy, which entails both different actors, different networks and different forms of rent extraction (Balogun, 2018). The Gulf of Guinea, on West Africa's southern coast, and Nigeria's coastal waters in particular, has become the world's most pirate-infested sea (Lopez-Lucia, 2015; Jacobsen and Nordby, 2015). The International Maritime Bureau reports that attacks on vessels at sea between Ivory Coast and Cameroon have grown dramatically since the early 2000s. Piracy has been common in Nigerian coastal waters over the last two decades with the region's booming oil theft and kidnapping-ransom economy, while in other piracy hotspots (Somalia, southeast Asia) piracy is in decline (*The Economist*, 2019). Niger Delta-based piracy has a historically long pedigree dating to the nineteenth century, but since the amnesty of 2009 pirates have the wind in their sails. Certainly, the number of attacks has ebbed and flowed this century, reaching an earlier peak in 2008 and 2013, but the current wave of violence is greater in scope and deadlier. The number of crew kidnapped in the Gulf of Guinea increased more than 50 percent from 78 in 2018 to 121 in 2019.[20] Currently, the Niger Delta region accounts for the vast majority of global maritime kidnappings: it equates to over 90 percent of global kidnappings reported at sea, with 64 crew members kidnapped across six incidents in the last quarter of 2019 alone. The region accounted for 64 incidents, including all four vessel hijackings that occurred in 2019, as well as ten out of eleven vessels that reported coming under fire (Lumpur, 2020). As in South-East Asia, pirates in Nigeria used to confine themselves to raiding oil-tankers to sell their cargo on the black market. When the oil price fell after 2014, they began copying their Somali counterparts and focused on kidnapping crews, though oil theft made a comeback in 2018 and 2019. Unlike the Somalis, West African pirates rarely retain the vessels or the workers. Instead, armed with AK-47s and knives, they storm a ship, round up some of the crew and return to land, where they hide their hostages.

Alternatively, if the prize is oil—and large quantities of oil that cannot be transshipped to the coast—then pirates engage in ship-to-ship oil transfer to a mother ship. Again, a different array of actors and rents are implicated. Pirates themselves do not have personal access to the networks with which to profit from the oil and typically deliver the oil to the principals for a flat sum. Once loaded to the tanker, the pirate groups are directed by the broker to deliver to specific locations along the West African coast and to oversee security while loading to tanks on shore. As Hastings and Philips (2015, p. 572) show, the boundary between licit and illicit has dissolved; the entities purported providing security are also involved in facilitating theft and providing protection: "the ship and cargo seizures are technically criminal activities, but at nearly every step of the way the pirates depend on the infrastructure (the ships, and storage and refining facilities) and the institutions (local brokerage, oil processing, and shipping companies, local and foreign buyers) of the formal oil economy." The visible and the invisible parts of the supply chain are in many respects indistinguishable. To add another layer of complexity, the kidnapping and piratical networks often overlap and intersect with other illicit maritime networks in the Gulf of Guinea especially drugs, human trafficking, and commodity smuggling (United Nations Office on Drugs and Crime, 2008; Ralby and Soud, 2018).

The After-life of Oil

What is the after-life of stolen crude? One answer is artisanal refining, locally known as "Kpo-fire." In virtually every community in the more isolated reaches of Niger Delta creeks and swamplands, households depend upon illegally refined fuels derived from stolen crude oil, typically selling at prices that undercut official fuel prices (see Garuba, 2010; Ikanone and Oyekan, 2014; Gelber, 2015). Plastic jerry cans of artisanal fuel (kerosene and petrol) are ubiquitous, retailing at roundabouts and markets even in large cities such as Port Harcourt or Warri. A small percentage of Nigerian crude is refined locally in state-owned refineries that are notoriously inefficient and typically lose vast quantities of money: over 12 months between June 2019 and 2020, the four state-owned refineries were idled and had operational losses of $367 (Smith, 2020). The year previously they operated at 13 percent of capacity. As a consequence, virtually all refined oil products are imported and then sold at subsidized prices ($0.48 per liter), a sort of vast "permit raj" that was exposed in a House of Representatives report in 2012 that entailed illicit activity totaling $6.9 billion, one of the most monumental cases of fraud in Nigeria's history (Mark, 2012; Sayne et al., 2015). A 200-page government inquiry revealed underhanded practices that fueled a sixfold increase in spending on oil handouts between 2009 and 2011. Fuel subsidies, part of a decades-old program meant to keep fuel prices low for millions of ordinary Nigerians, increased by 700 percent over three years. A report by a Nigerian House of Representatives committee identified the shadowy Nigerian National Petroleum Company—ranked the world's least transparent state oil firm—was single-handedly responsible for almost half of the siphoned subsidy funds and was "found not to be accountable to any body or authority." Seventy-two fuel importers, some with allegedly close links to senior government officials, were also singled out. In one case, payments totaling $6.4 million flowed from the state treasury 128 times within 24 hours to "unknown entities."

If the oil import business represents another massive tranche of the system of oil theft—in which traders and "briefcase" companies fight over the rents—fuel shortages nevertheless abound, especially in remote delta communities. Diesel and kerosene are in short supply and at a premium. In impoverished creek communities in which there is a sense that the state (through nationalization) has stolen "their oil," the oil theft business was able to facilitate the emergence of what has become over the last decade a major growth industry. Every year, security forces claim to have destroyed literally hundreds and in some cases thousands of illegal refining encampments dotted across the creeks in the oil-producing states[21]. A report estimated that by 2018, some 43,000 barrels of crude were refined locally each day from roughly 500 camps; in two states (Rivers and Bayelsa) it was estimated that between 2013 and 2018 the number of refineries increased five-fold (to 2,500) driven in part by national fuel scarcity and a growing demand for diesel and kerosene, and also by new forms of investment associated with "informal business associations improved information sharing and coordination of the supply chain" (Stakeholder Democracy Network, 2014, p. 11; 2018, p. 4). As profitability has increased, new investors are bankrolling the camps and the distribution system, and a greater share of stolen oil now ends up on the domestic black market (roughly 70 percent). The value of illegal oil products in these two states alone was almost $1 billion.

Illegal refining arose during the civil war (1967–70) among Biafran rebels cut off from fuel supply, but as oil theft began to proliferate in the 1990s and especially the early 2000s, so did artisanal refining. During 2003–04 as armed militia activity intensified, largely in response to state violence and the use by politicians and political Godfathers in the 2003 elections of armed youth groups and so-called cults to intimidate opponents. Competing non-state armed groups financed their activities increasingly through oil theft and refining. The leader of one important militant group (the Niger Delta People's Volunteer Force), Asari Dokubo, claimed that he had a tapped pipeline running to his compound and that his refined products—"Asari fuel"—were cheaper, better and more widespread in creek communities than commercial refined products. The bunkering territories are protected and indeed fought over while the security forces—the Navy and the Joint Task Force—simultaneously destroy illegal refineries while taxing their operations to ensure that the well-connected and wealthy refineries are protected. Especially since 2009, new refineries vastly outpace the rate at which refineries are destroyed.

Illegal refining depends on crude oil tapped from the tapping "unions" who deliver (and sell) the crude, often by Cotonou boats, to remote creeks' refining "camps." The distributors typically exclude middlemen and the vessels (and their work crews) can be owned by the tap owner, by larger refiners or by local transporters and vessel owners (Stakeholder Democracy Network, 2013). Distributors unload the crude either into open air pits or into so-called plastic Geepee tanks. An average camp might have ten to 20 people of all ages and genders; it requires capital investments (storage tanks, a "cooking oven," a cooling system and systems of hoses and drums). The refining process (dangerous for people and devastating for the environment) deploys a simplified version of fractional distillation in which crude oil is heated, condensed and separated. A camp operator (who might or might not be the owner) has workers, security, managers and "boatmen" in his employ. Tappers might earn $30 per day, boatmen $50–150 per day. Set-up costs for an average camp might be $5,000–6,000 and might generate $7,000–8,000 monthly in income.

The refining process uses a simplified version of fractional distillation (locally called "cooking"), in which crude oil is heated and condensed into separate petroleum products, aspects of which have been adapted from traditional gin and palm wine distillation. The illegal refining process yields diesel, petrol, kerosene, bitumen, and waste products.[22] The refining process begins when the "black" is heated in an "oven," burning crude oil to start the distillation, a process that releases dense black clouds into the camp, which, if not kept under control by spraying water onto the fire under the oven, can cause explosions. Distillation is kept cool through cold-water pumps and storage tanks, but the risks are substantial and the immediate impact on the environment catastrophic.

The illegally refined oil distributors typically represent yet another different network of actors and like tapping is one of the most profitable activities (in part because of the risks involved) in the oil theft assemblage. As the cost of buying stolen crude oil is a fraction of its true market price, the demand for cheap illegally refined products is considerable in both local and national markets. Most Nigerian crude oil grades are heavily diesel-rich but quality of refined products varies widely leading some refiners to purify diesel by mixing it with kerosene to improve the quality and launder the illegal product prior to distribution—much of which is sold in small quantities by women traders. Illegally refined diesel has become so intermixed with legal diesel distribution networks that it is impossible to say how far illegal products are being distributed, but there is a brisk trade in locally refined produce to other coastal states including Lagos. Locally, blended diesel is sold through pre-negotiated sales or along the roads or near filling stations and typically undercuts the official subsidized price of commercial fuels by fifteen percent or more. All movement and circulation operate under the cover of the police, the navy and military forces, and other security apparatuses.[23]

The major outlet for stolen crude is the international market. Barges of various sizes and conditions move the crude from the creeks where pipelines have been tapped—or in the case of theft at the export terminal simply add to the existing cargo in the tanker. Making their way downstream, pulled by tugboats, the barges meet awaiting tankers that, due to the topography of the Niger Delta, can anchor close to the places where the major rivers—the Benin, Escravos, Forcados and Ramos rivers—empty into the Atlantic. The vessels involved are typically in poor repair (but might cost from $50,000–75,000, far beyond the means of most local oil tappers) and might have been officially decommissioned. The chain from theft up to transference to oil tanker or local distribution is handled by the same gang but generally different units of the same group whereas the operation of the oil tankers and marketing of the stolen oil overseas appears to be handled by separate entities. While there are dedicated security forces devoted to surveillance and monitoring in order to apprehend bunkers, in practice few arrests are made and even fewer are prosecuted; in some cases, the tankers and their cargo mysteriously disappear. In 2003 Brigadier General Elias Zamani, then commanding a Delta peacekeeping force, was asked whether oil was being stolen by local people, the security forces, government officials, or an international element. His reply was: "All" (United Nations Office on Drugs and Crime, 2009, p. 22; see also Pérouse de Montclos, 2012).

Tracing stolen oil is virtually impossible for several reasons (Katsouris and Sayne, 2013). First, buyers of Nigerian oil load their cargoes onto tankers carrying crude from other oilfields, or even other countries—a process called "co-loading." For example, a trader might send a larger tanker to Nigeria to lift a 700,000 barrels cargo of Abo grade crude oil which then travels to the Forcados terminal, where it picks up an additional 300,000 barrels of Nigerian crude for delivery to Europe. Single tankers commonly carry multiple "parcels" of oil owned by different parties. The resulting full tanker-load of oil is called a "split cargo" and each parcel comes with its own bill of lading. Co-loaded and split cargoes, while perfectly legitimate, provide opportunities for bunkerers to disguise volumes of oil stolen at a terminal or in the field as a legal co-load. Mixed tanker-loads of stolen and legal oil are also rebranded as split cargoes by forging a separate bill of lading for the stolen portion. Second, complicated international delivery routes can hide stolen parcels. After leaving Nigerian waters, a mother ship carrying stolen crude can offload all of its cargo at a single refinery, offload parts of its cargo at different refineries, offload all or part of its cargo into storage, transfer all or part of its cargo STS to another vessel, or transfer all or part of its cargo STS to multiple vessels. Virtually all STS transfers of stolen oil probably take place further out at sea. Finally, export oil thieves blend stolen Nigerian crude with oil from other countries and with fuel oil produced in or outside Nigeria. A range of customers buy the adulterated goods once they are mixed onboard tankers or at sites onshore. Some is probably sold as bunker fuel for ships. And finally, there is the murky world of storage. Most traders place large amounts of oil into storage facilities around the world. This enables them to blend crudes or hold them until a particular market improves. Most oil storage is on land, but some floats at sea. Due diligence and reporting regulations vary by location. Selling crude oil into storage can allow sellers to disguise the oil's origins in future transactions. For example, an unscrupulous trader could receive a consignment of stolen oil into tanks it owns or rents, then blend or break it into smaller parcels. New bills of lading can be issued for each parcel when it is eventually sold, making less diligent buyers less likely to ask for an original bill of lading created in Nigeria.

In sum, diverted oil is also part of a transnational business—an oil mafia—linking the high-ranking military, politicians, business elites, security and regulatory forces, and domestic and foreign oil traders and shippers. The international oil companies have been an active part of this mix: local level employees often conspired with refiners and oil thieves, while corporate executives saw this rough and tumble supply chain and outright bribes as the price of doing business.

Making Oil Circulate: Political and Logistical Orders and the Invisible Supply Chain

Oil theft operations in Nigeria—as everywhere—entail a logistical and political order to tap, circulate, and distribute a variety of hydrocarbon products to local, regional, and international black markets. The dynamic shape of this assemblage—including elite political actors, youth groups, local and international and state-owned oil companies, shipping companies, insurgents, military, and much more—is secret and elusive yet in some respects conducted in broad daylight. The fact that the movement of tankers, or topping up, or illegal refining can often operate openly and indeed through formal channels speaks to the fact that the "invisible" (informal/illicit) supply chain operates through the same channels and with similar actors as the "visible" (formal/licit) global oil and gas supply chain. The same actors can be, and often are, involved in both sets of activities. The boundaries blur, the functions overlap and intersect. Furthermore, the invisible supply chain has its own formality. In the same way that the mafia constitutes a particular sort of order—a set of forms and conventions and relations to state powers—so too does the oil theft assemblage have its unions, taxes, dues, settlements, and returns. There are, too, enforcement (extra-economic) mechanisms; and like the formal gas supply chain, oil theft entrepreneurs and actors respond to market, security, and political signals. The oil theft industry has its own lexicon: foremen, tappers, sponsors, investors, buyers, and traders. If the illicit oil supply chain is in many respects co-terminus with the licit—with considerable porosity between the two—this observation questions the view that the resource curse is simply a reflection of the fact that at every step from extraction to final export, as Hastings and Phillips write, "oil firms are potentially subject to rents extracted by local political actors, both at the national and local levels, and must pay them off or establish informal understandings with them—often they must do both" (Hastings and Philips, 2015, p. 572). This is both true and incomplete since oil firms are not simply complicit but are active agents in not just the extraction of value through rents but in the reproduction of the entire system. The licit and illicit systems of petro-capitalism are deeply imbricated and mutually self-sustaining, feeding off each other and exhibiting remarkable durability over time even in the face of conflicts and violence.

Much of this shadow economy remains elusive and our understanding remains incomplete. It is elusive not only because of secrecy and complicities at the highest levels of the state and government, but also because of the incomplete picture of the oil theft enterprise. The fullest report claims that the oil-theft supply chain is more cellular than hierarchical (Sayne et al., 2015). If Nigerian politicians and the press speak of bunkering barons and kingpins, or describe oil-theft rings as mafias or syndicates, they argue that "most export operations are probably not run by one person, family, or ethnic group, and management tends to be more cooperative than based on command-and-control" (Sayne et al., 2015, p. 6). But these are surmises rather than conclusions since there seem to be mafia-like consortia, of differing degrees of complexity and organization, operating at multiple levels. But it is clear there is considerable heterogeneity across the cells' networks membership; they vary the size and location of operations, needs and political entanglements. Actor influence and positions might wax and wane (military commanders come and go) but there is "a common set of roles to fill...high-level opportunists, facilitators, operations, security, local transport, foreign transport, sales and low-level opportunists" (Sayne et al., 2015, p. 6; see also Balogun, 2018). If this sounds like a fractal landscape that constantly shape-shifts, that is for now at least as robust a generalization as we can make of this oil assemblage and the operations of capital.

Oil theft in my account is restricted to oil products stolen from nodes within the logistical infrastructure and various rents extracted around these operations. But theft is widespread in other hydrocarbon domains that are arguably of equal if not greater significance as regards illicit proceeds. One area pertains to illicit financial flows around the awarding of oil licenses and bonus payments through the leasing and tendering process. Licenses are assets that are traded among the political and business classes and represent one of the least transparent aspects of the industry, and the most corrupt (Sayne et al., 2017). Another is sales and so-called "first trades," namely NOC-buyer contracts and terms of trade (which I turn to next) (Extractive Industries Transparency Initiative, 2015; Longchamp and Perrot, 2017). And another is revenue collection and distribution (royalties, taxes and public financial management) and the public procurement contracts issued for oil and oil-related activities to oil-service companies, and so on (Organisation for Economic Co-operation and Development, 2016). These arenas are replete with all manner of value extractions—rents—of the sort I describe in my account of oil bunkering, and often on a vast scale.[24] In fact there is an entire industry and an edifice of regulatory authorities devoted to documenting the scale of the graft and theft associated with illicit flows in these other domains of the planetary oil assemblage, including the EITI, OECD, and advocacy organizations such as Global Witness and the Center for Research on Multinational Corporations Here, the assignment and use of property rights often resemble outright theft: oil prospecting and oil mining leases are acquired by members of the political class and are bought and sold as an asset class; massive bribes are paid to secure mega-engineering contracts; buyer-trader licenses are in effect licenses to print money. And not least, there is outright theft—pillage really—at the highest levels of leadership. During the late military period in Nigeria, the stolen assets sent out of the country by President Abacha to offshore financial centers were estimated at $5 billion, and the process has continued (especially in the period after 2009). Nigeria has no monopoly here. The infamous Bien Mal Acquis case affair involved a series of corruption scandals which emerged in oil-and-mineral-rich central African states in 2007. More recently, we saw the Luandagate affair[25] and revelations about so-called "oilygarchs" in the Paradise Papers and Wikileaks. While the theft involved in these instances turns on corrupt political elites, the role of the national oil companies—the black holes of any national oil sector—and international oil companies and trading houses is central to any understanding of the scale of financial hemorrhaging from the public purse.

Nigeria's universe of stolen oil returns us to Lefebvre's observations on global capitalism and space. First, oil theft is constituted through a myriad of overlapping, nested and intersecting spaces (the system is deeply territorialized): from bunkering territories, to oil concessions, to pipeline networks, to trade corridors, oil host community territories, military jurisdictions, and so on. All are more or less regulated and orderly; each has some form of quasi-sovereignty and is populated by its own petty sovereigns. It is a space of hypercomplexity overlaid with layered forms of sovereignty (in which state, corporate and forms of petty sovereign abound). Second, Lefebvre referred to a particular form of what he called state capitalism to understand the growth of post-war European capitalism and the complex spatial hypercomplexity. What is on offer in Nigeria and the licit/illicit value chain is less a version of *pur et dur* neoliberalism than a variant of oil-fueled state capitalism.

Two final points. Nigeria has no monopoly on oil theft: Russia, Colombia, Iraq, and the Caucasus are known to have significant losses, especially in downstream fuel theft. The fuel-smuggling trade is vibrant across the Turkish border to Syria. In the eastern Mediterranean, there is a flourishing smuggling of oil focused on Libya, Malta, and Cyprus. In 2018 a major oil theft occurred in a Shell refinery in Singapore, the company's largest refinery in the world. In Indonesia, there were 63 cases of oil theft from pipelines from one concession, the Rokan Block managed by Chevron Pacific Indonesia. And in Europe, pipeline theft grew from barely a few cases in 2010 to some 150 cases in 2015.

Mexico represents an intriguing case providing a sort of counterpoint to Nigeria both in terms of scale and organization. The country is a major oil producer and exporter of oil, accounting for fifteen percent of exports and twenty percent of state revenues, and like Nigeria has a large, complex national oil company (Petroleos Mexicanos, or PEMEX) controlling the upstream and downstream sectors. But oil theft (*robo de combustibles*), illegal oil traders (*huanicoleros*), and pipeline taps (*tomas clandestinas*) have grown from a cottage industry run by local gangs during the 1980s and the 1990s into a massive industry in the hands of cartels and specialized *huachicolero* syndicates who violently compete for control over the trade. Centered on two "Red Triangles" located in Puebla and Guanajuato, with secondary centers in Veracruz and Tamalaulipas, by 2019, 22 states had reported oil theft (Sullivan, 2012; Duhaukt, 2017). In 2006 there were 213 illegal pipeline taps; by 2016 they had grown to 6,873 (accounting for over $11.3 billion for the period 2009–2016). By 2018 the number of taps had almost doubled to a staggering 12,582 (Jones and Sullivan, 2019). Of the 1,533 pipeline taps reported in 2016, 1,071—or 70 percent of the total—were located along Highway 150D that parallels the trunk pipeline for refined products from Veracruz (and its refineries) to Mexico City. Not only was PEMEX itself in crisis, but oil theft and the violence it generated in a country marked by a pre-existing cascade of homicides (some 35,000 in 2019) reached crisis proportions. Oil theft in fact became one of, if not *the,* defining features of the first year on President Andres Manuel Lopez Obrado's *sexenio* following his landslide victory in July 2018.

Mexico's oil-theft assemblage reflects a rather different architecture to that of Nigeria, rooted as it is in the political history—and the political settlement—of post-revolutionary Mexico. Refined products (gasoline, diesel, kerosene) rather than crude represent the illicit commodities that are trafficked, and the focus is on a massive underground system of distribution (and to a degree larger scale refining) designed to undercut official fuel prices.[26] Energy reforms put in place by the Nieto administration (2012–18) permitted oil prices to rise and incentivized *huachicoleros* to undercut the formal market system. More crucially, while there are extremely porous boundaries between the military and security forces, the NOC and the oil thieves (like Nigeria), the central players in Mexico are transnational drug cartels that came to oil theft late in their institutional careers (the 2000s) on the backs of the deepening role of Mexico after the 1980s in the global cocaine, heroin and other narcotics wholesale trade. In part because of the anti-drug policies on both sides of the border and the changing markets structures for drugs, the cartels diversified and moved into oil theft, for which their national and transnational trade networks could be easily repurposed (Correa-Cabrera, 2017). The territorial natures of the cartels, their constant fragmentation and division as a result of the Mexican government's kingpin strategy which produced intra- and inter-cartel violence, and the geography of the PEMEX pipeline networks meant in practice a ferocious and violent struggle between cartels and other subsidiary or independent fuel traffickers to control the fuel business. And not least, the fuel cartels—currently the fuel trade is dominated by Cartel de Santa Rosa Lima, the Cartel Jalisco Nueva Generación, and the Los Zetas cartel (a splinter group of the Gulf Cartel)—used their pre-existing military capabilities to extort and threaten PEMEX workers (to access pipelines, refineries, liquefied natural gas storage tanks and even offshore rigs), secure protection from the military and the judiciary, and develop a national (and cross-border to the United States) tanker distribution system quite unlike the Nigeria domestic black market. The Mexican cartelized theft system is marked by extraordinary violence even by Nigerian standards: cities like Salamanca which houses a large refinery and the PEMEX Minatitlan Mexico City pipeline have been marked by extraordinary bloodletting and conflicts between the cartels and gangs and by period pipeline explosion involving hundreds of casualties.

Reading Mexico's oil theft history against that of Nigeria throws up some obvious parallels in terms of state capture and the complicities among the state oil companies, security forces, political classes, and oil thieves. But the actors, processes, and differing political histories and political settlements in each petro-state shape the specific forms in which the oil assemblage operates and reproduces (Hickey and Izama, 2017). Oil theft grew out of and was captured by drug cartels that were at the time a product of both the changing global character of the drug trade, the nature of the drug markets, and the declining powers of the then-ruling party, the Institutional Revolutionary Party (Correa-Cabrera, 2017). Nigeria's theft, by contrast, grew out of a political settlement in which an elite cartel presided over a provisioning system and a multi-ethnic federal system (Roy, 2017; Porter and Watts, 2017). For complex reasons, the systems fed popular resentments on the oilfields that resulted in the proliferation of armed non-state groups and ultimately an armed insurgency and amnesty. These histories color the oil-theft assemblages while retaining family resemblances. Interestingly, in a way that has no obvious parallel in Nigeria, the election of left populist President Obrador unleashed a major assault on the oil-theft sector both to stop the loss of revenues and also to stabilize a crippled PEMEX, reduce the extraordinary violence, and provide a better environment for investment by international oil companies. By mid-2020 it was reported that oil theft had decreased by 90 percent and in August "El Marro," the head of the Cartel de Santa Rosa Lima, was captured by Mexican security forces (Dalby, 2020).

Finally, oil and gas, like most supply chains, disclose the fact that "the trappings of logistical giants in one place actually hinge on logistical work in utterly deregulated zones elsewhere" (Schouten et al., 2019, p. 780). These are the circulation struggles—the imperatives to control place, space and territory and what moves through and across it—which do not produce a clean logistical space, a well-ordered supply chain in which place has been thinned out or eviscerated. Quite the reverse. Forces of calculability and order fulfill "disorderly" functions and vice versa. They are organically and dialectically related and constituted. These two faces of the oil life world—and the porosity between them—constitute an important expression of the oil cosmos but of so many global supply chains as the contributions show. Logistical orders can be and regularly are disrupted, blocked, diverted, and appropriated in novel and creative ways, and all of this points to the co-production of logistical and political orders. Making things move and circulate is both an expression of power and constructs and depends upon systems of public and private authority.

Oil theft reveals powerfully how the intersection of logistical/infrastructural and political orders constructs ontologies—what Julian Reid (2006) calls logistical life—within the vast oil assemblage. These logistical and infrastructural orders, as uneven and irregular as they are over space, create different opportunities and different kinds of space "because they create the thickenings of publics, and offer the possibility of assembling people or slowing them down" (De Boeck, 2012). Oil theft is built in and through oil infrastructures and represents what I call an "oil cosmos": not a circumscribed enclave of social thinness but an entire lived world. As a measure of this cosmos one only need note that the very presence of oil infrastructure (a wellhead, a pipeline) confers an existential status on communities: when present in a community territory a village or town or city neighborhood becomes an "oil host community" which confers particular rights, rents and identities. Of course if there is something of the entrepreneurial spirit at work in oil theft networks, and of resistance too (popular appropriation by those who see their oil resources as having been taken from them), it is a world of violence, conflict, subterfuge, and precarity. It is an ambient and combustible world, a vast provisioning machine in the business of shaping human experience and social identities. It is a sort of sensorium.

Logistics, Finance, and First Trades: Contract Theft and Commodity Traders[27]

The circulation of large quantities of stolen oil points to one of the great unexplored domains of the oil and gas assemblage: namely, so-called "first trade" between NOCs and buyer-traders, and specifically the role of the private commodity-traders, finance capital (on whom the traders rely) and not least the shady world of shipping and maritime movement. Value extraction and rent-seeking abound in this world, and I can only offer here the briefest of glimpses with the aim of shining light on the links among extraction, finance, and logistics and on the degree to which the planetary well turns on marginal, liminal, and frontier spaces that are not so much on the oilfields—frontiers though they might be—but in in the capitalist world of banks and offshore finance and in the great (and spectral) world of trading contracts. These are populated not only by the trading divisions of the international oil companies but more crucially by the likes of Glencore, Koch Industries, Mercuria, Trafigura, and large independent commodity traders.

Until recently, the trading system, and the relations between government sales and private buyers, have been strikingly absent from transparency measures and very little research has systematically focused on illicit flows arising from first trades. This interface is the key moment at which oil produced (that is to say the upstream sector) enters the global market (the midstream sector) with its price tag. First trade or equity oil is acquired by a considerable variety of buyers and traders—from international oil companies with their large trading desks to the large commodity trading houses, small independents, and even other national oil companies. Commodity-trading firms are essentially in the business of transforming commodities in space (logistics), time (storage), and form (processing). Their basic function is to perform physical "arbitrages" which enhance value through these various transformations.

The scale and scope of oil trading and their significance for oil-producing states and their treasuries are substantial. In 2014 the Berne Declaration—now known as Public Eye—and Swissaid analyzed the oil sale activities of the top ten oil exporters in SSA and found that from 2011 to 2013 the governments of these countries generated more than $250 billion in sales revenue, equaling 56 percent of their combined government revenues (Gillies *et al.*, 2014). A NRGI National Oil Company Database released in 2019 analyzed the oil, gas, and product sales by NOCs in 35 countries for which data is available and revealed that the first trades made by NOCs in these 35 countries to commodity traders and other buyers generated over $1.5 trillion in 2016 (22 percent of total state revenues). Updated oil sales data for 2016–18 available for 28 countries indicate that government revenue generated from commodity trading has risen significantly, from $1.4 trillion in 2016 to $2.1 trillion in 2018.

The scale of revenues generated from oil sales, coupled with the lack of regulation of how these sales are conducted, creates enormous opportunity for value extraction and rent-seeking. According to Global Financial Integrity, unrecorded oil sales amount to 17 billion annually (500,000 barrels per day) (McHugh, 2012). In 2016 OECD published a study that analyzed 131 corruption cases involving foreign public officials in the natural resources sector, including trading. Significantly, 26 of the cases (20 percent) appeared to involve commodity trading. These figures refer only to the number of cases, not to the sums of money misappropriated. If the latter were considered, then the scale of corruption in the trading phase, measured in terms of financial flows, would be greater still. Trade corruption involving Vitol, Philia, and Gunvor in the Democratic Republic of Congo, and Glencore in Kazakhstan, have been well documented (Public Eye, 2017). On February 26, 2020 the Swiss Federal Council published a report, *Supervision of commodity trading activities from the point of view of money laundering*, written in response to a postulate by the Council of States, that recognizes the high risk of corruption to which the commodity trading sector is exposed (Swiss Federal Council, 2020). The Money Laundering Communication Office (MROS) shows that over the past ten years, several thousand suspicious transactions related to trading have been reported. Two major international corruption scandals involving Brazilian and Venezuelan oil companies (Petrobras and PdVSA) alone resulted in more than 1,500 reports between 2015 and 2018. For the report, MROS evaluated a sample of 367 communications on suspicious transactions linked to trading between 2016 and 2018 (without taking into account Petrobras, PdVSA and other "laundromat cases"). These related to around 1.1 billion francs. MROS identified trading in fossil fuels as particularly risky, accounting for 85 percent of the samples examined. In addition to consulting companies and trust companies, the report mentions real estate companies and pension funds, in which profits are suspected of being of criminal origin, and which are said to have been reaped as part of raw materials trading activities. It concludes that "the Swiss financial center, given the size of the sector, is particularly exposed to the risk of money laundering linked to commodity trading, both through its banks and through traders established in Switzerland" (Swiss Federal Council, 2020, p. 9).

The menu of trading risks is broad, including not only the potential for tax evasion and money laundering associated with mis-invoicing but also the possibility of bribery, collusion, and below-market pricing associated with the largely opaque oil-backed loans and oil-for-product swap agreements. In Nigeria, for example, a number of beneficiaries of export allocations are nothing but letterbox companies whose sole function is that they are linked to high-ranking political officials or their entourages. Politically linked holders of "letterbox" or "briefcase" companies have, as the Nigerian Task Force explained, little or no commercial and financial capacity. In Nigeria, such fake entities represent a major part of the "oil market." As pointed out in a Chatham House report, only 25 percent to 40 percent of the holders of export allocations actually have the capacity or will to finance, ship, and sell their cargoes directly (Katsouris and Sayne, 2013, p. 8). The entire trading system attracts shadowy idle men, because these companies belong to individuals serving as fronts for the political class and power brokers.

First trades have been linked with a class of risks surrounding resource-backed loans (RBLs), namely loans provided to a government or a state-owned company, where repayment is either made directly in natural resources, (that is, in kind, or from a natural resource-related future income stream), or repayment is guaranteed by a natural resource-related income stream, or a natural resource asset serves as collateral (Mihalyi et al., 2020). RBLs are simply one set of transactions linking buyers and NOCs, but they carry significant risks because of their size and opacity. NRGI's analysis demonstrates that RBLs are remarkably opaque (only in a single case was the key contract document public) and they carry immense major public finance risks, as reflected in the fact that of 14 RBL recipient countries, ten experienced serious debt problems after the commodity price.

Although all commodity traders engage in transformation activities, they are tremendously diverse. Switzerland, which is the world's leading commodities trading hub with an estimated 35 percent share of the oil market, has over 500 trading companies, almost 90 percent of which are private; 42 percent had fewer than ten employees and ten percent more than 300 (Ascher et al., 2012; Pirrong, 2014). The five largest Swiss independent traders (Vitol, Glencore, Trafigura, Gunvor, and Mercuria) typically trade almost 18 million barrels per day, equivalent to about 20 percent of global demand. There is no common pattern in terms of the commodities they trade and transform, in the types of transformations they undertake, in their financing, and in their forms of ownership. Traders and sellers are often linked through complex financial and joint-venture agreements. The trading assemblage is diverse not only in virtue of the nature of the sale contracts and price negotiations, but also because of the relations and networks linking companies, buyers, finance capital, audit houses, and credit rating agencies. In engaging in these transformation activities, commodity traders face a wide array of risks, some of which can be managed by hedging, insurance, or diversification, but they face others that must be borne by the firms' owners. On a global canvas, much of the trading activity is centered on a cluster of global trading hubs (the UK, the Netherlands, Singapore, and Switzerland). Overall, the oil trading system is one of the most complex attributes of the global oil assemblage.

International oil companies—for example BP, Shell, Total—with their own trading desks sell their own production but also buy and sell third-party production. As "asset-based traders" they have access to and frequently use their own capital to fund trading activities. BP, Shell and Total alone traded 15 million barrels of crude a day in 2016. Independent international commodity traders are generally companies that have not traditionally engaged in production, are often privately held[28], and typically deploy bank-provided trade finance to fund their trading activities. Many of these companies are based in Switzerland, the UK, Dubai, and Singapore. Domestic buying companies operate only in the producing country and are heterogeneous in their scale and operations. Some are large and established, others are more akin to "middlemen" or "briefcase" companies acting as intermediaries between the NOC and other larger buyers. National Oil Companies whose subsidiaries trade in commodities either produced by themselves, or by third parties (e.g. China's Sinopec and Azerbaijan's SOCAR). Finally, there are the investment banks that trade commodities, such as Goldman Sachs and Citigroup—the number of which has fallen dramatically due to bank regulatory changes after 2008—and the end-users such as refiners, smelters, and processors (e.g. Sinopec, Japanese refiners such as JX Nippon Oil, and Energy Corp.). Traders generally do not have fully integrated supply chains and are lighter on assets, instead chartering vessels and entering into joint ventures with local counterparties. All traders depend upon liquidity and loans, typically through a suite of instruments to manage financing and settling of accounts that might include, for trader-NOC trades in particular, opaque and complexly structured crude-for-product swaps, oil-backed loans, or off-take agreements.

Viewed through the lens of the planetary well, the global oil trading system is intricate and byzantine, composed of varied assemblages of actors with contrasting interests and positions within a commodity system operating across multiple regulatory jurisdictions. In systemic terms, commodity trading is arguably one of the most complex and difficult to regulate within the oil and gas value chain. The trading system is dynamic, market prices are capricious, and risks are legion; this is not least because the architecture of the system has changed, and is changing, in relation to global capitalism in its recent financialized forms, and in response to market volatility and global competitive pressures. Over the last four decades, the system has experienced a thorough-going financialization (Gkanoutas-Leventis and Nesvetailova, 2015; Gkanoutas-Leventis, 2017). The 1980s liberalization and the institutional changes in the market triggered by the launch of commodity indexes by financial institutions in the early 1990s contributed to the growth of futures contracts and a raft of new actors. But recent market developments spurred by the introduction of permissive regulations in 2000 with the launch of the Commodities Future Modernisation Act in the United States opened the oil commodity markets to mutual funds, insurance institutions, and banks. Some of the largest investment banks, later known as "Wall Street Refiners," established specialized departments for trading in the oil market. By 2003 most of the biggest U.S. hedge funds were engaged in commodity markets, their involvement tripling between 2004 and 2007.

As oil became an increasingly popular asset class with investors, it widened the opportunities for hedging but also for financial speculation. Furthermore, the advance of financialization and the integration of financialized markets through indexification, produced endogenous dynamics in this market creating new sources of fragility and risk. Sometimes called "oil vega," this financialization of oil and the rise of paper trades made oil prices both volatile and largely independent of physical trades and market fundamentals. At the same time, despite the plethora of regulatory agencies in global finance, regulatory arbitrage is a defining quality of the global financial system, permitting commodities markets to thrive in between regulatory niches, capitalizing on permissive regulatory policies nationally, and exploiting unregulated spaces internationally (Gibbon, 2004). Most traders operate in and through trading hubs or offshore financial centers associated with favorable regulation and tax rates, strong capital markets, a deep tradition of trade and shipping and human capital resources (London, New York, Chicago, Houston, Calgary, Tokyo, Hong Kong, Geneva, Zug, and more recently the UAE and Singapore). Traders might be involved simultaneously in the buying, selling, transportation, storage, and refining of physical oil yet at the same time in value terms the overwhelming majority of trades are in so-called "paper trades" (the futures and derivative markets). In this hub-and-spoke network system, populated by a diverse suite of buyers, traders, and financiers, it is the *opacity* that presents such a challenge to anti–IFF measures.

The oil trading assemblage is not just complex, variegated, global and multiscalar in its operations. It exhibits a number of distinctive structural properties, three which are key for my purposes. First, the extent of operations that make use of offshore financial centers (OFCs) and subsidiaries that have ambiguous functions. Second, the lack of opacity in the trades themselves. The trading system seems to seek out, and even reproduce, opacity, operating in frontier-like (unregulated) spaces both within the oil producing states themselves but also in the trading hubs and OFCs. And third, increasingly the role of deregulated banking functions (KPMG, 2015). A blend of low commodity prices, deepening competition, capital requirements, and increased price transparency has eroded margins, reduced arbitrage opportunities and modified the players participating in this competitive arena. In addition, new banking regulations have also changed the financial architecture of the trading system. Large commodity trading houses have become active in the financial and credit markets, extending credit to economy and becoming part of the unregulated segment of the financial system, or the shadow banking system in large part because of the withdrawal of the larger investment banks as a consequence of regulatory changes in the financial sector most (notably Basel III, Dodd-Frank, MiFD II). Increasingly, smaller banks with a higher risk appetite are coming to the forefront, including Chinese banks looking to participate in syndicated facilities while hedge funds, private equity and specialized trade finance funds are incrementally being used by commodity traders as financing alternatives. All this makes for a world that lacks transparency, is shrouded in secrecy, and often operates in the shadows.

A stock-take of first trade transparency by EITI in 2018 revealed that of fourteen countries reviewed, over half did not provide core information by the seller, and virtually all buyers failed to disclose information on contracts, beneficial ownership, loading points, or buyer selection processes (Extractive Industries Transparency Initiative, 2018). The first pilot reports reveal both the limited impact of disclosure requirements and the contentiousness of the regulatory domain itself. Engagement by traders has in general been very low (and some of the Chinese and Russian buyers and financial houses are for the most part beyond the reach of EITI); the data is uneven in detail and quality and often inconsistent in what is measured. Contract disclosure is almost wholly absent, and beneficial ownership data is missing in cases such as Nigeria where there is a strong emphasis on local buyers. In Nigeria, 66 of 73 companies did not submit the reporting templates and as a result were unable to reconcile 81 percent of NNPC crude sales (Nigeria Extractive Industries Transparency Initiative, 2019).

Nigeria is, once again, a textbook example of value extraction through NOC-buyer contracts where the selection of buyers, the allocation of buyers' rights, and the negotiation of the terms of sale are shrouded in secrecy. The entire arena of contracts (for licensing and exploration and for oil-related engineering and service work) is the most opaque of sectors. The Nigerian case is so complex because virtually all sales are mediated through middlemen, and because the scale and size of the cargoes is vast. Conversely, in 2017 Nigeria sold 453 cargoes to 61 buyers totaling $13.2 billion, including Glencore, Trafigura, BP, Total, NNPC's trading subsidiaries (Duke Oil, Carlson Bermuda), and domestic buyers (Sahara Energy), including a number that are seemingly shell companies with no palpable operations, and foreign national oil companies (SONAR, SINOPEC). New research is gradually exposing not only the shady oil swap and oil-for-product deals and the key role of the major trading houses, but a much wider landscape of complicities between rogue Nigerian middlemen and enablers, and the diverse world of oil traders, speculators, and financiers (Berne Declaration, 2013; Gillies *et al.*, 2014; van Drunen *et al.*, 2020) Politically exposed letterbox companies, secret calls for tender, opaque and shady partnerships, and links between Nigerian importers of refined fuels and Swiss trading houses (making use of the fuel subsidies discussed earlier) are all part and parcel of the normalized operations of the trading houses.

The intensive use of Offshore Financial Centers, particularly by independent traders, combined with the complexity of corporate holdings through OFC jurisdictions, weakens the system of corporate governance at the same time as the relatively light touch and willful laxity of oversight by public authorities is a key point of attraction for corporate managers. However, banking regulations and the withdrawal of formerly dominant international banks from directly financing trades has seen the rise of local banks and traders and joint venture arrangements—a trend that has increased the relative difficulty of establishing the *bona fides* and identity of counterparties to the deal. In turn, this is reported to be weakening the effectiveness of corporate governance protocols where incentives exist for them to be applied. The oil-trading system in this sense has its own *differentia specifica* compared to other sectors of the oil and gas value chain.

Frontiers Across Planetary Oil

> Capitalism is a frontier process.
>
> *Jason Moore,* Capitalism in the Web of Life, *2015, p. 107*

I want to conclude with a word about commodity frontiers and planetary oil. To the petro-geologist, the frontier has a set of technical meanings. It is a geological province which becomes a working petroleum system, a play with its own unique reservoir properties, particular temperatures, flow characteristics, viscosity, and so on. In another sense, as a geological formation located in space—and therefore located within the supply chain—these plays are often at the margins and fringes of the global value chain opening and closing with the shifting horizons of the exploration and production process of the oil industry. Much of my account focuses on these frontiers in the rough-and-tumble terrains, the "fragile and conflicted" petro-states of the Global South marked by "poor governance." To this extent, the oil frontier represents a particular sort of social space. Frontiers are sites within the global supply chain "beyond the sphere of routine actions of centrally located violence producing enterprises….[populated] by classes specialized in expediency whose only commitment [is] to preserve the order that made possible the profitable utilization of such expediency" (Baretta and Markoff, 2006, pp. 36, 51). Frontiers are social spaces at the limits of central power where authority—and indeed the rule of law and its forms of enforcement and oversight—is neither secure nor non-existent. The key attribute here is institutional patchiness or unevenness, or what James Ron usefully distinguishes as weakly institutionalized spaces not tightly integrated into adjacent core states (Ron, 2015, p. 7). Oil frontiers in this sense do not necessarily conform to Tsing's (2005) much-cited view that frontiers are unpredictable, free for all, not yet mapped, unstable. In my view this is not quite right: frontiers can stably reproduce, and their dynamics frighteningly predictable and ordered. As Grandin (2019) says of the frontier, the state often precedes it; authority, power, and institutions of all sorts are present in complex and differentiated ways. Put differently, the characteristic of frontiers everywhere is the circumvention of infrastructural and administrative grids of the formalized economy.

The world of oil theft and invisible-visible supply chains shows how across the space of planetary oil are all multiple frontiers some of which are located at the other end of the oil assemblage, in offshore financial centers populated by shell and dormant companies and consolidated and encased by law, financial institutions, audit companies, and the like. As the world of oil trading shows so clearly, *that* part of planetary oil is no less subject to opacity and lack of transparency than oil theft or the byzantine operations of national oil companies. The trading houses and trading desks, offshore financial centers in the Cayman Islands, Luxembourg, Bermuda, Hong Kong, the Netherlands, Ireland, the Bahamas, Singapore, Belgium, the British Virgin Islands, and Switzerland are also part of a great cosmos of oil theft and hyper-extraction. In some of these frontier settings, extraordinarily capable expertise and regulatory capabilities are brought to bear to limit the reach of public authority. It is not only in the oil producing states that "old margins are becoming new frontiers" but also in places like Singapore and Zurich "where mobile, globally competitive capitals find minimally regulated zone in which to vets its operations" (Mezzadra and Neilson, 2019, p. 123).

In planetary, hyper-extractive systems frontiers open and close over time and space. Frontier phenomena, which populate supply chains everywhere, are marked by institutional patchiness, by overlapping and complex nested forms of power and authority. Sometimes frontiers might throw up alternatives—counter-logistics or even emancipatory political orders—but as often as not they are precarious, violent, and illicit. Frontiers point to the fact that the infrastructural and political orders that operate across and through planetary oil might operate in close proximity to, or in conjunction with, the state or they might exist largely outside of it. Perhaps this is what Bertolt Brecht (1927) had in mind when he wrote in his poem *700 Intellectuals Pray to an Oil Tanker* that "God has descended again in the form of an oil tanker."

Notes

1 I would like to thank Doug Porter for many conversations on these topics and the recent OECD research teams on illicit finance and the oil sector. Judith Shapiro provided much-needed input and editorial guidance.

2 The issue of oil and gas data—or more properly epistemology—is extremely murky. Figures produced by state, corporate, and advocacy organizations on the number of spills, their cause, and the volume of spillage, to say nothing of output and revenues, vary enormously. It is emblematic of larger questions about transparency and the degree to which the most basic information—for example, wellhead and export terminal metering systems—are either non-existent or are inadequate and/or tampered with. For an industry marked by technological complexity and sophistication, the appalling quality—to say nothing of the elasticity—of the most basic data is striking (see Watts and Zalik, 2020).

3 Spatial complexity is matched by various oil 'temporalities' (pertaining to the duration/ longevity of the reservoir, time to first or peak oil, length of concession and so on) and 'verticalities' (pertaining to petro-geology, well depth, land or sub-sea based reservoirs and the like). See Lunning (2018), Woegink (2018), and Yusoff (2017).

4 Materials are taken to include biomass, fossil fuels, metal ores, and non-metallic minerals. Primary materials refer to those sourced from mining and extraction activities in their raw form, such as mineral ores. Secondary materials refer to materials that have been used previously (i.e. recycling). See UNEP 2019 Global Resources Outlook: 2019. Nairobi: United Nations Environment Program, p. 42.

5 Resource extraction and processing make up about half of the total global greenhouse gas emissions and more than 90 percent of land- and water-related impacts (biodiversity loss and water stress).

6 In his memoir *A Promised Land* (2020) President Barack Obama emphasizes how the novelty and enormity of the Deepwater disaster shook him. Until then, Obama had maintained a "fundamental confidence" that he "could always come up with a solution through sound process and smart choices." But those plumes of oil—"emanations from hell"—rushing out of a cracked earth and into the sea's ghostly depths" seemed of another order, unassimilable to his generally imperturbable worldview.

7 They pull upon the work of Lauren Benton (2010) to emphasize the forms of quasi- or partial sovereignties, and the world of non-state world of petty sovereigns, to expose the fragmented and uneven complexities of contemporary capitalism. In a somewhat different register, albeit more sensitive to racialized extraction, Macarena Gomez-Barris (2017, p. 1) offers a decolonial theoretical account "foregrounding submerged perspectives" anchored in "anarcho-feminist indigenous critique."

8 The proliferation of these rents means not only that they are the basis of capitalist expansion but are the objects of contest and struggle over, for example, which group elites receive import licenses, which ethnic groups are awarded mining leases, and who benefits from corporate community development projects.

9 A research project involving myself, Alexander Arroyo, Arthur Mason, and Berit Kristofferson entitled "The Digital Arctic" is currently in progress.

10 Almost 5 million producing oil wells puncture the surface of the earth (77,000 were drilled in 2019, 4,000 offshore); 3,300 are subsea. There are by some estimates over 40,000 oil fields in operation, more than 2 million kilometers of pipelines blanket the globe in a massive trunk-network and another 75,000 kilometers of lines transport oil and gas along the seafloor.

11 Subsea exploration in the Barents Sea has been challenged in Norway on legal and environmental grounds (www.nytimes.com/2020/11/05/world/europe/norway-supreme-court-climate-change.html), but the new Arctic has also meant explosive commercial competition among littoral powers (www.nytimes.com/2020/11/12/us/russia-military-alaska-arctic-fishing.html).

12 The assets of the largest ten oil and gas companies is roughly $3 trillion.

13 The latest trade data to estimate the magnitude of import and export trade misinvoicing—one of the largest components of measurable illicit financial flows—among 135 developing countries and 36 advanced economies shows that by industrial sector, mineral fuels exhibited the second largest value gap ($113 billion, representing 16 percent of total trade) between 2008 and 2017 (see Global Financial Integrity, 2020).

14 On 29 August, 2019 the Ad-Hoc Committee of the National Economic Council on Crude Theft disclosed that Nigeria lost about 22 million barrels in the first six months of 2019. This loss was later put at $1.35 billion. This amount is already about 5 percent of the entire year's budget. Also, it is more than the capital allocations for education, healthcare, defense, and agriculture combined.

15 Cited in www.bloomberg.com/news/articles/2019-06-05/nigeria-s-oil-thieves-roar-ba ck-even-as-militants-kept-in-check.

16 In 2018, four pipelines—the NCTL, the Trans Forcados pipeline, the Trans Niger pipeline and the Obagi flowstation—accounted for 600,000 barrels of lost crude oil.

17 In this section of oil theft, I make use of fieldwork conducted in the Nigeria delta over the last 15 years and the following: Katsouris and Sayne, 2013; UNODC, 2009; Oyefusi, 2014; Fiennes, 2020; Naanen and Tolani, 2015: Schultze-Kraft, 2017; Ugor, 2013; Balogun, 2018.

18 A manifold is a more complex arrangement of piping or valves designed to control, distribute, and typically monitor oil and are often configured for specific functions that require a higher degree of control and instrumentation. A flowstation is usually the first stop for hydrocarbon fluids coming from crude oil and gas wells. Its purpose is to separate the hydrocarbon into liquid and vapor phases, reduce turbulence, and pass on the liquid to the next facility.

19 *Pipeline Surveillance Contracts in the Niger Delta*, Policy Brief (Port Harcourt: Stakeholder Democracy Network) describes an archetypical surveillance contract or as follows: "through the network of relationships the pipeline surveillance contractor maintains across communities, he is able to neutralize such opposition by distributing 'royalties' to 'settle' with chiefs, elders, young people, and women's groups before work begins. These demands typically amount to 10–15 percent of the value of the work to be done. He keeps track of the total amount of money that he distributes and that the international oil company reimburses him, depositing that amount into his bank account, using payments euphemistically known as 'local content'" (Stakeholder Democracy Network, 2019:1).

20 In 2019, over 90 percent of globally reported kidnappings and hostage-taking at sea took place in the Gulf of Guinea, and the vast majority of attacks are launched on shipping from within Nigerian territorial waters: www.hstoday.us/channels/global/global-con cern-grows-as-gulf-of-guinea-piracy-attacks-increase-in-number-and-violence.

21 According to the Nigerian Navy, 2,287 refineries were destroyed between 2015 and 2019 with a peak of 1,218 in 2017: https://guardian.ng/news/nigeria/nigerian-na vy-destroy-2287-illegal-refineries.

22 After refining 30 drums of crude, a refiner might produce 25 drums of diesel worth N250,000; six drums of kerosene are worth N30,000 in the black market.

23 A recent Transparency International (2019: 4–5) report documented military personnel demanding payments, regular and scheduled, from illegal refineries in exchange for allowing them to operate have continued to surface. Interviewees in Bayelsa state, for example, reported that after an illegal refinery failed to meet a deadline to pay an "operational fee" of 4 million Nigerian naira ($11,000), military officers arrived on the site and opened fire, allegedly killing one person and demanding an extra 200,000 Nigerian naira ($550) for the delay. On the next day, 1.7 million Nigerian naira ($5,000) was delivered to military personnel with a promise to pay the balance of 2.3 million Nigerian naira ($6,000) later. Standard "tax" payments for each drum of product was 1,000 to 2,000 Nigerian naira ($3 to $6), and retailers of illegal oil products spend an average of 60,000 Nigerian naira ($167) on transportation "settlements" for different security personnel, including the military, and police at road checkpoints if trucks were deployed to move the oil.

24 In 2008, Albert J. Stanley, a former executive with a Halliburton subsidiary (KBR), pleaded guilty to charges that he conspired to pay $182 million in bribes to Nigerian officials in return for contracts to build a $6 billion liquefied natural gas complex.

25 Isabel dos Santos, the daughter of former Angolan president José Eduardo dos Santos and Africa's richest woman, has a reported fortune of over $2 billion. According to the Luanda Leaks, as well as reports from Maka Angola and other sources, Ms dos Santos and her husband earned some of their money thanks to public contracts approved by her father's government and suspicious deals struck with state-owned companies. Ms. dos Santos was appointed as head of Angola's national oil company, Sonangol, by her father in June 2016 and remained in place until she was removed by the current President in November 2017 (see www.icij.org/investigations/luanda-leaks).

26 As in Nigeria with some of the militants who steal and refine oil and supply local communities, the oil cartels have often gained popular support by provisioning cheap fuel, making gifts of fuels to poor communities to celebrate holidays, and by developing a *huachicolero* subculture that involves the adoption of Catholic saints.

27 Parts of this section draws upon research I conducted for an OECD project on Illicit Financial Flows and Oil Trading. I am grateful for the insight provided by the research teams including Catherine Anderson, Doug Porter, Alexandra Gillies, Joe Williams, Anastasia Nesvetailova, Ronen Palan, and Phil Culbert.

28 The exception in the top five trading houses is Glencore (revenues of $219 billion in 2018; 158,000 employees), which is also now a major extractive company, following its merger with Xstrata, in the mining sector and owns limited upstream assets in the oil sector.

References

Adunbi, O. (2015) *Oil Wealth and Insurgency in Nigeria*. Indianapolis, IN: Indiana University Press.

Amnesty International. (2013) *Bad Information: Oil Spill Investigations in the Niger Delta*. London: Amnesty International Publications.

Amnesty International. (2015) 'Niger Delta: Shell's Manifestly False Claims About Oil Pollution Exposed, Again'. Available at: www.amnesty.org/en/latest/news/2015/11/shell-fa lse-claims-about-oil-pollution-exposed.

Amnesty International. (2018) 'Negligence in the Niger Delta: Decoding shell and Eni's poor record on oil spills'.

Andreucci, D., García-Lamarca, M., Wedekind, J., and Swyngedouw, E. (2017) '"Value grabbing": A political ecology of rent', *Capitalism Nature Socialism*, 28 (3), pp. 28–47.

Appel, H. (2019) *The Licit Life of Capitalism*. Durham, NC: Duke University Press.

Appel, H., Mason, A., and Watts, M. (eds.) (2015) *Subterranean Estates: Life Worlds of Oil and Gas*. Ithaca, NY: Cornell University Press.

Arboleda, M. (2020) *Planetary Mine: Territories of Extraction Under Late Capitalism*. London: Verso.

Arroyo, A. (In progress) 'Designing a digital ocean: Speculative oceanographies in the New Arctic,' PhD Dissertation, Berkeley: University of California.

Ascher, J., Lazlo, P., and Quiviger, G. (2012) Commodity trading at a strategic crossroad, Working Paper on Risk #39. London: McKinsey & Company.

Bala-Gbogbo, E. (2019) 'Nigeria's oil thieves roar back as militants kept in check'. Bloomberg. Available at: www.bloomberg.com/news/articles/2019-06-05/nigeria-s-oil-thie ves-roar-back-even-as-militants-kept-in-check.

Balogun, W. (2018) 'Crude Oil Theft, Petro–Piracy and Illegal Trade in Fuel', PhD Dissertation, Lancaster University.

Baretta, S. and Markoff, J. 2006 'Civilization and Barbarism' in Coronil, F. and Skurski, J. (eds.) *States of Violence*. Detroit, MI: University of Michigan Press.

Barry, A. (2006) 'Technological zones', *European Journal of Social Theory*, 9 (2), pp. 239–253.

Berne Declaration. (2013) *Swiss Traders Opaque Deals in Nigeria*. Zurich.

Boyd, W., Prudham, W.S., and Schurman, R.A. (2001) 'Industrial dynamics and the problem of nature', *Society and Natural Resources*, 14 (7), pp. 555–570.

Brecht, B. (1927) '700 Intellectuals Pray to an Oil Tanker', *Revolutionary Democracy*. Available at: https://revolutionarydemocracy.org/rdv9n2/poemsbb.htm.

Bridge, G. (2008) 'Global production networks and the extractive sector: Governing resource–based development', *Journal of Economic Geography*, 8 (3), pp. 389–419.

Bridge, G. (2015). 'The Hole World: Scales and Spaces of Extraction', *Scenario Journal*. Available at: https://scenariojournal.com/article/the-hole-world.

Chalfin, B. (2014) 'Public things, excremental politics, and the infrastructure of bare life in Ghana's city of Tema', *American Ethnologist*, 41 (106).

Christophers, B. (2019) 'The problem of rent', *Critical Historical Studies*, 6 (2), pp. 308–309.

Christophers, B. (2020). *Rentier Capitalism*. London: Verso.

Clark, M. (2014) 'Nigeria's oil thieves drive Shell To distraction as company plans pipeline sale', *IB Times*, 27 February, 2014.

Correa-Cabrera, G. (2017) *Los Zetas Inc.: Criminal corporations, energy, and civil war in Mexico*. Austin, TX: University of Texas Press.

Cowen, D. (2010) 'A geography of logistics: Market authority and the security of supply chains', *Annals of the Association of American Geographers*, 100 (3): pp. 600–620.

Dalby, C. (2020) '*Three takeaways from the capture of 'El Marro' in Mexico*'. *Insight Crime*. Available at: www.insightcrime.org/news/analysis/three-takeaways-el-marro-mexico.

De Boeck, F. (2012) Infrastructure: Commentary from Filip De Boeck. Curated Collections, Cultural Anthropology Online. Available at: https://journal.culanth.org/index.php/ca/infrastructure-filip-de-boeck.

Duhaukt, A. (2017) *Looting Fuel Pipelines in Mexico*. Houston, TX: Baker Institute, Rice University.

Extractive Industries Transparency Initiative. (2015). The EITI, NOCs and the first trade. Oslo: EITI Secretariat.

Extractive Industries Transparency Initiative. (2018) Commodity trading transparency stocktake. Oslo: EITI Secretariat.

Ferguson, J. (2005) 'Seeing like an oil company', *American Anthropologist*, 107 (3), pp. 377–382.

Ferguson, J. (2006) *Global shadows: African in the Neoliberal World Order*. Durham, NC and London: Duke University Press.

Foreign Corrupt Practices Act Clearing House. (2019) Key Statistics. Stanford Law School. Available at: http://fcpa.stanford.edu/industry.html.

Garuba, D.S. (2010). 'Trans-border economic crimes, illegal oil bunkering and economic reforms in Nigeria', *Policy brief series*, 15.

Gelber, E. (2015). 'Rogue pipelines, oil and amnesty: The social life of infrastructure in the Niger Delta' (Doctoral dissertation, Columbia University).

Gereffi, G., Humphrey, J., and Sturgeon, T. (2005). 'The governance of global value chains', *Review of International Political Economy*, 12 (1), pp. 78–104.

Gibbon, P. (2004) 'Trading Houses during and since the Great Commodity Boom: Financialization, productivization or...?', DIIS Working Paper, Copenhagen, 2014(12).

Gillies, A., Gueniat, M., and Kummer, L. (2014) 'Big spenders: Swiss trading companies, African oil and the risk of opacity'. Zurich, Berne Declaration.

Gkanoutas–Leventis, A. (2017) *Spikes and Shocks: The Financialization of the Oil Market from 1980 to the Present Day*. London: Palgrave.

Gkanoutas–Leventis, A. and Nesvetailova, A. (2015) 'Financialisation, oil and the Great Recession', *Energy Policy*, 86, pp. 891–902.

Global Financial Integrity. (2013) *Illicit Financial Flows and the Problem of Net Resource Transfers from Africa: 1980–2009*. New York, NY: Global Financial Integrity.

Grandin, G. (2019) *The End of the Myth: From the Frontier to the Border Wall in the Mind of America*. New York, NY: Metropolitan Books.

Harvey, D. (2007) *The Limits to Capital*. London: Verso.

Hastings, J.V. and Phillips, S.G. (2015) 'Maritime piracy business networks and institutions in Africa', *African Affairs*, 114 (457), p. 573.

Hausmann, R. (1981) 'State landed property, oil rent, and accumulation in Venezuela: an analysis in terms of social relations'.

Henely, K. (2012) 'Review: The Forgotten Space', *Slant Magazine*. Available at: www.slantmagazine.com/film/the-forgotten-space.

Hertog, S. (2010) 'Defying the resource curse: Explaining successful state–owned enterprises in rentier states', *World Politics*, 62, pp. 261–301.

Hickey, S. and Izama, A. (2017) 'The politics of governing oil in Uganda: Going against the grain?', *African Affairs*, 116 (463), pp. 163–185.

Ikanone, C.E.O. and Oyekan, P.O. (2014). 'Effect of boiling and frying on the total carbohydrate, vitamin C and mineral contents of Irish (*Solanun tuberosum*) and sweet (*Ipomea batatas*) potato tubers', *Nigerian Food Journal*, 32 (2), pp. 33–39.

Jacobsen, K.L. and Nordby, J.R. (2015) *Maritime Security in the Gulf of Guinea*. Copenhagen: Royal Danish Defence College Publishing House.

Jensen, C. and Morita, S. (2015) 'Infrastructures as ontological experiments', *Engaging Science, Technology, and Society*, 1, pp. 81–87.

Jingzhong, Y., van der Ploeg, J.D., Schneider, S., and Shanin, T. (2020) 'The incursions of extractivism: Moving from dispersed places to global capitalism', *The Journal of Peasant Studies*, 47 (1), pp. 155–183,

Jones, N.P. and Sullivan, J.P. (2019) 'Huachicoleros: Criminal cartels, fuel theft, and violence in Mexico', *Journal of Strategic Security*, 12 (4), pp. 1–24.

Katsouris, C. and Sayne, A. (2013) *Nigeria's Criminal Crude: International Options to Combat the Export of Stolen Oil*. London: Chatham House.

Klinger, J.M. (2018) *Rare Earth Frontiers: From Terrestrial Subsoils to Lunar Landscapes*. Ithaca, NY: Cornell University Press.

KPMG. (2015) Clarity on commodities trading.

Labban, M. (2014) 'Deterritorializing extraction: Bioaccumulation and the planetary mine', *Annals of the Association of American Geographers*, 104 (3), pp. 560–576.

Lapavitsas, C. (2009) 'Financialised capitalism: Crisis and financial expropriation', *Historical Materialism*, 17 (2), pp. 114–148.

Larkin, B. (2013) 'The politics and poetics of infrastructure', *Annual Review of Anthropology*, 42, pp. 327–343.

Lefebvre, H. (2005) *The Production of Space*. Oxford: Blackwell.

Lombardi, M. (2003) *Global Networks*. New York, NY: Independent Curators.

Longchamp, O. and Perrot, N. (2017) *Trading in Corruption: Evidence and Mitigation Measures for Corruption in the Trading of Oil and Minerals*. U4 Anti–Corruption Centre.

Lopez-Lucia, E. (2015) *Fragility, Conflict and Violence in the Gulf of Guinea* (rapid literature review). Birmingham: GSDRC, University of Birmingham.

Lumpur, K. (2020) Unprecedented number of crew kidnappings in the Gulf of Guinea despite drop in overall global numbers. International Chamber of Commerce. Available at: https://iccwbo.org/media-wall/news-speeches/unprecedented-number-of-crew-kidnappings-in-the-gulf-of-guinea-despite-drop-in-overall-global-numbers/#:~:text=The%20number%20of%20crew%20kidnapped,last%20quarter%20of%202019%20alone.

Luning, S. (2018) 'Mining temporalities: Future perspectives', *The Extractive Industries and Society*, 5 (2), pp. 281–286.

Mark, M. (2012) 'Nigeria fuel subsidy scheme hit by corruption', *The Guardian*. Available at: www.theguardian.com/world/2012/apr/19/nigeria–fuel–subsidy–scheme–corruption.

Mason, A. (ed.) (Forthcoming) *Arctic Abstractive Industry*. Oxford and New York, NY: Berghan.

Mazzucato, M. (2018) *The Value of everything: Making and taking in the global economy*. London: Allen Lane.

McHugh, L. (2012). *The Threat of Organised Crime to the Oil Industry*. Future Directions International.

Mezzadra, S. and Neilson, B. (2017) 'On the multiple frontiers of extraction: Excavating contemporary capitalism', *Cultural Studies*, 31 (2–3), pp. 185–204.

Mezzadra, S. and Neilson, B. (2019) *The Politics of Operations: Excavating Contemporary Capitalism*. Durham, NC: Duke University Press.

Mihalyi, D., Adam, A., and Hwang, J. (2020) Resource-backed loans: pitfalls and potential. New York, NY: Natural Resource Governance Institute.

Mining Review Africa. (2019) 'The digital mine: How miners are turning a vision into reality', *Mining Review Africa*. Available at: www.miningreview.com/health–and–safety/the–digital–mine–how–miners–are–turning–a–vision–into–reality.

Mommer, B. (1990) 'Oil Rent and rent Capitalism: The example of Venezuela', *Fernand Braudel Center*, 14, pp. 417–437.

Moore, J.W. (2015) *Capitalism in the Web of Life: Ecology and the Accumulation of Capital*. London: Verso Books.

Murray, P. (2004) 'The Social and Material Transformation of Production by Capital: Formal and real subsumption in Capital' in Bellofiore, R. and N. Taylor (eds.) *The Constitution of Capital: Essays on Volume 1 of Marx's Capital*. Cham: Springer. pp. 243–272.

Nasir, J., (2020) 'Analysis: "N556bn spent on oil pipeline repairs since 2015"—Nigeria losing trillions to vandalism'. Petrobarometer. Available at: https://petrobarometer.theca ble.ng/2020/01/22/analysis-n556bn-spent-on-oil-pipeline-repairs-since-2015-nigeria -losing-trillions-to-vandalism/.

National Oil Spills Detection and Response Agency. (2020) Nigerian Oil Spill Monitor. Available at: https://oilspillmonitor.ng.

Ngada, T. and Bowers, K. (2018) 'Spatial and temporal analysis of crude oil theft in the Niger Delta', *Security Journal*, 31 (2), pp. 501–523.

Nigerian Bulletin. (2020) 'Nigeria lost N4.57tn revenue to crude oil theft in 4 years – maritime security review'. Available at: www.nigerianbulletin.com/threads/nigeria–lost–n4–57tn–re venue–to–crude–oil–theft–in–4–years–%E2%80%93–maritime–security–review.413541.

Nigeria Extractive Industries Transparency Initiative. (2019) *Nigeria's Oil and Gas Trading Pilot Report*. Oslo/Abuja.

Nigeria National Petroleum Corporation. (2008–2018). *Annual Statistical Reports*. Annual Statistics Bulletin. Available at: www.nnpcgroup.com/Public-Relations/Oil-and-Gas-Sta tistics/Pages/Annual-Statistics-Bulletin.aspx.

Nigeria Natural Resources Charter. (2018) NNRC Assesses the impact of crude oil theft on Nigerians. Available at: www.nigerianrc.org/nnrc–assesses–the–impact–of–crude–oil–theft.

Nwajiaku-Dahou, K. (2012). 'The political economy of oil and "rebellion" in Nigeria's Niger Delta', *Review of African Political Economy*, 39 (132), pp. 295–313.

Obama, B. (2020) *A Promised Land*. New York, NY: Crown Books.

Obi, C. and Rustad, S.A. (Eds.) (2011). *Oil and Insurgency in the Niger Delta: Managing the Complex Politics of Petro-violence*. London: Zed Books.

Ogunde, A. (2012) 'Nigeria loses over $500 million monthly due to crude bunkering – Shell'. *Business News*. Available at: http://businessnews.com.ng/2012/02/07/nigeria-lose s-over-500-million-monthly-due-to-crude-bunkering-shell.

Oil and Gas IQ. (2019) 'What does digital transformation in oil and gas look like?', *Oil & Gas IQ*. Available at: www.oilandgasiq.com/oil–gas/news/what–is–digital–transformation.

Olawuyi, D. and Tubodenyefa, Z. (2018) *Review of the Environmental Guidelines and Standards for the Petroleum Industry of Nigeria*. Ado Ekiti: OGEES Institute, Afe Babalola University.

Omeje, K.C. (2006) *High Stakes and Stakeholders: Oil Conflict and Security in Nigeria*. London: Ashgate Publishing.

Organisation for Economic Co-operation and Development. (2014) *OECD foreign bribery report: an analysis of the crime of bribery of foreign public officials*. Paris: OECD Publishing.

Organisation for Economic Co-operation and Development. (2015) *Material resources, productivity and the environment*. Paris: OECD Publishing.

Organisation for Economic Co-operation and Development. (2016) *Corruption in the extractive value chain: Typology of risks, mitigation measures and incentives*. Paris: OECD Publishing, pp. 11–12.

Organisation for Economic Co-operation and Development. (2016) *Corruption in the Extractive Value Chain: Typology of Risks, Mitigation Measures and Incentives*. Paris: OECD Publishing.

Organisation for Economic Co-operation and Development. (2018) *Global Material Resources Outlook to 2060*. Paris.

Østensen, Å.G. and Stridsman, M. (2017) 'Shadow Value Chains: Tracing the link between corruption, illicit activity and lootable natural resources from West Africa', *U4 Issue*.

Pérouse de Montclos, M.A. (2012) 'Maritime piracy in Nigeria: Old wine in new bottles?', *Studies in Conflict & Terrorism*, 35(7–8), pp. 531–541.

Piketty, T. (2014) *Capital in the Twenty–first Century*. Cambridge, MA: Belknap Press of Harvard University Press.

Pirrong, C. (2014) *The Economics of Commodity Trading Firms*. Zurich: Trafigura.

Porter, D. and Watts, M. (2017) 'Righting the resource curse: Institutional politics and state capabilities in Edo State, Nigeria', *The Journal of Development Studies*, 53 (2), pp. 249–263.

Public Eye. (2017) *Gunvor in Congo: Oil, Cash and Misappropriation of a Swiss Trader in Brazzaville*. Zurich.

Ralby, I. and Soud, D. (2018) 'Oil on the water: Illicit hydrocarbons activity in the maritime domain', Atlantic Council.

Reinhart, L.B. (2014) 'The aftermath of Mexico's fuel–theft epidemic: Examining the Texas black market and the conspiracy to trade in stolen condensate', *St. Mary's Law Journal*, 45 (49), pp. 749–786.

Rexer, J. (2019) *Black Market Crude*. Philadelphia, PA: Kleinman Center for Energy Policy, University of Pennsylvania.

Ron, J. (2005) *Frontiers and Ghettoes*. Berkeley, CA: University of California Press.

Roy, P. (2017) Anti-Corruption in Nigeria: A political settlements analysis.

Sayne, A., Gillies, A., and Katsouris, C. (2015) *Inside NNPC Oil Sales: A Case for Reform in Nigeria*. New York, NY: Natural Resources Governance Institute.

Sayne, A., Gillies, S., and Watkins, A. (2017) *Twelve Red Flags: Corruption Risks in the Award of Extraction Licenses and Contracts*. New York, NY: Natural Resource Governance Institute.

Schmitt, C. (2003). *The Nomos of the Earth in the International Law of the Jus Publicum Europaeum* (trans. G.L. Ulmen). New York, NY: Telos.

Schouten, P., Stepputat, F., and Bachmann, J. (2019) 'States of circulation: Logistics off the beaten path', *Environment and Planning D: Society and Space*, 37 (5), p.780.

Signé, L., Sow, M., and Madden, P. (2020) *Illicit Financial Flows in Africa*. Washington, DC: The Brookings Institution.

Smith, M. (2020) 'Losses mount at idle Nigerian refineries', *Petroleum Economist*. Available at: www.petroleum–economist.com/articles/midstream–downstream/refining–marketing/ 2020/losses–mount–at–idle–nigerian–refineries.

Stakeholder Democracy Network. (2013) *Communities not criminals: Illegal oil refining in the Niger delta*. Port Harcourt: Stakeholder Democracy Network Publications.

Stakeholder Democracy Network. (2014) *Communities not Criminals: Illegal Oil Refining in the Niger Delta*. London and Port Harcourt: International Secretariat Development House.

Stakeholder Democracy Network. (2017) *Communities Not Criminals: Illegal Oil Refining in the Niger Delta*. Port Harcourt: Stakeholder Democracy Network.

Stakeholder Democracy Network. (2018) *More Money, More Problems: Economic Dynamics of The Artisanal Oil Industry in The Niger Delta Over Five Years*. Port Harcourt: Stakeholder Democracy Network.

Stakeholder Democracy Network. (2019) *Pipeline Surveillance Contracts in the Niger Delta*. Port Harcourt: Stakeholder Democracy Network.

Standing, G. (2016) *The Corruption of Capitalism: Why Rentiers Thrive and Work Does Not Pay*. London: Biteback.

Sullivan, J. (2012) *From Drug Wars to Criminal Insurgency: Mexican Cartels, Criminal Enclaves and Criminal Insurgency in Mexico and Central America*. Paris: Fondation Maison des Sciences de l'Homme.

Swiss Federal Council. (2020) Supervision of commodity trading activities from the point of view of money laundering.

Szeman, I. (2017) 'On the politics of extraction', *Cultural Studies*, 31(2–3), pp. 440–447.

The Economist. (2019) 'The Gulf of Guinea is now the world's worst piracy hotspot', June 20, 2019. Available at: www.economist.com/international/2019/06/29/the-gulf-of-gui nea-is-now-the-worlds-worst-piracy-hotspot.

Tichý, L. (2019) 'The Islamic State oil and gas strategy in North Africa', *Energy Strategy Reviews*, 24, pp. 254–260.

Tsing, A. (2009). 'Supply chains and the human condition', *Rethinking Marxism*, 21 (2), pp. 148–176.

United Nations Development Programme. (2006) *Niger Delta Human Development Report.* Abuja.

United Nations Economic Commission for Africa. (2018) *Base Erosion and Profit Shifting in Africa: Reforms to Facilitate Improved Taxation of Multinational Enterprises.* Addis Ababa: Economic Commission for Africa.

United Nations Environment Programme. (2019) Global Resources Outlook 2019. Nairobi.

United Nations Office on Drugs and Crime. (2009) *Transnational Trafficking and the Rule of Law in West Africa.* Vienna.

Van Drunen, S., Hartlief, I., Bassey, C., and Henshaw, K. (2020) *Big Business, Low Profile: Shedding Light on Oil Trader Vitol's Operations in Nigeria.* Amsterdam: SOMO.

Vienneast. (2016) *An Investigation into Oil Smuggling and Revenue Generation by Islamic State.* London: Vienneast.

Watts, M. (2007). 'Petro-insurgency or criminal syndicate? Conflict & violence in the Niger Delta', *Review of African Political Economy*, 34 (114), pp. 637–660.

Watts, M. (2015) 'Spaces of Insurgency: power, place and spectacle in Nigeria' in Merrill, H. and Hoffman, L. (eds.) *Geographies of Power: Re-cognizing the Present Moment of Danger.* Athens, GA: University of Georgia Press, pp. 191–227.

Watts, M. (2018) 'Authority, precarity and conflict at the edge of the state' in Engels, B. and Dietz, K. (eds.) *Climate Change in Africa: Social and Political Impacts, Conflicts and Strategies.* Berlin: Peter Lang, pp. 167–206.

Weizman, E. (2007) *Hollow Land: Israel's Architecture of Occupation.* London: Verso Books.

Whanda, S., Adekola, O., Adamu, B., Yahaya, S., and Pandey, P.C. (2017) 'Geo–spatial analysis of oil spill distribution and susceptibility in the Niger delta region of Nigeria', *Journal of Geographic Information Systems*, 8, pp. 438–456.

Wiegink, N. (2018) 'Imagining booms and busts: Conflicting temporalities and the extraction "Development" nexus in Mozambique', *The Extractive Industries and Society*, 5 (2), pp. 245–252.

Yates, D.A. (1996) *The Rentier State in Africa: Oil Rent Dependency and Neocolonialism in the Republic of Gabon.* Trenton, NJ: Africa World Press.

Ye, J., van der Ploeg, J.D., Schneider, S., and Shanin, T. (2020). 'The incursions of extractivism: moving from dispersed places to global capitalism', *The Journal of Peasant Studies*, 47 (1), pp. 155–183.

Yusoff, K. (2017) 'Geosocial strata', *Theory, Culture & Society*, 34(2–3), pp. 105–127.

INDEX

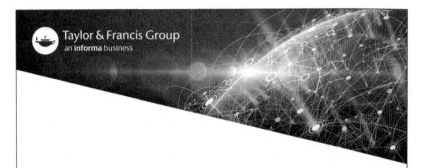